GPS for
Land Surveyors

Fourth Edition

Accessing the E-book edition

Using the VitalSource® ebook

Access to the VitalBook™ ebook accompanying this book is via VitalSource® Bookshelf — an ebook reader which allows you to make and share notes and highlights on your ebooks and search across all of the ebooks that you hold on your VitalSource Bookshelf. You can access the ebook online or offline on your smartphone, tablet or PC/Mac and your notes and highlights will automatically stay in sync no matter where you make them.

1. **Create a VitalSource Bookshelf account at** https://online.vitalsource.com/user/new or log into your existing account if you already have one.

2. **Redeem the code provided in the panel below to get online access to the ebook.** Log in to Bookshelf and click the **Account** menu at the top right of the screen. Select **Redeem** and enter the redemption code shown on the scratch-off panel below in the **Code To Redeem** box. Press **Redeem**. Once the code has been redeemed your ebook will download and appear in your library.

GPS for Land Surveyors, Fourth Edition

DOWNLOAD AND READ OFFLINE

To use your ebook offline, download BookShelf to your PC, Mac, iOS device, Android device or Kindle Fire, and log in to your Bookshelf account to access your ebook:

On your PC/Mac

Go to *http://bookshelf.vitalsource.com/* and follow the instructions to download the free **VitalSource Bookshelf** app to your PC or Mac and log into your Bookshelf account.

On your iPhone/iPod Touch/iPad

Download the free **VitalSource Bookshelf** App available via the iTunes App Store and log into your Bookshelf account. You can find more information at *https://support. vitalsource.com/hc/en-us/categories/200134217- Bookshelf-for-iOS*

On your Android™ smartphone or tablet

Download the free **VitalSource Bookshelf** App available via Google Play and log into your Bookshelf account. You can find more information at *https://support.vitalsource.com/ hc/en-us/categories/200139976-Bookshelf-for-Android- and-Kindle-Fire*

On your Kindle Fire

Download the free **VitalSource Bookshelf** App available from Amazon and log into your Bookshelf account. You can find more information at *https://support.vitalsource.com/ hc/en-us/categories/200139976-Bookshelf-for-Android- and-Kindle-Fire*

N.B. The code in the scratch-off panel can only be used once. When you have created a Bookshelf account and redeemed the code you will be able to access the ebook online or offline on your smartphone, tablet or PC/Mac.

SUPPORT

If you have any questions about downloading Bookshelf, creating your account, or accessing and using your ebook edition, please visit *http://support.vitalsource.com/*

GPS for
Land Surveyors

Fourth Edition

Jan Van Sickle

CRC Press
Taylor & Francis Group
Boca Raton   London   New York

CRC Press is an imprint of the
Taylor & Francis Group, an **informa** business

CRC Press
Taylor & Francis Group
6000 Broken Sound Parkway NW, Suite 300
Boca Raton, FL 33487-2742

International Standard Book Number-13: 978-1-4665-8310-8 (Pack - Book and Ebook)

Visit the Taylor & Francis Web site at
http://www.taylorandfrancis.com

and the CRC Press Web site at
http://www.crcpress.com

Printed and bound in Great Britain by
TJ International Ltd, Padstow, Cornwall

Contents

Preface

Global Positioning System (GPS) is now a part of a growing international context—the Global Navigation Satellite System (GNSS). One definition of GNSS embraces any constellation of satellites providing signals from space that facilitate autonomous positioning, navigation, and timing on a global scale. Currently, there is one other system besides GPS that satisfies this definition. It is the Russian Federation's GLONASS, an acronym for Globalnaya Navigationnaya Sputnikovaya Sistema. There are several other satellite systems in development that will likely reach similar capability soon and will take their place within the GNSS definition. In fact, some of them are considered by many to already fall under the GNSS label. These include the Galileo system administered by the European Union, the Chinese BeiDou Navigation Satellite System (BDS), and regional systems such as the Indian Regional Navigation Satellite System (IRNSS) and the Japanese Quasi-Zenith Satellite System (QZSS) are often included as well. Much of the information presented here applies equally to both GPS and the larger GNSS context, and GNSS is mentioned throughout this book. The emphasis on GPS modernization and GNSS is most prominent in Chapter 8.

The changes in GPS hardware, software, and procedures are accelerating. Keeping up is a real challenge. My hope in offering this fourth edition of *GPS for Land Surveyors* is to contribute some small assistance in that process.

This book has been written to find a middle ground. It is intended to be neither simplistic nor overly technical but an introduction to the concepts needed to understand and use GPS, not a presentation of the latest research in the area. An effort has been made to explain the progression of the ideas at the foundation of satellite positioning and delve into some of the particulars.

This is a practical book, a guide to some of the techniques used in GPS, from their design through observation, processing, real-time kinematic, and real-time networks. Today, some of the aspects of satellite navigation are familiar and some are not. This book is about making them all familiar.

1 Global Positioning System (GPS) Signal

GPS SIGNAL STRUCTURE

GPS AND TRILATERATION

GPS can be compared to trilateration. Both techniques rely exclusively on the measurement of distances to fixed positions. One of the differences between them, however, is that the distances, called ranges in GPS, are not measured to control points on the surface of the Earth. Instead, they are measured to satellites orbiting in nearly circular orbits at a nominal altitude of about 20,000 km above the Earth.

Passive System

The ranges are measured with signals that are broadcast from the GPS satellites to the GPS receivers in the microwave part of the electromagnetic spectrum; this is sometimes called a passive system. GPS is passive in the sense that only the satellites transmit signals; the users simply receive them. As a result, there is no limit to the number of GPS receivers that may simultaneously monitor the GPS signals. Just as millions of television sets may be tuned to the same channel without disrupting the broadcast, millions of GPS receivers may monitor the satellite's signals without danger of overburdening the system. This is a distinct advantage, but as a result, GPS signals must carry a great deal of information. A GPS receiver must be able to gather all the information it needs to determine its own position from the signals it collects from the satellites.

Time

Time measurement is essential to GPS surveying in several ways. For example, the determination of ranges, like distance measurement in a modern trilateration survey, is done electronically. In both cases, distance is a function of the speed of light, an electromagnetic signal of stable frequency and elapsed time. In a trilateration survey, frequencies generated within an electronic distance measuring (EDM) device can be used to determine the elapsed travel time of its signal because the signal bounces off a reflector and returns to where it started. However, the signals from a GPS satellite do not return to the satellite; they travel one way, to the receiver. The satellite can mark the moment the signal departs, and the receiver can mark the moment it arrives, and because the measurement of the ranges in GPS depends on the measurement of the time it takes a GPS signal to make the trip, the elapsed time must be determined by decoding the GPS signal itself.

Control

Both GPS surveys and trilateration surveys begin from control points. In GPS the control points are the satellites themselves; therefore, knowledge of the satellite's position is critical. Measurement of a distance to a control point without knowledge of that control point's position would be useless. It is not enough that the GPS signals provide a receiver with information to measure the range between itself and the satellite. That same signal must also communicate the position of the satellite at that very instant. The situation is complicated somewhat by the fact that the satellite is always moving with respect to the receiver at a speed of approximately 4 km/s. In a GPS survey, as in a trilateration survey, the signals must travel through the atmosphere. In a trilateration survey, compensation for the atmospheric effects on an EDM signal, estimated from local observations, can be applied at the signal's source. This is not possible in GPS. The GPS signals begin in the virtual vacuum of space, but then, after hitting the Earth's atmosphere, they travel through much more of the atmosphere than EDM signals.

Therefore, the GPS signals must give the receiver some information about needed atmospheric corrections.

It takes more than one measured distance to determine a new position in a trilateration survey or in a GPS survey. Each of the several distances used to define one new point must be measured to a different control station. For trilateration, three distances are adequate for each new point. For a GPS survey, the minimum requirement is a measured range to each of at least four GPS satellites.

Just as it is vital that every one of the three distances in a trilateration is correctly paired with the correct control station, the GPS receiver must be able to match each of the signals it tracks with the satellite of its origin. Therefore, the GPS signals themselves must also carry a kind of satellite identification. To be on the safe side, the signal should also tell the receiver where to find all the other satellites as well.

To sum up, a GPS signal must somehow communicate to its receiver: (1) what time it is on the satellite, (2) the instantaneous position of a moving satellite, (3) some information about necessary atmospheric corrections, and (4) some sort of satellite identification system to tell the receiver where it came from and where the receiver may find the other satellites. How does a GPS satellite communicate all that information to a receiver? It uses codes.

Codes

GPS codes are binary, zeroes and ones, the language of computers. There are three basic legacy codes in GPS that have been around since the beginning of the system. They are the Precise code, or P(Y) code, the Coarse/Acquisition (C/A) code, and the Navigation (NAV) code.

There are also new codes now being broadcast by some of the GPS satellites. Among these new codes are the M code, the L2C code, and the L1C code. These new codes have not yet reached full operational capability.

All these codes contain the information GPS receivers need to function, but they must travel from the GPS satellites to the receivers to deliver it. Carrier waves provide the conveyance the codes need to make the trip.

In order to transport information a carrier wave has to have at least one characteristic that can be changed or *modulated*. The characteristic can be the carrier waves phase, amplitude, or frequency. For example, the information, music, or speech received from an AM radio station is encoded onto the carrier wave by amplitude modulation, and the information on the signal from an FM radio station is there because of frequency modulation. GPS carriers use phase modulation to do the job. Looking at some of the characteristics of waves is a good start toward understanding how that works.

Wavelength and Frequency

A wavelength with duration of 1 s, known as 1 cycle per second, has a frequency of 1 Hertz (Hz) in the International System of Units (SI). A frequency of 1 Hz is rather low. The lowest sound human ears can detect has a frequency of about 25 Hz. The highest is about 15,000 Hz, or 15 kilohertz (KHz). Most of the modulated carriers used in EDMs and all those in GPS have frequencies that are measured in units of a million cycles per second, or megahertz, MHz.

The three fundamental frequencies assigned to GPS carriers come from a part of the electromagnetic spectrum known as the L-band. The L-band includes high frequencies from approximately 390 to 1550 MHz with wavelengths from 15 to 30 cm. The two legacy GPS carriers are L1 at 1575.42 MHz, which is just a little above the L-band range, and L2 at 1227.60 MHz. There is also a third, relatively new, carrier being broadcast by some GPS satellites. It is known as L5, and its frequency is 1176.45 MHz.

NAV Messages

Getting back to codes and the information they transport, one of the most important codes is known as the NAV code. It is the primary vehicle for communicating the NAV message to GPS receivers. The NAV message is also known as the GPS message. It includes some of the information the receivers need to determine positions (Figure 1.1). Today, there are several NAV messages being broadcast by GPS satellites, but we will look at the oldest of them first.

The legacy NAV message continues to be one of the mainstays on which GPS relies. The NAV code is broadcast at a low frequency of 50 Hz on both the L1 and the L2 GPS carriers. It carries information about the location of the GPS satellites called the ephemeris and data used in both time conversions and offsets called clock corrections. Both GPS satellites and receivers have clocks on board. The NAV also communicates the health of the satellites on orbit and information about the ionosphere. The ionosphere is a layer of atmosphere through which the GPS signals must travel to get to the user. It includes data called almanacs that provide a GPS receiver with enough little snippets of ephemeris information to calculate the coordinates of all the satellites in the constellation with an approximate accuracy of a couple of kilometers.

Here are some of the parameters of the design of the legacy NAV message.

The master frame of the NAV message contains 25 frames, or pages, of data. Each of the 25 frames is 1500 bits long and is divided into five subframes. Each of the five subframes contains 10 words, and each word is comprised of 30 bits. In other words,

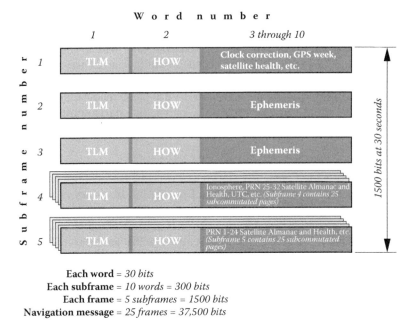

Word   number

1   *2*   *3 through 10*

Each word = *30 bits*
Each subframe = *10 words = 300 bits*
Each frame = *5 subframes = 1500 bits*
Navigation message = *25 frames = 37,500 bits*

FIGURE 1.1   One frame of the legacy NAV message.

each of the five subframes has 300 bits and because there are five of them in each of the 25 frames, the entire NAV message contains 37,500 bits. At a rate of 50 bits per second it takes 12.5 min to broadcast and to receive the entire message. However, it only takes 6 seconds to collect a single subframe and 30 s to pick up the whole frame.

The first two words in every subframe are the telemetry (TLM) word and the handover word (HOW). The first 8 bits in the telemetry word contain synchronization information that is used by the receiver to integrate itself with the NAV message and decode it. The handover word contains the truncated Z-count. The Z-count is one of the primary units for GPS Time and can be considered a counter that is incremented at the beginning of each consecutive subframe and thereby provides the receiver with the instant at the leading edge of the next data subframe.

Frame to frame, the information contained in subframes 1, 2, and 3 is repeated every 30 s. The information in subframes 4 and 5 changes. These two subframes 4 and 5 are comprised of 25 subcommutated pages. That means that each of these pages of subframe 4 and 5 contain different information, and a receiver must collect all 25 frames to get all the data. The receiver gets the first of the 25 pages on the first time through the five subframes. It gets the second of the 25 pages on the second time through the subframes and so on. On the good side, because subframes 4 and 5 contain the information for all satellites, a receiver need only lock on to one satellite to be able to pick up the almanac data for the whole constellation.

Unfortunately, the accuracy of some aspects of the information included in the NAV message deteriorates with time. Translated into the rate of change in the three-dimensional position of a GPS receiver, it is about 4 cm/min. Therefore, mechanisms are in place to prevent the message from getting too old. For example, every two hours the data in subframes 1, 2, and 3, the ephemeris and clocks parameters, are updated. The data in subframes 4 and 5, the almanacs, are renewed every six days. These updates are provided by the government uploading facilities around the world that are known, along with their tracking and computing counterparts, as the Control Segment. The information sent to each satellite from the Control Segment makes its way through the satellites and back to the users in the NAV message.

There are new NAV messages coming into play. There are four of them. The content and format of the three new civil messages (L2-CNAV, CNAV-2, L5-CNAV) and one military message (MNAV) are improved compared with the legacy NAV. In general, these NAV messages are more flexible and robust. They are also transmitted at a higher rate than the legacy NAV.

P AND COARSE/ACQUISITION CODES

Like the NAV code, the P and C/A codes are designed to carry information from the GPS satellites to the receivers. Also like the NAV code, they are modulated onto carrier waves as are the new M, L2C, and L1C codes. However, unlike the NAV code, these codes are not vehicles for broadcasting information that has been uploaded by the Control Segment; instead, they carry the raw data from which GPS receivers derive their position and time measurements.

Pseudorandom Noise

The P and C/A codes are complicated; so complicated that they appear to be noise at first. In fact, they are known as pseudorandom noise (PRN) codes. Actually, they are carefully designed. They have to be. They must be capable of repetition and replication.

P Code

The P code is called the Precise code. It is a particular series of ones and zeroes generated at a rate of 10.23 million bits per second. It is carried on both L1 and L2, and it is very long, 37 weeks (2×10^{14} bits in code). Each GPS satellite is assigned a part of the P code all its own and then repeats its portion every seven days. This assignment of one particular week of the 37 week long P code to each satellite helps a GPS receiver distinguish one satellite's transmission from another. For example, if a satellite is broadcasting the fourteenth week of the P code, it must be Space Vehicle 14. The encrypted P code is called the P(Y) code.

There is a flag in subframe 4 of the NAV message that tells a receiver when the P code is encrypted into the P(Y) code. This security system has been activated by the Control Segment since January 1994. It is done to prevent spoofing from working. Spoofing is generation of false transmissions masquerading as the Precise code. This countermeasure called anti-spoofing (AS) is accomplished by the modulation of a W code to generate the more secure Y code that replaces the P code. Commercial GPS receiver manufacturers are not authorized to use the P(Y) code directly. Therefore, most have developed proprietary techniques both for carrier wave and pseudorange measurements on L2 indirectly. Dual-frequency GPS receivers must also overcome AS.

C/A Code

The C/A code is also a particular series of ones and zeroes, but the rate at which it is generated is 10× slower than the P(Y) code. The C/A code rate is 1.023 million bits per second. Here satellite identification is quite straightforward. Not only does each GPS satellite broadcast its own completely unique 1023 bit C/A code, but it also repeats its C/A code every millisecond. The legacy C/A code is broadcast on L1 only. It used to be the only civilian GPS code but no longer. The legacy C/A has been joined by a new civilian signal known as L2C that is carried on L2.

Standard Positioning Service and Precise Positioning Service

Still the C/A code is the vehicle for the Standard Positioning Service (SPS), which is used for most civilian surveying applications. The P(Y) code, on the other hand, provides the same service for Precise Positioning Service (PPS). The idea of SPS and PPS was developed by the Department of Defense many years ago. SPS was designed to provide a minimum level of positioning capability considered consistent with national security, ±100 m, 95 percent of the time, when intentionally degraded through Selective Availability (SA).

Selective Availability, the intentional dithering of the satellite clocks by the Department of Defense, was instituted in 1989 because the accuracy of the C/A point positioning as originally rolled out was too good! As mentioned above, the accuracy was supposed to be ±100 m horizontally, 95 percent of the time with a vertical accuracy of about ±175 m. However, in fact, it turned out that the C/A code point positioning gave civilians access to accuracy of about ±20 to ±40 m. That was not according to plan, so the satellite clocks' accuracy was degraded on the C/A code. The good news is that the intentional error source called SA is gone now. It was switched off on May 2, 2000, by presidential order. Actually, SA never did hinder the surveying applications of GPS (more about that in Chapter 3). However, satellite clock errors, the unintentional kind, still contribute error to GPS positioning.

PPS is designed for higher positioning accuracy and was originally available only to users authorized by the Department of Defense; that has changed somewhat and will be discussed in this chapter. It used to be that the P(Y) code was the only military code. That is no longer the case. It has been joined by a new military signal called the M code.

GPS Time

We will return to codes and how they are modulated onto the carrier waves in a moment, but to help in understanding code generation, it is useful first to cover an aspect at the foundation of that process, time. Actually, it is fundamental to the whole Global Positioning System.

For example, there is time-sensitive information is the NAV message in both subframes 1 and 4. The information in subframe 4 helps a receiver relate two different time standards to one another. One of them is GPS Time and the other is Coordinated Universal Time (UTC).

GPS Time is the time standard of the GPS system. It is also known as GPS System Time (GPST). UTC is the time standard for the world. The rates of these two standards are virtually the same. Specifically, the rate of GPS Time is kept within 1 μs, and usually less than 25 ns, of the rate of UTC as it is determined by the U.S. Naval Observatory Master Clock (USNO MC). The exact difference is in two constants, A0 and A1, in the NAV message that give the time difference and rate of system time against UTC.

The rate of UTC itself is carefully determined. It is steered by about 65 timing laboratories and hundreds of atomic clocks around the world and is remarkable in its stability. In fact, it is more stable than the rotation of the Earth itself, such that UTC and the rotation gradually get out of sync with one another. Therefore, in order to keep the discrepancy between UTC and the Earth's actual motion under 0.9 seconds corrections of 1 second, called leap seconds, are periodically introduced into UTC. In other words, the rate of UTC is consistent and stable all the time, but the numbers denoting the moment of time changes whenever a 1 second leap second is introduced.

However, leap seconds are not used in GPS Time. It is a continuous time scale. Nevertheless, there was a moment when GPS Time was identical to UTC. It was midnight, January 6, 1980. Since then many leap seconds have been added to UTC, but none have been added to GPS Time. So even though their rates are virtually identical, the numbers expressing a particular instant in GPS Time are different by some seconds from the numbers expressing the same instant in UTC. For example, GPS Time was 16 s ahead of UTC on July 1, 2012.

Information in subframe 4 of the NAV message includes the relationship between GPS Time and UTC, and it also notes future scheduled leap seconds. In this area, subframe 4 can accommodate 8 bits, 255 leap seconds, which should suffice until about 2330. The NAV message also contains information the receiver needs to come close to correlating its clock with that of the clock on the satellite. However, because the time relationships in GPS are changing constantly, they can only be partially defined in these subframes. It takes more than a portion of the NAV message to define those relationships to the necessary degree of accuracy.

Subframe 1 contains some information on the health of the satellite, an estimate of the signal dispersion due to the ionosphere, and the GPS week number (more details in Satellite Health). Most importantly, subframe 1 contains coefficients that are used to estimate the difference between the satellite clock and GPS Time.

Satellite Clocks

Designs of GPS satellites have changed over the years, but within each generation of satellites there have also been similar characteristics. Those generations are called blocks. For example, the early satellites were known as Block I. The next satellites on orbit were Blocks II and IIA. Even amid the changes in satellite design, one of the common elements from the very beginning of the system is that each GPS satellite has carried its own onboard clocks. These clocks are also known as time or frequency standards. They are in the form of very stable and accurate atomic clocks that are now and have in the past been regulated by the vibration frequencies of the atoms of two elements, cesium and rubidium. There were three rubidium clocks on the first three GPS satellites. At the time, cesium clocks were performing a bit better than rubidium so subsequent satellites had four clocks, three rubidium and one cesium. As mentioned, these early satellites were known as Block I. The Block II and IIA satellites carry four clocks as well, but there are two rubidium and two cesium.

On the subsequent Block IIR, IIR-M, and IIF satellites, there are three improved Rubidium Atomic Frequency Standard (RAFS) clocks and they are all regulated by rubidium. These improved clocks have proven to be excellent performers. There has been some consideration of equipping the upcoming Block IIF satellites with hydrogen maser clocks, which will have even higher precision than the rubidium clocks.

The operation of the clocks in any one satellite is completely independent from the operation of the clock in any other satellite, so the government tracking stations of the Control Segment monitor all the satellite clocks. The Control Segment could continuously adjust so all the satellite clocks would stay in lockstep with the GPS Time standard, but that is not what is done. Constantly tweaking the satellite's onboard clocks would tend to reduce their useful life. Instead, each clock is allowed to drift up to 1 ms from GPS Time and their individual drifts are recorded and uploaded into subframe 1 of each satellite's NAV message. These deviations from GPS Time available in the NAV message are known as the broadcast clock correction. Given the broadcast clock correction a GPS receiver can relate the satellite's clock to GPS Time. This is part of the solution to the problem of directly relating the receiver's own clock to the satellite's clock. However, the receiver will need to rely on other aspects of the GPS signal for a complete time correlation. Further, the drift of each satellite's clock is not constant. Nor can the broadcast clock correction updates, which occur about every two hours, be frequent enough to completely define the drift. Therefore, one of the 10 words included in subframe 1 provides a definition of the reliability of the broadcast clock correction. This is called IODC, or Issue of Data Clock. It gives the issue number of the data set for the clock correction and provides a way to identify the currency of the correction.

GPS Week

Subframe 1 contains information on the GPS week. It is worthwhile to mention that in GPS weeks are counted consecutively. The first GPS week, GPS week 0, began at 00 hour UTC on January 6, 1980, and ended on January 12, 1980. It was followed by GPS week 1, GPS week 2, and so on. In Table 1.1 the first and second GPS weeks in January 2018 are shown as calculated from this initial second the column heading GPS Week from 1/6/1980.

However, about 19.6 years later, at the end of GPS week 1023 (August 15–21, 1999), it was necessary to start the numbering again at 0. This necessity accrued from the fact that the following week, August 22–28, 1999, would have been GPS week 1024 and that would have been beyond the capacity of the GPS week field in the legacy NAV message. The GPS week field was only 10 bits, and the largest week count a 10-bit field can accommodate is 1024. Therefore, a rollover was required at 00 hour on August 22, 1999. However, in UTC, taking into account leap seconds, the moment was 23:59:47 on August 21, 1999. In any case, the GPS week consecutive numbering began again at GPS week 0. In Table 1.1 the first and second GPS weeks in January 2018 are shown as calculated from this second beginning under the column heading GPS Week from 8/21/1999.

To alleviate this problem in the future, the modernized messages L2-CNAV, CNAV-2, L5-CNAV, and MNAV have 13-bit field for the GPS week count. This means that the GPS week will not need to roll over again for about 157 years.

TABLE 1.1
Seconds, Days, and Weeks

January 2018

Gregorian Date	Day	GPS Week from 01/06/1980	Day of Week	GPS Week from 08/21/1999	Seconds from Week at Midnight on that Day	Day of the Year (Julian Day)	Modified Julian Day	Julian Day (CE)
1	Monday	1982	1	958	86400	1	58119	2458119.5
2	Tuesday	1982	2	958	172800	2	58120	2558120.5
3	Wednesday	1982	3	958	259200	3	58121	2458121.5
4	Thursday	1982	4	958	345600	4	58122	2458122.5
5	Friday	1982	5	958	432000	5	58123	2458123.5
6	Saturday	1982	6	958	518400	6	58124	2458124.5
7	Sunday	1983	0	959	0	7	58125	2458125.5
8	Monday	1983	1	959	86400	8	58126	2458126.5
9	Tuesday	1983	2	959	172800	9	58127	2458127.5
10	Wednesday	1983	3	959	259200	10	58128	2458128.5

Julian Date

Here is a little more concerning dates. It is usual in GPS practice to define particular dates in a sequential manner from the first of the year. For example, most practitioners of GPS use the term *Julian date* to mean the day of the year counted consecutively from January 1 of the current year. The day of the year is also known as the ordinal date. With this method, January 1 is day 1 and December 31 is day 365, excepting in leap years. In Table 1.1 the first 10 days of January 2018 are illustrated. As expected, the days of the year are shown as 1 through 10. However, if the Julian dates were taken literally, they would be counted from noon on January 1, 4713 BC. Those are the dates on the far right in Table 1.1 under the heading Julian Day (CE).

The modified Julian date is sometimes referenced. It is in the column adjacent and is found by simply subtracting 2400000.5 from the Julian day.

Broadcast Ephemeris

Another example of time-sensitive information is found in subframes 2 and 3 of the NAV message. They contain information about the position of the satellite with respect to time. This is called the satellite's ephemeris. The ephemeris that each satellite broadcasts to the receivers provides information about its position relative to the Earth. Most particularly, it provides information about the position of the satellite antenna's phase center. The ephemeris is given in a right ascension (RA) system of coordinates. There are six orbital elements; among them are the size of the orbit, that is its semimajor axis a, and its shape, that is the eccentricity e. However, the orientation of the orbital plane in space is defined by other things, specifically the RA of its ascending node, Ω, and the inclination of its plane, i. These parameters along with the argument of the perigee, ω, and the description of the position of the satellite on the orbit, known as the true anomaly, provides all the information the user's computer needs to calculate Earth-centered, Earth-fixed, World Geodetic System 1984, GPS Week 1762 (WGS84 [G1762]) coordinates of the satellite at any moment. See Figure 1.2.

The broadcast ephemeris, however, is far from perfect. It is expressed in parameters named for the seventeenth century German astronomer Johann Kepler. The ephemerides may appear Keplerian, but in this case, the orbits of the GPS satellites deviate from nice smooth elliptical paths because they are unavoidably perturbed by gravitational and other forces. Therefore, their actual paths through space are found in the result of least squares, curve-fitting analysis of the satellite's orbits.

The accuracies of both the broadcast clock correction and the broadcast ephemeris deteriorate with time. As a result, one of the most important parts of this portion of the NAV message is called IODE. IODE is an acronym that stands for Issue of Data Ephemeris, and it appears in both subframes 2 and 3. See Figure 1.3.

Atmospheric Correction

Subframe 4 addresses atmospheric correction. As with subframe 1, the data there offer only a partial solution to a problem. The Control Segment's monitoring stations find the apparent delay of a GPS signal caused by its trip through the ionosphere through an analysis of the different propagation rates of the carrier frequencies

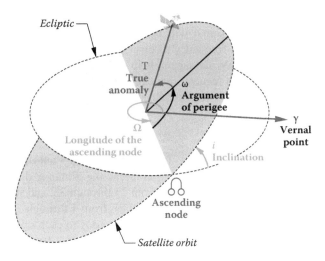

FIGURE 1.2   GPS ephemeris elements.

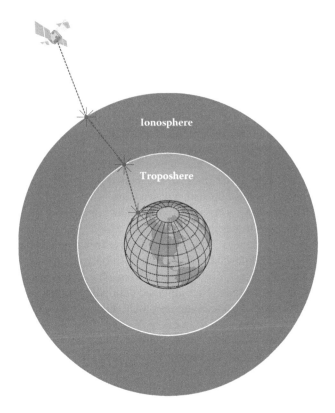

FIGURE 1.3   Atmospheric delay.

broadcast by GPS satellites, L1, L2, and L5. These frequencies and the effects of the atmosphere on the GPS signal will be discussed in this book. For now, it is sufficient to say that a single-frequency receiver depends on the ionospheric correction in subframe 4 of the NAV message to help remove part of the error introduced by the atmosphere, whereas a receiver that can track more than one carrier has a bit of an advantage by comparing the differences in the frequency-dependent propagation rates.

Almanac

As mentioned briefly, the almanac data in subframes 4 and 5 tell the receiver where to find all the GPS satellites. Subframe 4 contains the almanac data for satellites with pseudorandom noise (PRN) numbers from 25 through 32, and subframe 5 contains almanac data for satellites with PRN numbers from 1 through 24. The Control Segment generates and uploads a new almanac every day to each satellite.

While a GPS receiver must collect a complete ephemeris from each individual GPS satellite to know its correct orbital position, it is convenient for a receiver to be able to have some information about where all the satellites in the constellation are by reading the almanac from just one of them. The almanacs are much smaller than the ephemerides because they contain coarse orbital parameters and incomplete ephemerides, but they are still accurate enough for a receiver to generate a list of visible satellites at power-up. They, along with a stored position and time, allow a receiver to find its first satellite.

On the one hand, if a receiver has been in operation recently and has some left over almanac and position data in its nonvolatile memory from its last observations, it can begin its search with what is known as a warm start, which is known as a normal start. In this condition the receiver might begin by knowing the time within about 20 s and its position within 100 km or so. This approximate information helps the receiver estimate the range to satellites. For example, it will be able to restrict its search for satellites to those likely above its horizon rather than wasting time on those below it. Limiting the range of the search decreases the time to first fix (TTFF). It can be as short as 30 s with a warm start.

On the other hand, if a receiver has no previous almanac or ephemeris data in its memory, it will have to perform a cold start, which is also known as a factory start. Without previous data to guide it, the receiver in a cold start must search for all the satellites without knowledge of its own position, velocity, or the time. When it does finally manage to acquire the signal from one, it gets some help and can begin to download an almanac. That almanac data will contain information about the approximate location of all the other satellites. The period needed to receive the full information is 12.5 min.

The TTFF is longest at a cold start, less at warm, and least at hot. A receiver that has a current almanac, a current ephemeris, time, and position can have a hot start. A hot start can take from 0.5 to 20 s.

Estimating how long each type of start will actually take is difficult. Overhead obstructions interrupting the signal from the satellites and the GPS signals reflecting from nearby structures can delay the loading of the ephemeris necessary to lock onto the satellite's signals.

Satellite Health

Subframe 1 contains information about the health of the satellite the receiver is tracking when it receives the NAV message and allows it to determine if the satellite is operating within normal parameters. Subframes 4 and 5 include health data of all the satellites, data that are periodically uploaded by the Control Segment. These subframes inform users of any satellite malfunctions before they try to use a particular signal. The codes in these bits may convey a variety of conditions. They may tell the receiver that all signals from the satellite are good and reliable or that the receiver should not currently use the satellite because there may be tracking problems or other difficulties. They may even tell the receiver that the satellite will be out of commission in the future; perhaps it will be undergoing a scheduled orbit correction. GPS satellites health is affected by a wide variety of breakdowns, particularly clock trouble. That is one reason they carry multiple clocks.

Telemetry and Handover Words

As mentioned, each of these five subframes begins with the same two words: the telemetry (TLM) word and the handover word (HOW). Unlike nearly everything else in the NAV message, these two words are generated by the satellite itself. As shown in the column headed Seconds of the Week at Midnight on that Day in Table 1.1, GPS Time restarts each Sunday at midnight (0:00 o'clock). These data contain the time since last restart of GPS Time on the previous Sunday 0:00 o'clock.

The TLM is the first word in each subframe. It indicates the status of uploading from the Control Segment while it is in progress and contains information about the age of the ephemeris data. It also has a constant unchanging 8-bit preamble of 10001011, and a string helps the receiver reliably find the beginning of each subframe.

The HOW provides the receiver information on the time of the GPS week (TOW) and the number of the subframe, among other things. For example, the HOW's Z count (an internally derived 1.5 s epoch) tells the receiver exactly where the satellite stands in the generation of positioning codes. In fact, the handover word actually helps the receiver go from tracking the C/A code to tracking the P(Y) code, the primary GPS positioning codes. It is used by military receivers.

PRODUCTION OF A MODULATED CARRIER WAVE

All the codes mentioned come to a GPS receiver on a modulated carrier; therefore, it is important to understand how a modulated carrier is generated. The signal created by an electronic distance meter (EDM) in a total station is a good example of a modulated carrier.

EDM Ranging

As mentioned, an EDM only needs one frequency standard because its electromagnetic wave travels to a retroprism and is reflected back to its point of origin. The EDM is both the transmitter and the receiver of the signal. Therefore, in general terms, the instrument can take half the time elapsed between the moment of transmission and the moment of reception, multiply by the speed of light, and find the distance between itself and the retroprism (Distance = Elapsed Time × Rate).

Illustrated in Figure 1.4, the fundamental elements of the calculation of the distance measured by an EDM, ρ, are the time elapsed between transmission and reception of the signal, Δt, and the speed of light, c.

$$\text{distance} = \rho$$

$$\text{elapsed time} = \Delta t$$

$$\text{rate} = c$$

GPS Ranging

The one-way ranging used in GPS is more complicated. It requires the use of two clocks. The broadcast signals from the satellites are collected by the receiver, not reflected. Nevertheless, in general terms, the full time elapsed between the instant a GPS signal leaves a satellite and arrives at a receiver, multiplied by the speed of light, is the distance between them.

Unlike the wave generated by an EDM, a GPS signal cannot be analyzed at its point of origin. The measurement of the elapsed time between the signal's transmission by the satellite and its arrival at the receiver requires two clocks, one in the satellite and one in the receiver. This complication is compounded because to correctly represent the distance between them, these two clocks need to be perfectly synchronized with one another. Because perfect synchronization is physically impossible, the problem is addressed mathematically.

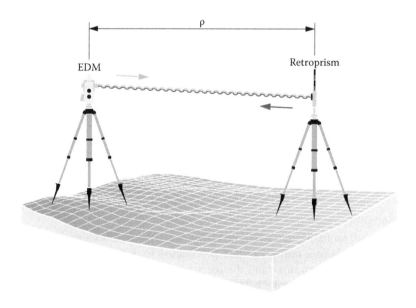

FIGURE 1.4   Two-way ranging.

In Figure 1.5, the basis of the calculation of a range measured from a GPS receiver to the satellite, ρ, is the multiplication of the time elapsed between a signal's transmission and reception, Δt, by the speed of light, c.

A discrepancy of 1 μs, 1 millionth of a second from perfect synchronization, between the clock aboard the GPS satellite and the clock in the receiver can create a range error of 300 m, far beyond the acceptable limits for nearly all surveying work.

Oscillators

The time measurement devices used in both EDM and GPS measurements are clocks only in the most general sense. They are more correctly called oscillators, or frequency standards. In other words, rather than producing a steady series of audible ticks, they keep time by chopping a continuous beam of electromagnetic energy at extremely regular intervals. The result is a steady series of wavelengths and the foundation of the modulated carrier.

For example, the action of a shutter in a movie projector is analogous to the modulation of a coherent beam by the oscillator in an EDM. Consider the visible beam of light of a specific frequency passing through a movie projector. It is interrupted by the shutter, half of a metal disk rotating at a constant rate that alternately blocks and uncovers the light. In other words, the shutter chops the continuous beam into equal segments. Each length begins with the shutter closed and the light beam entirely blocked. As the shutter rotates open, the light beam is gradually uncovered. It increases to its maximum intensity, and then decreases again as the shutter gradually closes. The light is not simply turned on and off; it gradually increases and decreases. In this analogy the light beam is the carrier, and it has a wavelength much shorter than the wavelength of the modulation of that carrier produced by the shutter. This modulation can be illustrated by a sine wave.

The wavelength begins when the light is blocked by the shutter. The first minimum is called a 0° phase angle. The first maximum is called the 90° phase angle and occurs when the shutter is entirely open. It returns to minimum at the 180° phase angle when the shutter closes again, but the wavelength is not yet complete. It

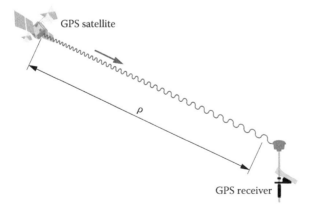

FIGURE 1.5   One-way ranging.

continues through a second shutter opening, 270°, and closing, 360°. The 360° phase angle marks the end of one wavelength and the beginning of the next one. The time and distance between every other minimum, that is from the 0° to the 360° phase angles, is a wavelength and is usually symbolized by the Greek letter lambda, λ.

As long as the rate of an oscillator's operation is very stable, both the length and elapsed time between the beginning and end of every wavelength of the modulation will be the same.

CHAIN OF ELECTROMAGNETIC ENERGY

GPS oscillators are sometimes called clocks because the frequency of a modulated carrier, measured in hertz, can indicate the elapsed time between the beginning and end of a wavelength, which is a useful bit of information for finding the distance covered by a wavelength. The length is approximately

$$\lambda = \frac{c_a}{f}$$

where
    λ = length of each complete wavelength in meters
    c_a = speed of light corrected for atmospheric effects
    f = frequency in hertz

For example, if an EDM transmits a modulated carrier with a frequency of 9.84 MHz and the speed of light is approximately 300,000,000 m/s (a more accurate value is 299,792,458 m/s, but the approximation 300,000,000 m/s will be used here for convenience), then

$$\lambda = \frac{c_a}{f}$$

$$\lambda = \frac{300,000,000 \text{ m/s}}{9,840,000 \text{ Hz}}$$

$$\lambda = 30.49 \text{ m}$$

The modulated wavelength would be about 30.49 m long, or approximately 100 feet.

The modulated carrier transmitted from an EDM can be compared to a Gunter's chain constructed of electromagnetic energy instead of wire links. Each full link of this electromagnetic chain is a wavelength of a specific frequency. The measurement between an EDM and a reflector is doubled with this electronic chain because, after it extends from the EDM to the reflector, it bounces back to where it started. The entire trip represents twice the distance and is simply divided by 2, but like the surveyors who used the old Gunter's chain, one cannot depend on a particular measurement ending conveniently at the end of a complete link (or wavelength). A measurement is much more likely to end at some fractional part of a link (or wavelength). The question is, Where?

Phase Shift

With the original Gunter's chain, the surveyor simply looked at the chain and estimated the fractional part of the last link that should be included in the measurement. Those links were tangible. Because the wavelengths of a modulated carrier are not, the EDM must find the fractional part of its measurement electronically. Therefore, it does a comparison. It compares the phase angle of the returning signal to that of a replica of the transmitted signal to determine the *phase shift*. That phase shift represents the fractional part of the measurement. This principle is used in distance measurement by both EDM and GPS systems.

How does it work? First, it is important to remember that points on a modulated carrier are defined by phase angles, such as 0°, 90°, 180°, 270°, and 360° (see Figure 1.6). When two modulated carrier waves reach exactly the same phase angle at exactly the same time, they are said to be *in phase, coherent,* or *phase locked*. However, when two waves reach the same phase angle at different times, they are *out of phase* or *phase shifted*.

For example, in Figure 1.7, the sine wave shown by the dashed line has returned to an EDM from a reflector. Compared with the sine wave shown by the solid line, it is out of phase by one-quarter of a wavelength. The distance between the EDM and the reflector ρ is then

$$p = \frac{(N\lambda + d)}{2}$$

where
    N = number of full wavelengths the modulated carrier completed
    d = fractional part of a wavelength at the end that completes the doubled distance

In this example, d is three-quarters of a wavelength because it lacks its last quarter, but how would the EDM know that? It knows because at the same time an external carrier wave is sent to the reflector, the EDM keeps an identical internal reference wave at home in its receiver circuits. In Figure 1.8, the external beam returned from the reflector is compared to the reference wave, and the difference in phase between the two can be measured.

Both EDM and GPS ranging use the method represented in this illustration. In GPS, the measurement of the difference in the phase of the incoming signal and the phase of the internal oscillator in the receiver reveals the small distance at the end of a range. In GPS, the process is called *carrier phase ranging*. As the name implies, the *observable* is the carrier wave itself.

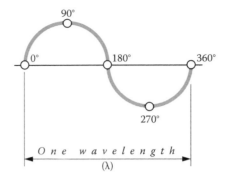

FIGURE 1.6   0° to 360° = one wavelength.

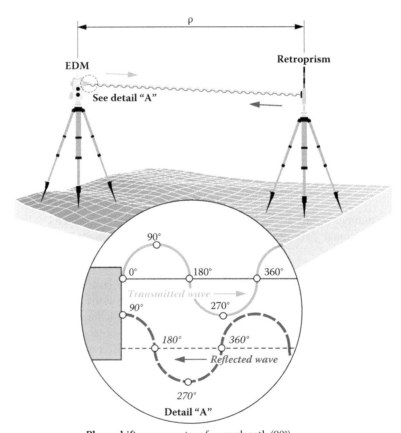

Phase shift—one quarter of a wavelength (90°)

FIGURE 1.7   An EDM measurement.

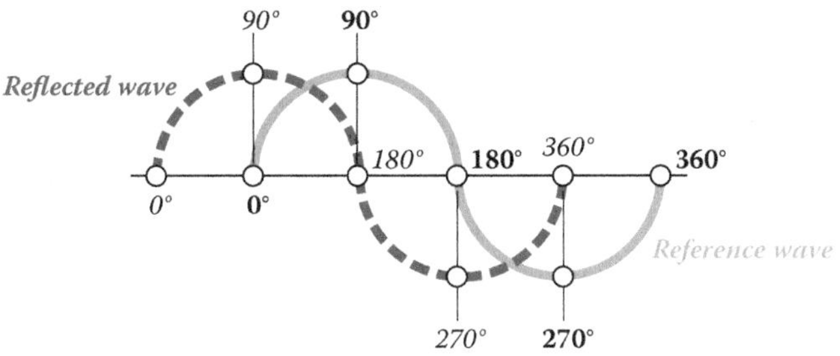

FIGURE 1.8   Reference and reflected waves.

TWO OBSERVABLES

The word observable is used throughout GPS literature to indicate the signals whose measurement yields the range or distance between the satellite and the receiver. The word is used to draw a distinction between the thing being measured, the observable, and the measurement, the observation.

In GPS, there are two types of observables: the *pseudorange* and the *carrier phase*. The latter, also known as the *carrier beat phase*, is the basis of the techniques used for high-precision GPS surveys. On the other hand, the pseudorange can serve applications when virtually instantaneous point positions are required or relatively low accuracy will suffice.

These basic observables can also be combined in various ways to generate additional measurements that have certain advantages. It is in this latter context that pseudoranges are used in many GPS receivers as a preliminary step toward the final determination of position by a carrier phase measurement. The foundation of pseudoranges is the correlation of code carried on a modulated carrier wave received from a GPS satellite with a replica of that same code generated in the receiver.

Most of the GPS receivers used for surveying applications are capable of code correlation. That is, they can determine pseudoranges from the C/A code or the P(Y) code. These same receivers are usually capable of determining ranges using the unmodulated carrier as well. However, first let us concentrate on the pseudorange.

ENCODING BY PHASE MODULATION

A carrier wave can be modulated in various ways. Radio stations use modulated carrier waves. The radio signals are AM, amplitude modulated, or FM, frequency modulated. When your radio is tuned to 105 FM, you are not actually listening to 105 MHz despite the announcer's assurances; 105 MHz is well above the range of human hearing. That frequency, 105 MHz, is the frequency of the carrier wave that is being modulated. It is when those modulations occur that make the speech and music intelligible. They come to you at a much slower frequency than the carrier wave.

The GPS carriers L1, L2, and L5 could have been modulated in a variety of ways to carry the binary codes, the 0s and 1s that are the C/A and P(Y) codes. Neither amplitude nor frequency modulation are used in GPS. It is the alteration of the phases of the carrier waves that encodes them. It is phase modulation that allows them to carry the codes from the satellites to the receivers.

One consequence of this method of modulation is that the signal can occupy a broader bandwidth than would otherwise be possible. The GPS signal is said to have a *spread spectrum* because of its intentionally increased bandwidth. In other words the overall bandwidth of the GPS signal is much wider than the bandwidth of the information it is carrying. In other words, while L1 is centered on 1575.42 MHz, L2 on 1227.60 MHz, and L5 on 1176.45 MHz, the width of these signals takes up a good deal of space on each side of these frequencies than might be expected. For example, the C/A code signal is spread over a width of 2.046 MHz or so, the P(Y) code signal is spread over a width about 20.46 MHz on L1, and the coming L1C signal will be spread over 4.092 MHz as shown in Figure 1.9.

This characteristic offers several advantages. It affords better signal-to-noise ratio, more accurate ranging, less interference, and increased security. However, spreading the spectral density of the signal also reduces its power so that the GPS signal is sometimes described as a 25-watt light bulb seen from 10,000 miles away. Clearly, the weakness of the signal makes it somewhat difficult to receive under cover of foliage.

In any case the most commonly used spread spectrum modulation technique is known as *binary phase shift keying* (BPSK). This is the technique used to create the NAV Message, the P(Y) code, and the C/A code. The *binary biphase* modulation is the switching from 0 to 1 and from 1 to 0 accomplished by phase changes of 180° in the carrier wave. Put another way at the moments when the value of the code must change from 0 to 1, or from 1 to 0, the change is accomplished by an instantaneous reverse of the phase of the carrier wave. It is flipped 180°. And each one of these flips occurs when the phase of the carrier is at the zero crossing. Each 0 and 1 of the binary code is known as a *code chip*. The 0 represents the normal state, and the 1 represents the *mirror image* state.

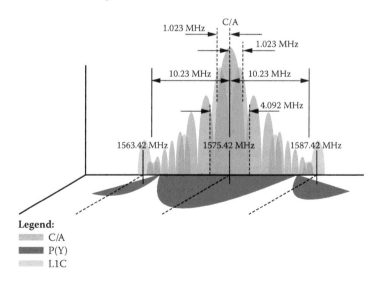

FIGURE 1.9   Spread spectrum on L1.

The rate of all of the components of GPS signals are multiples of the standard rate of the oscillators. The standard rate is 10.23 MHz. It is known as the *fundamental clock rate* and is symbolized F_0. For example, the GPS carriers are $154 \times F_0$, or 1575.42 MHz, $120 \times F_0$, or 1227.60 MHz, and $115 \times F_0$, or 1176.45 MHz. These represent L1, L2, and L5 respectively. Figure 1.10 illustrates the code modulation of the L1 carrier.

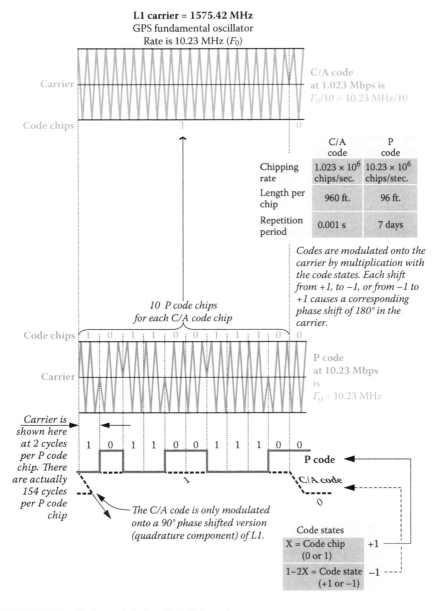

FIGURE 1.10   Code modulation of the L1 carrier.

The codes are also based on F_0: 10.23 code chips of the P(Y) code, 0s or 1s, occur every microsecond. In other words, the *chipping rate* of the P(Y) code is 10.23 million bits per second (Mbps), exactly the same as F_0, 10.23 MHz.

The chipping rate of the C/A code is 10× slower than the P(Y) code. It is one-tenth of F_0, 1.023 Mbps. Ten P(Y) code chips occur in the time it takes to generate one C/A code chip. This is a reason why a P(Y) code derived pseudorange is more precise than a C/A code pseudorange and why the C/A code is known as the *Coarse/Acquisition code*.

Even though both codes are broadcast on L1, they are distinguishable from one another by their transmission in *quadrature*. That means that the C/A code modulation on the L1 carrier is phase shifted 90° from the P(Y) code modulation on the same carrier.

PSEUDORANGING

Strictly speaking, a pseudorange observable is based on a time shift. This time shift can be symbolized by $d\tau$ and is the time elapsed between the instant a GPS signal leaves a satellite and the instant it arrives at a receiver. The concept can be illustrated by the process of setting a watch from a time signal heard over a telephone.

PROPAGATION DELAY

Imagine that a recorded voice said, "The time at the tone is 3 hours and 59 minutes." If a watch was set at the instant the tone was heard, the watch would be wrong. Supposing that the moment the tone was broadcast was indeed 3 hours and 59 minutes; the moment the tone is heard must be a bit later. It is later because it includes the time it took the tone to travel through the telephone lines from the point of broadcast to the point of reception. This elapsed time would be approximately equal to the length of the circuitry traveled by the tone divided by the speed of the electricity, which is the same as the speed of all electromagnetic energy, including light and radio signals. In fact, it is possible to imagine measuring the actual length of that circuitry by doing the division.

In GPS, that elapsed time is known as the propagation delay and it is used to measure length. The measurement is accomplished by a combination of codes. The idea is somewhat similar to the strategy used in EDMs, but where an EDM generates an internal replica of its *modulated carrier wave* to correlate with the signal it receives by reflection, a pseudorange is measured by a GPS receiver using a replica of a portion of the code that is modulated onto the carrier wave. The GPS receiver generates this replica itself, and it is used to compare with the code that is coming in from the satellite.

CODE CORRELATION

To conceptualize the process, one can imagine two codes generated at precisely the same time and identical in every regard: one in the satellite and one in the receiver. The satellite sends its code to the receiver, but on its arrival the codes do not line up even though they are identical. They do not correlate until the replica code in the receiver is time shifted a little bit. Once that is done, the receiver generated replica code fits the received satellite code. It is this time shift that reveals the propagation delay. The propagation delay is the time it took the signal to make the trip from the satellite to the receiver, $d\tau$. It is the same idea described above as the time it took the tone to travel through the telephone lines, except the GPS code is traveling through space and atmosphere. Once the time shift of the replica code is accomplished, the two codes match perfectly and the time the satellite signal spends in TRANSIT has been measured, well, almost.

It would be wonderful if that time shift could simply be divided by the speed of light and yield the true distance between the satellite and the receiver at that instant, and it is close. However, there are physical limitations on the process that prevent such a perfect relationship.

Autocorrelation

As mentioned, the almanac information from the NAV message of the first satellite a GPS receiver acquires tells it which satellites can be expected to come into view. With this information the receiver can load up pieces of the C/A codes for each of those satellites. Then the receiver tries to line up the replica C/A codes with the signals it is actually receiving from the satellites. The time required for correlation to occur is influenced by the presence and quality of the information in the almanac.

Actually lining up the code from the satellite with the replica in the GPS receiver is called *autocorrelation* and depends on the transformation of code chips into *code states*. The formula used to derive code states (+1 and −1) from code chips (0 and 1) is

$$\text{code state} = 1 - 2x$$

where x is the code chip value. For example, a normal code state is +1, and corresponds to a code chip value of 0. A mirror code state is −1 and corresponds to a code chip value of 1.

The function of these code states can be illustrated by asking three questions: First, if a tracking loop of code states generated in a receiver does not match code states received from the satellite, how does the receiver know? In that case, for example, the sum of the products of each of the receiver's 10 code states, with each of the code states from the satellite, when divided by 10, does not equal 1. Second, what does the receiver do when the code states in the receiver do not match the code states received from the satellite? It shifts the frequency of its search a little bit from the center of the L1 1575.42 MHz. This is done to accommodate the inevitable Doppler shift of the incoming signal since the satellite is always either moving toward or away from the receiver. The receiver also shifts its piece of code in time. These iterative small shifts in both time and frequency continue until the receiver code states do, in fact, match the signal from the satellite. Third, how does the receiver know when a tracking loop of replica code states does match code states from the satellite? In the case illustrated in Figure 1.11 the sum of the products of each code state of the receiver's replica 10, with each of the 10 from the satellite, divided by 10, is exactly 1.

In Figure 1.11, before the code from the satellite and the replica from the receiver are matched, the sum of the products of the code states is not 1. Following the

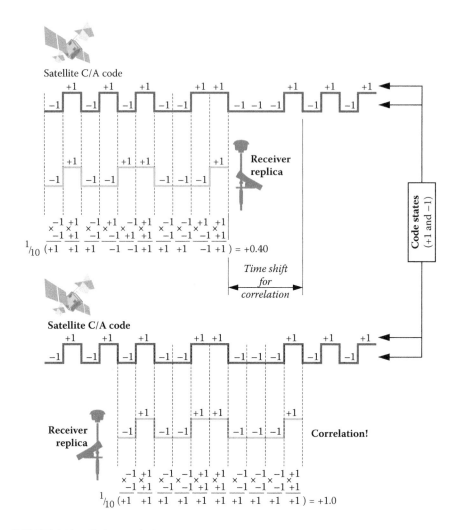

FIGURE 1.11   Code correlation.

correlation of the two codes the sum of the code states is exactly 1, and the receiver's replica code fits the code from the satellite like a key fits a lock.

CORRELATION PEAK

As shown in the somewhat idealized triangular plot in Figure 1.12, the condition of maximum correlation has been achieved at 1. That is when the receiver generated replica matches the code being received from the GPS satellite. In practice, the correlation peak is actually a bit rounded rather than being so emphatically triangular, but in any case, once the correlation peak has been reached it is then maintained by continual adjustment of the receiver generated code as described in this chapter.

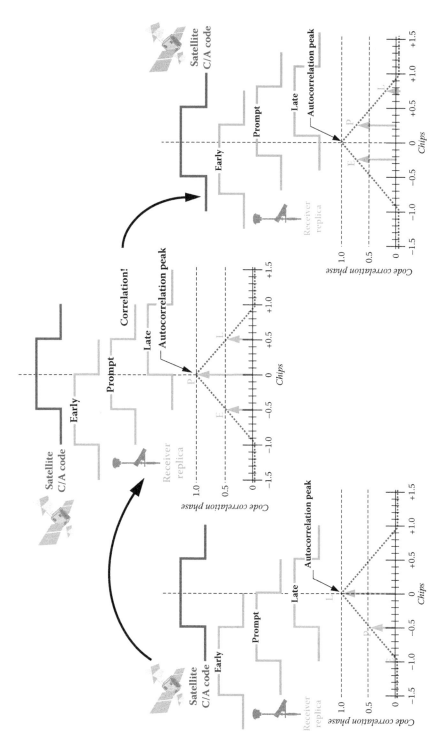

FIGURE 1.12 Correlation peak.

It is usual to find three correlators used for tracking. One is at the prompt (P), punctual, or on-time position with the other two symmetrically located somewhat early (E) and late (L). The separation between the early, prompt, and late code phases is always symmetrical. The separation is often 1 chip as illustrated here, but in a *narrow correlator* they may be separated by as little as 1/10th of a chip. Whatever the size of the separation, the symmetry is important to their purpose. That purpose is giving the receiver the information it needs to perform the shifts, the continual adjustments, alluded to above.

Because there is an equal distance between them when the receiver's replica code is aligned with the code from the satellite, the early and late correlators will have exactly the same amplitude on the triangle as shown in the middle of Figure 1.12. On the other hand, when the codes are not aligned, the early and late correlator amplitudes will be different. And, it is important to note that the symmetry ensures that the difference in their amplitudes will be unequal in proportion to the amount that the codes are out of phase with one another. Using this information the receiver can not only calculate the amount of the correlation error, but also whether the receiver's replica code is ahead (early) or behind (late) the incoming satellite code. Using this information it can correct the replica code generation to match the satellite's signal. Using this feedback loop correlation typically takes about 1 to 5 ms and does not exceed 20 ms. Please note that the same early, prompt, and late correlator approach is also utilized to align the receiver's replica carrier phase with the incoming satellite carrier phase.

LOCK AND THE TIME SHIFT

Once correlation of the two codes is achieved with a *delay lock loop* (DLL), it is maintained by a *correlation channel* within the GPS receiver, and the receiver is sometimes said to have *achieved lock* or to be *locked on* to the satellites. If the correlation is somehow interrupted later, the receiver is said to have *lost lock*. However, as long as the lock is present, the NAV message is available to the receiver. Remember that one of its elements is the broadcast clock correction that relates the satellite's onboard clock to GPS Time, and a limitation of the pseudorange process comes up.

Imperfect Oscillators

One reason the time shift $d\tau$ found in autocorrelation cannot quite reveal the true range ρ of the satellite at a particular instant is the lack of perfect synchronization between the clock in the satellite and the clock in the receiver. Recall that the two compared codes are generated directly from the fundamental rate F_0 of those clocks, and because these widely separated clocks, one on Earth and one in space, cannot be in perfect lockstep with one another, the codes they generate cannot be in perfect synch either. Therefore, a small part of the observed time shift $d\tau$ must always be due to the disagreement between these two clocks. In other words, the time shift not only contains the signal's TRANSIT time from the satellite to the receiver, it contains clock errors, too.

In fact, whenever satellite clocks and receiver clocks are checked against the carefully controlled GPS Time, they are found to be drifting a bit. Their oscillators are imperfect. It is not surprising that they are not quite as stable as the more than 150 atomic clocks around the world that are used to define the rate of GPS Time. They are subject to the destabilizing effects of temperature, acceleration, radiation, and other inconsistencies. As a result, there are two clock offsets that bias every satellite to receiver pseudorange observable. That is one reason it is called a pseudorange (see Figure 1.13).

A Pseudorange Equation

Clock offsets are only one of the errors in pseudoranges. Their relationship can be illustrated by the following equation (Fotopoulos 2000):

$$p = \rho + d_\rho + c(dt - dT) + d_{ion} + d_{trop} + \varepsilon_{mp} + \varepsilon_p$$

where
p = pseudorange measurement
ρ = true range
d_ρ = satellite orbital errors

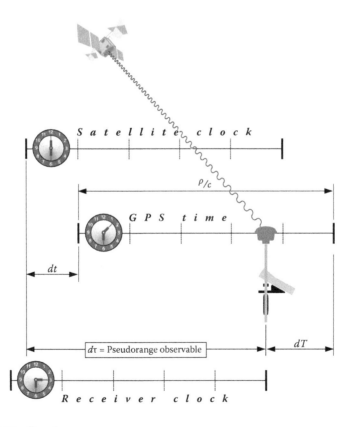

FIGURE 1.13   Pseudorange.

c = speed of light
dt = satellite clock offset from GPS Time
dT = receiver clock offset from GPS
d_{ion} = ionospheric delay
d_{trop} = tropospheric delay
ε_{mp} = multipath
ε_p = receiver noise

Please note that the pseudorange p and the true range ρ cannot be made equivalent without consideration of clock offsets, atmospheric effects, and other biases that are inevitably present. This discussion of time can make it easy to lose sight of the real objective, which is the position of the receiver. Obviously, if the coordinates of the satellite and the coordinates of the receiver were known perfectly, then it would be a simple matter to determine time shift $d\tau$ or find the true range ρ between them.

In fact, receivers placed at known coordinated positions can establish time so precisely they are used to monitor atomic clocks around the world. Several receivers simultaneously tracking the same satellites can achieve resolutions of 10 ns or better. Also, receivers placed at known positions can be used as base stations to establish the relative position of receivers at unknown stations, a fundamental principle of most GPS surveying.

It can be useful to imagine the true range term ρ, also known as the *geometric range*, actually includes the coordinates of both the satellite and the receiver. However, they are hidden within the measured value, the pseudorange p, and all of the other terms on the right side of the equation. The objective then is to mathematically separate and quantify these biases so that the receiver coordinates can be revealed. Clearly, any deficiency in describing, or *modeling*, the biases will degrade the quality of the final determination of the receiver's position, but even if the biases were modeled completely, there would be a limit to how correctly a pseudorange could represent the range between a satellite and a receiver.

One Percent Rule of Thumb

Here is a convenient approximation. The maximum resolution available in a pseudorange is about 1 percent of the chipping rate of the code used, whether it is the P(Y) code or the C/A code. In practice, positions derived from these codes are rather less reliable than described by this approximation; nevertheless, it offers a basis to evaluate pseudoranging, in general, and compare the potentials of P(Y) code and C/A code pseudoranging, in particular.

A P(Y) code chip occurs every 0.0978 of a microsecond. In other words, there is a P(Y) code chip about every tenth of a microsecond. That's one code chip every 100 ns. Therefore, a P(Y) code based measurement can have a maximum precision of about 1 percent of 100 ns, or 1 ns. What is the length of 1 ns? Multiplied by the speed of light it's approximately 30 cm, or about a foot. So, just about the very best you can do with a P(Y) pseudorange is a foot or so.

Because its chipping rate is 10× slower, the C/A code based pseudorange is 10× less precise. Therefore, 1 percent of the length of a C/A code chip is 10 × 30 cm, or 3 m. Using the rule of thumb, the maximum resolution of a C/A code pseudorange is nearly 10 feet. Actually, this is a bit optimistic. The actual positional accuracy of a single C/A code receiver was about ±100 m with *Selective Availability* (SA) turned on. It was ±30 m with SA turned off in May 2000 and is a bit better today.

The carrier phase observable provides substantially better precision than the pseudorange, and this can be illustrated with the same rule of thumb. First, the length of a single wavelength of each carrier is calculated using

$$\lambda = \frac{c_a}{f}$$

where

λ = length of each complete wavelength in meters
c_a = speed of light corrected for atmospheric effects
f = frequency in hertz

The L1-1575.42 MHz carrier transmitted by GPS satellites has a wavelength of approximately 19 cm.

$$\lambda = \frac{c_a}{f}$$

$$\lambda = \frac{300 \times 10^6 \text{ m/s}}{1575.42 \times 10^6 \text{ Hz}}$$

$$\lambda = 0.19 \text{ m}$$

The L2-1227.60 MHz frequency carrier transmitted by GPS satellites has a wavelength of approximately 24 cm.

$$\lambda = \frac{c_a}{f}$$

$$\lambda = \frac{300 \times 10^6 \text{ m/s}}{1227.60 \times 10^6 \text{ Hz}}$$

$$\lambda = 0.24 \text{ m}$$

The L5-1176.45 MHz frequency carrier transmitted by GPS satellites has a wavelength of approximately 25 cm.

$$\lambda = \frac{c_a}{f}$$

$$\lambda = \frac{300 \times 10^6 \text{ m/s}}{1176.45 \times 10^6 \text{ Hz}}$$

$$\lambda = 0.25 \text{ m}$$

Therefore, using the wavelength of any GPS carrier as the observable, the measurement resolved to 1 percent of the wavelength would be about 2 mm. It is no surprise then that the carrier phase observable is preferred for the higher precision work most surveyors have come to expect from GPS.

CARRIER PHASE RANGING

THE CYCLE AMBIGUITY PROBLEM

Even though the carrier phase is observable at the center of high accuracy surveying applications of GPS, it introduces a difficulty that needs to be overcome. It is called the cycle ambiguity problem. It is similar to the hurdle encountered by the EDM.

As you know, by comparing the phase of the signal returned from the reflector with the reference wave it kept at home, an EDM can measure how much the two are out of phase with one another. However, this measurement can only be used to calculate a small part of the overall distance. It only discloses the length of a fractional part of a wavelength used. This leaves a big unknown, namely, the number of full wavelengths of the EDM's modulated carrier between the transmitter and the receiver at the instant of the measurement. This cycle ambiguity is symbolized by N. Fortunately, the cycle ambiguity can be solved in the EDM measurement process. The key is using carriers with progressively longer wavelengths. For example, the submeter portion of the overall distance can be resolved using a carrier with the wavelength of a meter. This can be followed by a carrier with a wavelength of 10 m, which provides the basis for resolving the meter aspect of a measured distance. This procedure may be followed by the resolution of the tens of meters using a wavelength of 100 m. The hundreds of meters can then be resolved with a wavelength of 1000 m, and so on.

Such a method works in the EDM's two-way ranging system, but the GPS one-way ranging makes use of the same strategy impossible. GPS ranging must use an entirely different strategy for solving the cycle ambiguity problem because the satellites broadcast only three carriers, currently. The venerable L1 and L2 are now joined by L5 from some of the satellites. Those carriers have constant wavelengths, and they only propagate from the satellites to the receivers, in one direction. Therefore, unlike an EDM measurement, the wavelengths of these carriers in GPS cannot be periodically changed to resolve the cycle ambiguity problem.

CARRIER PHASE COMPARISONS

The unmodulated L1, L2, and L5 are the observables used in the carrier phase solution rather than the P and C/A codes. This fact has always offered the good news that the user of the carrier phase solution was immune from the effects of Selective Availability (SA). Selective Availability was the intentional degradation of the SPS, the standard positioning service, available through the C/A code, but because carrier phase observations do not use codes, they were never affected by SA. Fortunately, it was turned off in May 2000 so the point is moot, that is unless SA should be reinstituted.

Understanding carrier phase is perhaps a bit more difficult than the pseudorange, but the basis of the measurements has some similarities. As you know, the foundation of a pseudorange measurement is the correlation of the codes received from a GPS satellite with replicas of those codes generated within the receiver. The foundation of the carrier phase measurement is the combination of the unmodulated carrier itself received from a GPS satellite with a replica of that carrier generated within the receiver.

As in the EDM example, it is the phase difference between the incoming signal and the internal reference that reveals the fractional part of the carrier phase measurement in GPS. The incoming signal is from a satellite rather than a reflector, of course, but like an EDM measurement, the internal reference is derived from the receiver's oscillator and the number of complete cycles is not immediately known.

Beat

The carrier phase observable is sometimes called the *reconstructed carrier phase* or *carrier beat phase* observable. In this context, a *beat* is the pulsation resulting from the combination of two waves with different frequencies. An analogous situation occurs when two musical notes of different pitch are sounded at the same time. Their two frequencies combine and create a third note called the beat. Musicians can tune their instruments by listening for the beat that occurs when two pitches differ slightly. This third pulsation may have a frequency equal to the difference or the sum of the two original frequencies.

The beat phenomenon is by no means unique to musical notes; it can occur when any pair of oscillations with different frequencies is combined. In GPS, a beat is created when a carrier generated in a GPS receiver and a carrier received from a satellite are combined (see Figure 1.14).

At first, that might not seem sensible. How could a beat be created by combining two absolutely identical unmodulated carriers? There should be no difference in frequency between an L1 carrier generated in a satellite and an L1 carrier generated in a receiver. They both should have a frequency of 1575.42 MHz. If there is no difference in the frequencies, how can there be a beat? But there is a slight difference between the two carriers. Something happens to the frequency of the carrier on its trip from a GPS satellite to a receiver; its frequency changes. The phenomenon is described as the Doppler Effect.

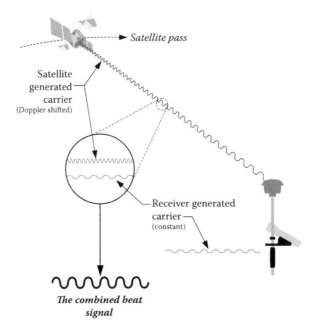

FIGURE 1.14   Carrier beat phase.

DOPPLER EFFECT

Sound provides a model for the explanation of the phenomenon. An increase in the frequency of a sound is indicated by a rising pitch; a lower pitch is the result of a decrease in the frequency. A stationary observer listening to the blasting horn on a passing train will note that as the train gets closer, the pitch rises, and as the train travels away, the pitch falls. Furthermore, the change in the sound, clear to the observer standing beside the track, is not heard by the engineer driving the train. He hears only one constant, steady pitch. The relative motion of the train with respect to the observer causes the apparent variation in the frequency of the sound of the horn.

In 1842, Christian Doppler described this frequency shift now named for him. He used the analogy of a ship on an ocean with equally spaced waves. In his allegory, when the ship is stationary, the waves strike it steadily, one each second. However, if the ship sails into the waves, they break across its bow more frequently. If the ship then turns around and sails with the waves, they strike less frequently across its stern. The waves themselves have not changed; their frequency is constant, but to the observer on the ship, their frequency seems to depend on his motion.

GPS and the Doppler Effect

From the observer's point of view, whether it is the source, the observer, or both that are moving, the frequency increases while they move together and decreases while they move apart. The Doppler Effect is inversely proportional to the wavelength of the signal. Therefore, if a GPS satellite is moving toward an observer, its carrier wave comes into the receiver with a higher frequency than it had at the satellite. If a GPS satellite is moving away from the observer, its carrier wave comes into the receiver with a lower frequency than it had at the satellite. Because a GPS satellite is virtually always moving with respect to the observer, any signal received from a GPS satellite is Doppler shifted.

Carrier Phase Approximation

The carrier phase observation in cycles is often symbolized by ϕ in GPS literature. Other conventions include superscripts to indicate satellite designations and subscripts to define receivers. For example, in the following equation ϕ_r^s is used to symbolize the carrier phase observation between satellite s and receiver r. The difference that defines the carrier beat phase observation is (Wells 1986)

$$\phi = \phi_r^s = \phi^s(t) - \phi_r(T)$$

where $\phi^s(t)$ is the phase of the carrier broadcast from the satellite s at time t. Please note that the frequency of this carrier is the same, nominally constant frequency that is generated by the receiver's oscillator. The value $\phi_r(T)$ is its phase when it reaches the receiver r at time T.

A description of the carrier phase observable to measure range can start with the same basis as the calculation of the pseudorange, travel time. The time elapsed between the moment the signal is broadcast, t, and the moment it is received, T, multiplied by the speed of light, c, will yield the range between the satellite and receiver, ρ:

$$(T - t)c \approx \rho$$

Even using this simplified equation, it is possible to get an approximate idea of the relation between time and range for some assumed, nominal values. These values could not be the basis of any actual carrier phase observation because, among other reasons, it is not possible for the receiver to know when a particular carrier wave left the satellite. However, for the purpose of illustration, suppose a carrier left a satellite at 00 hours 00 minutes 00.000 seconds and arrived at the receiver 67 milliseconds later:

$$(T - t)c \approx \rho$$

$$(00h:00m:00.067s - 00h:00m:00.000s)300,000 \text{ km/s} \approx \rho$$

$$(0.067)300,000 \text{ km/s} \approx \rho$$

$$20,100 \text{ km} \approx \rho$$

This estimate indicates that if the carrier broadcast from the satellite reaches the receiver 67 milliseconds later, the range between them is approximately 20,100 km.

Carrying this example a bit farther, the wavelength λ of the L1 carrier can be calculated by dividing the speed of light, c, by the L1 frequency f:

$$\lambda = \frac{c}{f}$$

$$\lambda = \frac{300,000,000 \text{ m/s}}{1,575,420,000 \text{ Hz}}$$

$$\lambda = 0.1904254104 \text{ m}$$

Dividing the approximated range ρ by the calculated L1 carrier wavelength λ yields a rough estimation of the carrier phase in cycles, ϕ:

$$\frac{\rho}{\lambda} \approx \phi$$

$$\frac{20,100,000 \text{ m}}{0.1904154104 \text{ m}} \approx \phi$$

$$105,553,140 \text{ cycles} \approx \phi$$

The 20,100 km range implies that the L1 carrier would cycle through approximately 105,553,140 wavelengths on its trip from the satellite to the receiver.

Of course, these relationships are much simplified. However, they can be made fundamentally correct by recognizing that ranging with the carrier phase observable is subject to all of the same biases and errors as the pseudorange. For example, terms such as the receiver clock offset may be incorporated, again symbolized by dT as it was in the pseudorange equation. The imperfect satellite clock can be included; its error is symbolized by dt. The tropospheric delay d_{trop}, the ionospheric delay d_{ion}, and multipath-receiver noise ε_ϕ are also added to the range measurement. The ionospheric delay will be negative here. With these changes the simplified travel time equation can be made a bit more realistic.

$$[(T + dT) - (t + dt)]c \approx \rho - d_{\text{ion}} + d_{\text{trop}} + \varepsilon_\phi$$

This more realistic equation can be rearranged to isolate the elapsed time $(T - t)$ on one side by dividing both sides by c and then moving the clock errors to the right side (Wells et al. 1986).

$$\frac{[(T + dT) - (t + dt)]c}{c} = \frac{\rho - d_{\text{ion}} + d_{\text{trop}} + \varepsilon_\phi}{c}$$

$$[(T + dT) - (t + dt)] = \frac{\rho - d_{\text{ion}} + d_{\text{trop}} + \varepsilon_\phi}{c}$$

$$(T - t + dT - dt) = \frac{\rho - d_{\text{ion}} + d_{\text{trop}} + \varepsilon_\phi}{c}$$

$$(T - t + dT - dt) + (dt - dT) = \frac{\rho - d_{\text{ion}} + d_{\text{trop}} + \varepsilon_\phi}{c} + (dt - dT)$$

$$T - t = dt - dT + \frac{\rho - d_{\text{ion}} + d_{\text{trop}} + \varepsilon_\phi}{c}$$

This expression now relates the travel time to the range. However, in fact, a carrier phase observation cannot rely on the travel time for two reasons. First, in a carrier phase observation the receiver has no codes with which to tag any particular instant on the incoming continuous carrier wave. Second, because the receiver cannot distinguish one cycle of the carrier from any other, it has no way of knowing the initial phase of the signal when it left the satellite. In other words, the receiver cannot know the travel time, and, therefore, it is hard to see how it can determine the number of complete cycles between the satellite and itself. This unknown quantity is called the *cycle ambiguity*.

Remember, the approximation of 105,553,140 wavelengths in this example is for the purpose of comparison and illustration only. In actual practice, a carrier phase observation must derive the range from a measurement of phase at the receiver not from a known travel time of the signal.

The missing information is the number of complete phase cycles between the receiver and the satellite at the instant that the tracking began. The critical unknown integer, symbolized by N, is the cycle ambiguity, and it cannot be directly measured by the receiver. The receiver can count the complete phase cycles it receives from the moment it starts tracking until the moment it stops. It can also monitor the fractional phase cycles, but the cycle ambiguity N is unknown.

Illustration of the Cycle Ambiguity Problem

The situation is somewhat analogous to an unofficial technique used by some nine-teenth century contract surveyors on the Great Plains. The procedure can be used as a rough illustration of the cycle ambiguity problem in GPS.

It was known as the *buggy wheel* method of chaining. Some of the lines of the public land system that crossed open prairies were originally surveyed by loading a wagon with stones or stakes and tying a cloth to a spoke of the wheel. One man drove the team, another kept the wagon in line with a compass, and a third counted the revolutions of the flagged wheel to measure the distance. When there had been enough turns of the improvised odometer to measure half a mile, they set a stone or stake to mark the corner and then rolled on, counting their way to the next corner.

A GPS receiver is like the man assigned to count the turns of the wheel. He is supposed to begin his count from the moment the crew leaves the newly set corner. Suppose he jumps into the wagon, gets comfortable, and takes an unscheduled nap. When he wakes up, the wagon is on the move. Trying to make up for his laxness, he immediately begins counting. However, at that moment the wheel is at a half turn, a fractional part of a cycle. He counts the subsequent half turn and then, back on the job, he intently counts each and every full revolution. His tally grows as the cycles accumulate, but he is in trouble and he knows it. He cannot tell how far the wagon has traveled; he was asleep for the first part of the trip. He has no way of knowing how far they had come before he woke up and started counting. He is like a GPS receiver that cannot know how far it is from the satellite when it starts counting phase cycles. They can tell it nothing about how many cycles stood between itself and the satellite when the receiver was locked on and began tracking. Those unknown cycles are the cycle ambiguity.

The 360° cycles in the carrier phase observable are wavelengths λ, not revolutions of a wheel. Therefore, the cycle ambiguity included in the complete carrier phase equation is an integer number of wavelengths, symbolized by λN (Fotopoulos 2000). So, the complete carrier phase observable equation can now be stated as

$$\Phi = \rho + d_\rho + c(dt - dT) + \lambda N - d_{ion} + d_{trop} + \varepsilon_{m\Phi} + \varepsilon_\Phi$$

where

Φ = carrier phase measurement
ρ = true range
d_ρ = satellite orbital errors
c = speed of light
dt = satellite clock offset from GPS Time
dT = receiver clock offset from GPS
λ = carrier wavelength
N = integer ambiguity in cycles
d_{ion} = ionospheric delay
d_{trop} = tropospheric delay
$\varepsilon_{m\Phi}$ = multipath
ε_Φ = receiver noise

EXERCISES

1. What is the function of the information in subframe 5 of the NAV message?
   a. Once a receiver is locked onto a satellite, it helps the receiver determine the position of the satellite that is transmitting the NAV message.
   b. Once a receiver is locked onto a satellite, it helps the receiver correct the part of the delay of the signal caused by the ionosphere.
   c. Once a receiver is locked onto a satellite, it helps the receiver correct the received satellite time to GPS Time.
   d. Once a receiver is locked onto a satellite, it helps the receiver acquire the signals of the other satellites.

2. Which height most correctly expresses a nominal altitude of GPS satellites above the Earth?
   a. 20,000 miles
   b. 35,420 km
   c. 20,183,000 m
   d. 108,000 nautical miles

3. Which of the following statements about the clocks in GPS satellites is not correct?
   a. Signal for each satellite is independent from the other satellites and is generated from its own onboard clock.
   b. Clocks in GPS satellites may also be called oscillators or frequency standards.
   c. Every GPS satellite is launched with very stable atomic clocks onboard.
   d. Clocks in any one satellite are allowed to drift up to 1 ns from GPS Time before they are tweaked by the Control Segment.

4. Global Positioning System is known as a passive system. What does that mean?
   a. Ranges are measured with signals in the microwave part of the electromagnetic spectrum.
   b. Only the satellites transmit signals; the users receive them.
   c. A GPS receiver must be able to gather all the information it needs to determine its own position from the signals it bounces off the satellites.
   d. Signals from a GPS receiver return to the satellite.

5. Which comparison of EDM and GPS processes is correct?
   a. EDM and GPS signals are both reflected back to their sources.
   b. EDM measurements require atmospheric correction; GPS ranges do not.
   c. EDMs and GPS satellites both transmit modulated carriers.
   d. Phase differencing is used in EDM measurement but not in GPS.

6. What information is critical to defining the relationship between GPS Time and UTC?
   a. Multipath
   b. Broadcast ephemeris
   c. Anti-spoofing flag
   d. Leap seconds

7. The P(Y) code and the C/A code are created by shifts from 0 to 1 or 1 to 0 known as code chips. There is a corresponding change of $180°$ in the GPS carrier waves. What changes?
   a. Fundamental clock rate
   b. Frequency
   c. Phase
   d. Amplitude

8. The P(Y) code and the C/A code are both broadcast on L1. How can a GPS receiver distinguish between them on that carrier?
   a. Chipping rate for the C/A code is 10× slower than the chipping rate for the P(Y) code.
   b. They are broadcast in binary biphase modulation.
   c. The fundamental clock rate is 10.23 MHz.
   d. They are broadcast in quadrature.

9. Which of the following is the most correct description of the Doppler Effect?
   a. Distortion of electromagnetic waves due to the density of charged particles in the Earth's upper atmosphere.
   b. Systematic changes that occur when a moving object or light beam pass through a gravitational field.
   c. Systematic changes that occur when a moving object approaches the speed of light.
   d. Shift in frequency of an acoustic or electromagnetic radiation emitted by a source moving relative to an observer.

10. When Selective Availability was switched off, how did it affect surveying applications of GPS?
    a. Accuracy of most surveying applications of GPS doubled.
    b. P(Y) code was not encrypted.
    c. Cycle ambiguity problem was eliminated.
    d. There was no significant change.

11. Which of the following numbers describe the frequencies of the L1, L2, and L5 carrier signals?
    a. 1755.42, 1227.60, 1176.45
    b. 1575.42, 1227.60, 1176.45
    c. 1175.42, 1226.70, 1576.45
    d. 1542.77, 1260.52, 1167.85

ANSWERS AND EXPLANATIONS

1. Answer is (d)
   Explanation: Subframe 5 contains the ephemerides of up to 24 satellites. The purpose of subframe 5 is to help the receiver acquire the signals of the other satellites. The receiver must first lock onto a satellite to have access to the NAV message, of course, but once the NAV message can be read the positions of all of the other satellites can be computed.

2. Answer is (c)
   Explanation: The GPS constellation is still evolving, but the orbital configuration is fairly well settled. Each GPS satellite's orbit is nearly circular and has a nominal altitude of 20,183 km or 20,183,000 m.
   The orbit is approximately 12,500 statute miles or 10,900 nautical miles above the surface of the planet. GPS satellites are higher than the usual orbit assigned to most other satellites including the space shuttle. Their high altitude allows each GPS satellite to be viewed simultaneously from a large portion of the Earth at any given moment. However, GPS satellites are well below the height required for the sort of geosynchronous orbit used for communications satellites, 35,420 km.

3. Answer is (d)

Explanation: Government tracking facilities monitor the drift of each satellite clock's deviation from GPS Time. The drift is allowed to reach a maximum of one millisecond before it is adjusted. Until the difference reaches that level, it is contained in the broadcast clock correction in subframe 1 of the NAV message.

4. Answer is (b)

Explanation: Some of the forerunners of GPS were NAVal systems that involved transmissions from the users but not GPS. The military designed the system to exclude anything that would reveal the location of the GPS receiver. If GPS had been built as a two-way system, it would have been much more complicated especially as the number of users grew. Therefore, it is a passive system, meaning that the satellites transmit and the users receive.

5. Answer is (c)

Explanation: An unmodulated carrier carries no information and no code. While GPS receivers can and do make phase measurements on the unmodulated carrier waves, they also make use of the code information available on the modulated carrier transmitted by GPS satellites to determine pseudoranges. GPS uses a one-way system. The modulated carrier with code information travels from the satellite to the receiver where it is correlated with a reference. This one-way method requires a frequency standard in both the satellite and the receiver.

An EDM's measurement is also based on a modulated carrier. However, the EDM uses a two-way system. The modulated carrier is transmitted to a retroprism. It is reflected and returns to the EDM. The EDM can then determine the phase delay by comparing the returned modulated wave with a reference in the instrument. This two-way method requires only one frequency standard in the EDM since the modulated carrier is reflected.

6. Answer is (d)

Explanation: GPS Time is calculated using the atomic clock at the Master Control Station at the Schriever Air Force Base, formerly known as Falcon Air Force Station near Colorado Springs, Colorado. GPS Time is kept within 1 millisecond of UTC. However, UTC is adjusted for leap seconds, and GPS Time is not. Conversion of GPS Time to UTC requires knowledge of the leap seconds applied to UTC since January 1980. This information is available from the U.S. Naval Observatory time announcements.

7. Answer is (c)

Explanation: Phase modulations are used to mark the divisions between the code chips. Whenever the C/A code or the P(Y) code switches from a binary 1 to a binary 0 or vice versa, its L1 or L2 carriers have a sharp mirror-image shift in phase. About every millionth of a second the phase of the C/A code carrier can shift. However, the P(Y) code carrier can have a phase shift about every ten millionth of a second.

8. Answer is (d)

   Explanation: Even though both codes are broadcast on L1, they are distinguishable from one another by their transmission in quadrature. That means that the C/A code modulation on the L1 carrier is phase shifted 90° from the P(Y) code modulation on the same carrier.

9. Answer is (d)

   Explanation: If the source of radiation is moving relative to an observer, there is a difference between the frequency perceived by the observer and the frequency of the radiation at its source. The shift is to higher frequencies when the source moves toward the observer and to lower frequencies when it moves away. This effect has been named for its discoverer, C. J. Doppler.

10. Answer is (d)

    Explanation: When Selective Availability (SA) was switched off on May 2, 2000, by presidential order, the effect on surveying applications of GPS was negligible. SA means that the GPS signals were transmitted from the satellites with intentional clock errors added. The stability of the atomic clocks onboard was deliberately degraded, and, therefore, the NAV message was degraded, too. While SA was removed by Precise Positioning Service, P(Y) code, users with decryption techniques, civilian users of the Standard Positioning Service C/A code were not able to remove these errors.

    SA was on since the first launch of the Block II satellites in 1989. It was turned off briefly to allow coalition forces in the Persian Gulf to use civilian GPS receivers in 1990 but was turned on again immediately.

    For the code phase receiver owner limited to pseudorange positioning, SA was a problem. A single pseudorange receiver could only achieve positional accuracy of about ±100 m, 95 percent of the time. This sort of single point positioning is not a usual surveying application of GPS. However, with a second code phase receiver as a base station on a known position and using differential data processing techniques, submeter accuracy was possible even with SA turned on. The carrier phase receivers used in most surveying applications have always been, for all practical purposes, immune from the effects of SA.

11. Answer is (b)

    Explanation: The two legacy GPS carriers are L1 at 1575.42 MHz, which is just a little above the L-band range, and L2 at 1227.60 MHz. There is also a third, relatively new, carrier being broadcast by some GPS satellites. It is known as L5, and its frequency is 1176.45 MHz.

2 Biases and Solutions

BIASES

A Look at the Error Budget

The understanding and management of errors is indispensable for finding the true geometric range ρ from either a pseudorange or a carrier phase observation.

$$p = \rho + d_\rho + c(dt - dT) + d_{ion} + d_{trop} + \varepsilon_{mp} + \varepsilon_p \text{ (pseudorange)}$$

$$\phi = \rho + d_\rho + c(dt - dT) + \lambda N - d_{ion} + d_{trop} + \varepsilon_{m\phi} + \varepsilon_\phi \text{ (carrier phase)}$$

Both equations include environmental and physical limitations called *range biases.*

Atmospheric errors are among the biases; two are the ionospheric effect d_{ion} and the tropospheric effect d_{trop}. The tropospheric effect may be somewhat familiar to total station and electronic distance measuring (EDM) device users, even if the ionospheric effect is not. Other biases, clock errors symbolized by $(dt - dT)$ and receiver noise (ε_p and ε_ϕ), multipath (ε_{mp} and $\varepsilon_{m\phi}$), and orbital errors d_ρ are unique to satellite surveying methods. As you can see, each of these biases comes from a different source. They are each independent of one another, but they combine to obscure the true geometric range. The objective here is to discuss each of them separately.

User Equivalent Range Error and User Range Error

The summary of the total error budget affecting a pseudorange is called the *user equivalent range error* (UERE). This expression, often used in satellite surveying literature, is the square root of the sum of the squares of the individual biases (see Figure 2.1). Differential correction techniques limit the effect of these errors.

Some of the biases that make up the UERE such as those attributable to the atmosphere d_{ion}, d_{trop} and satellite orbits d_ρ increase and decrease with the length of the baselines between receivers. Differential correction techniques can often limit the effect of these errors. Others included in the UERE such as those due to receiver noise, ε_p and ε_ϕ, multipath, ε_{mp} and $\varepsilon_{m\phi}$, do not increase and decrease with the length of the baselines between receivers.

When the estimated error considered only includes the biases attributed to atmosphere d_{ion}, d_{trop}, satellite orbits d_ρ, and the satellite clocks $(dt - dT)$, the term *user range error* (URE) is used. When it includes the total error budget, it is called the UERE.

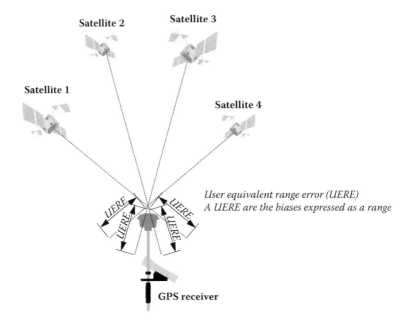

FIGURE 2.1   User equivalent range error.

IONOSPHERIC EFFECT d_{ion}

One of the largest errors in GPS positioning is attributable to the atmosphere. The long relatively unhindered travel of the GPS signal through the virtual vacuum of space changes as it passes through the Earth's atmosphere. Through both refraction and diffraction, the atmosphere alters the apparent speed and, to a lesser extent, the direction of the signal. This causes an apparent delay in the signals transit from the satellite to the receiver.

Ionized Plasma

The ionosphere is ionized plasma comprised of negatively charged electrons that remain free for long periods before being captured by positive ions. It extends from about 50 to 1000 km above the Earth's surface and is the first part of the atmosphere that the signal encounters as it leaves the satellite.

The magnitude of these delays is determined by the state of the ionosphere at the moment the signal passes through, so it's important to note that its density and stratification varies. The Sun plays a key role in the creation and variation of these aspects. Also, the daytime ionosphere is rather different from the nighttime ionosphere.

Ionosphere and the Sun

When gas molecules are ionized by the Sun's ultraviolet radiation, free electrons are released. As their number and dispersion varies, so does the electron density in the ionosphere. This density is often described as *total electron content* (TEC), a measure of the number of free electrons in a column through the ionosphere with a

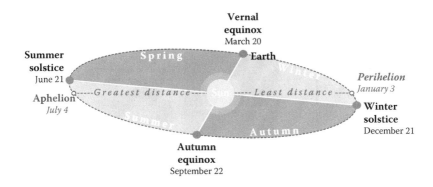

FIGURE 2.2   Earth's orbit.

cross-sectional area of 1 m^2: 10^{16} is one TEC unit. The higher the electron density, the larger the delay of the signal, but the delay is by no means constant.

The ionospheric delay changes slowly through a daily cycle. It is usually least between midnight and early morning and most around local noon or a little after. During the daylight hours in the midlatitudes the ionospheric delay may grow to be as much as 5× greater than it was at night, but the rate of that growth is seldom more than 8 cm/min. It is also nearly 4× greater in November, when the Earth is nearing its *perihelion*, its closest approach to the Sun, than it is in July near the Earth's *aphelion*, its farthest point from the Sun. The effect of the ionosphere on the GPS signal usually reaches its peak in March, about the time of the vernal equinox (see Figure 2.2).

Ionospheric Stratification

The ionosphere has layers sometimes known as the mesosphere and thermosphere that are themselves composed of D, E, and F regions (see Figure 2.3). Neither the boundaries between these regions nor the upper layer of the ionosphere can be defined strictly. Here are some general ideas on the subject. The lowest detectable layer, the D region, extends from about 50 to 90 km. It has almost no effect on GPS signals and virtually disappears at night. The E region, also a daytime phenomenon, is between 90 and 120 km. Its effect on the signal is slight, but it can cause the signal to scintillate. The layer that affects the propagation of electromagnetic signals the most is the F region. It extends from about 120 to 1000 km. The F region contains the most concentrated ionization in the atmosphere. In the daytime, the F layer can be further divided into F1 and F2. F2 is the most variable. F1, the lower of the two, is most apparent in the summer. These two layers combine at night. Above the F layer is fully ionized. It is sometimes known as the photosphere or the H region.

The ionosphere is also not homogeneous. Its behavior in one region of the Earth is liable to be unlike its behavior in another. For example, ionospheric disturbances can be particularly harsh in the polar regions. However, the highest TEC values and the widest variations in the horizontal gradients occur in the band of about 60° of *geomagnetic latitude*. That band lies 30° north and 30° south of the Earth's magnetic equator.

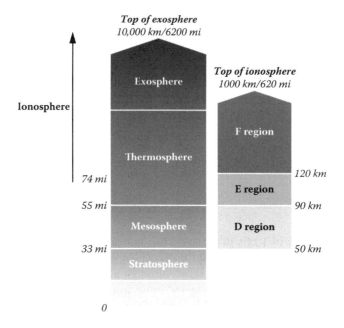

FIGURE 2.3   Atmospheric model.

Satellite Elevation and Ionospheric Effect

The severity of the ionosphere's effect on a GPS signal depends on the amount of time that signal spends traveling through it. A signal originating from a satellite near the observer's horizon must pass through a larger amount of the ionosphere to reach the receiver than does a signal from a satellite near the observer's zenith. In other words, the longer the signal is in the ionosphere, the greater the ionosphere's effect on it.

Magnitude of the Ionospheric Effect

The error introduced by the ionosphere can be very small, but it may be large when the satellite is near the observer's horizon, the vernal equinox is near, and/or sunspot activity is severe. For example, the TEC is maximized during the peak of the 11-year solar cycle. It also varies with magnetic activity, location, time of day, and even the direction of observation.

Group Delay and Phase Delay

The ionosphere is *dispersive*, which means that the apparent time delay contributed by the ionosphere depends on the frequency of the signal. This dispersive property causes the codes, the modulations on the carrier wave, to be affected differently than the carrier wave itself during the signal's trip through the ionosphere. The P code, the C/A code, the Navigation message, and all the other codes appear to be delayed, or slowed, affected by what is known as the *group delay*. However, the carrier wave itself appears to speed up in the ionosphere. It is affected by what is known as the *phase delay*. It may seem odd to call an increase in speed a delay. It is sometimes called *phase advancement*. In any case, it is governed by the same properties of electron content as the group delay; phase delay just increases negatively. Please note that the algebraic sign of d_{ion} is negative in the carrier phase equation and positive in the pseudorange equation. In other words, a range from a satellite to a receiver determined by a code observation will be a bit too long and a range determined by a carrier observation will be a bit too short.

Different Frequencies Are Affected Differently

Another consequence of the dispersive nature of the ionosphere is that the apparent time delay for a higher frequency carrier wave is less than it is for a lower frequency wave. That means that L1, 1575.42 MHz, is not affected as much as L2, 1227.60 MHz, and L2 is not affected as much as L5, 1176.45 MHz.

This fact provides one of the greatest advantages of a dual-frequency receiver over the single-frequency receivers. The separations between the L1 and L2 frequencies (347.82 MHz), the L1 and L5 frequencies (398.97 MHz), and even the L2 and L5 frequencies (51.15 MHz) are large enough to facilitate estimation of the ionospheric group delay. Therefore, by tracking all the carriers, a multiple-frequency receiver can model and remove not all, but a significant portion of, the ionospheric bias. There are now several possible combinations, L1/L2, L1/L5, and L2/L5. It is even possible to have a triple-frequency combination to help ameliorate this bias.

The frequency dependence of the ionospheric effect is described by the following expression (Brunner and Welsch 1993):

$$v = \frac{40.3}{cf^2} \cdot \text{TEC}$$

where

      v = ionospheric delay
      c = speed of light in meters per second
      f = frequency of the signal in Hz
   TEC = quantity of free electrons per square meter

As the formula illustrates, the time delay is inversely proportional to the square of the frequency; in other words, the higher the frequency, the less the delay. For example, the ionospheric delay at 1227.60 (L2) is 65 percent larger than at 1575.42 MHz (L1), and at 1176.45 MHz (L5) it is 80 percent larger than 1575.42 (L1).

Broadcast Correction

A predicted total UERE is provided in each satellite's Navigation message as the user range accuracy (URA), but it is minus ionospheric error. To help remove some of the effect of the ionospheric delay on the range derived from a single-frequency receiver, there is an ionospheric correction available in another part of the Navigation message, subframe 4. However, this broadcast correction should not be expected to remove more than about three-quarters of the error, which is most pronounced on long baselines. Where the baselines between the receivers are short, the effect of the ionosphere can be small, but as the baseline grows, so does the significance of the ionospheric bias.

SATELLITE CLOCK BIAS *dt*

One of the largest errors can be attributed to the satellite clock bias. It can be quite large especially if the broadcast clock correction is not used by the receiver to bring the time signal acquired from a satellite's onboard clock in line with GPS Time. As time is a critical component in the functioning of GPS, it is important to look closely at the principles behind this bias.

Relativistic Effects on the Satellite Clock

Albert Einstein's special and general theories of relativity apply to the clocks involved here. At 3.874 km/s the clocks in the GPS satellites are traveling at great speed, and that makes the clocks on the satellites appear to run slower than the clocks on Earth by about 7 μs a day. However, this apparent slowing of the clocks in orbit is counteracted by the weaker gravity around them. The weakness of the gravity makes the clocks in the satellites appear to run faster than the clocks on Earth by about 45 μs a day. Therefore, on balance, the clocks in the GPS satellites in space appear to run faster by about 38 μs a day than the clocks in GPS receivers on Earth. So, to ensure the clocks in the satellites will actually produce the correct fundamental frequency of 10.23 MHz in space, their frequencies are set to 10.22999999543 MHz before they are launched into space.

There is yet another consideration, the eccentricity of the orbit of GPS satellites. With an eccentricity of 0.02 this effect on the clocks can be as much as 45.8 ns. Fortunately, the offset is eliminated by a calculation in the GPS receiver itself, thereby avoiding what could be ranging errors of about 14 m. The receiver is moving, too, of course, so an account must be made for the motion of the receiver due to the rotation of the Earth during the time it takes the satellites signal to reach it. This is known as the Sagnac effect, and it is 133 ns at its maximum. Luckily, these relativistic effects can be accurately computed and removed from the system.

Satellite Clock Drift

Clock drift is another matter. As discussed in Chapter 1, the onboard satellite clocks are independent of one another. The rates of these rubidium and cesium oscillators are more stable if they are not disturbed by frequent tweaking and adjustment is kept to a minimum. While GPS Time itself is designed to be kept within 1 μs, or one-millionth of a second, of UTC, excepting leap seconds, the satellite clocks can be allowed to drift up to 1 ms, or one-thousandth of a second, from GPS Time.

There are three kinds of time involved here. The first is UTC per the U.S. Naval Observatory (USNO). The second is GPS Time. The third is the time determined by each independent GPS satellite.

Their relationship is as follows. The Master Control Station (MCS) at Schriever (formerly Falcon) Air Force Base near Colorado Springs, Colorado, gathers the GPS satellite's data from monitoring stations around the world. After processing, this information is uploaded back to each satellite to become the broadcast ephemeris, broadcast clock correction, and so forth. The actual specification for GPS Time demands that its rate be within 1 μs of UTC as determined by USNO, without consideration of leap seconds. Leap seconds are used to keep UTC correlated with the actual rotation of the Earth, but they are ignored in GPS Time. In GPS Time it is as if no leap seconds have occurred at all in UTC since 24:00:00, January 5, 1980. In practice, the rate of GPS Time is much closer than 1 μs of the rate of UTC; it is usually within about 25 ns, or 25 billionths of a second, of the rate of UTC.

By constantly monitoring the satellites clock error dt, the Control Segment gathers data for its uploads of the broadcast clock corrections. You will recall that clock corrections are part of the Navigation message.

RECEIVER CLOCK BIAS dT

The third largest error that can be caused by the receiver clock is its oscillator. Both a receiver's measurement of phase differences and its generation of replica codes depend on the reliability of this internal frequency standard.

Typical Receiver Clocks

GPS receivers are usually equipped with quartz crystal clocks, which are relatively inexpensive and compact. They have low power requirements and long life spans. For these types of clocks, the frequency is generated by the piezoelectric effect in an oven-controlled quartz crystal disk, a device sometimes symbolized by *OCXO*. Their reliability ranges from a minimum of about 1 part in 108 to a maximum of about 1 part in 1010, a drift of about 0.1 ns in 1 s. Even at that, quartz clocks are not as stable as the atomic standards in the GPS satellites and are more sensitive to temperature changes, shock, and vibration. Some receiver designs augment their frequency standards by also having the capability to accept external timing from cesium or rubidium oscillators.

ORBITAL BIAS d_ρ

Orbital bias has the potential to be the fourth largest error. It is addressed in the broadcast ephemeris.

Forces Acting on the Satellites

The orbital motion of GPS satellites is not only a result of the Earth's gravitational attraction; there are several other forces that act on the satellite. The primary disturbing forces are the nonspherical nature of the Earth's gravity, the attractions of the sun and the moon, and solar radiation pressure. The best model of these forces is the actual motion of the satellites themselves and the government facilities distributed around the world, known collectively as the *Control Segment*, *ground segment*, or the *Operational Control System* (OCS), continuously track them for that reason, among others (see Chapter 3).

TROPOSPHERIC EFFECT d_{trop}

The fifth largest UERE can be attributed to the effect of the troposphere.

Troposphere

The troposphere is that part of the atmosphere closest to the Earth. It extends from the surface to about 9 km over the poles and about 16 km over the equator, but in this work it will be combined with the tropopause and the stratosphere, as it is in much of GPS literature. Therefore, the following discussion of the tropospheric effect will include the layers of the Earth's atmosphere up to about 50 km above the surface.

Tropospheric Effect Is Independent of Frequency

The troposphere and the ionosphere are by no means alike in their effect on the satellite's signal. While the troposphere is refractive, its refraction of a GPS satellite's signal is not related to its frequency. The refraction is tantamount to a delay in the arrival of a GPS satellite's signal. It can also be conceptualized as a distance added to the range the receiver measures between itself and the satellite. The troposphere is part of the electrically neutral layer of the Earth's atmosphere, meaning it is not ionized. The troposphere is also nondispersive for frequencies below 30 GHz or so. Therefore, L1, L2, and L5 are equally refracted. This means that the range between a receiver and a satellite will be shown to be a bit longer than it actually is.

However, as it is in the ionosphere, density affects the severity of the delay of the GPS signal as it travels through the troposphere. For example, when a satellite is close to the horizon, the delay of the signal caused by the troposphere is maximized. The tropospheric delay of the signal from a satellite at zenith, directly above the receiver, is minimized.

Satellite Elevation and Tropospheric Effect

The situation is analogous to atmospheric refraction in astronomic observations; the effect increases as the energy passes through more of the atmosphere. The difference in GPS is that it is the delay, not the angular deviation, caused by the changing density of the atmosphere that is of primary interest. The GPS signal that travels the shortest path through the troposphere will be the least delayed by it. So, even though the delay at an elevation angle of 90° at sea level will only be about 2.4 m, it can increase to about 9.3 m at 75° and up to 20 m at 10°. There is less tropospheric delay at higher altitudes.

Modeling

Modeling the troposphere is one technique used to reduce the bias in GPS data processing, and it can be up to 95 percent effective. However, the residual 5 percent can be quite difficult to remove. Several *a priori* models have been developed, for example, the Saastamoinen model and the Hopfield models, which perform well when the satellites are at reasonably high elevation angles. However, it is advisable to limit GPS observations to those signals above 15 percent or so to ameliorate the effects of atmospheric delay.

Dry and Wet Components of Refraction

Refraction in the troposphere has a dry component and a wet component. The dry component, which contributes most of the delay, perhaps 80–90 percent, is closely correlated to the atmospheric pressure. The dry component can be more easily

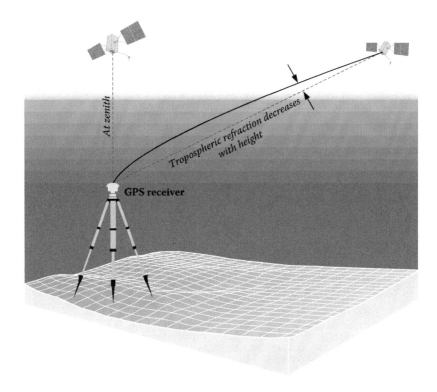

FIGURE 2.4 Tropospheric effect.

estimated than the wet component. It is fortunate that the dry component contributes the larger portion of range error in the troposphere because the size of the delay attributable to the wet component depends on the highly variable water vapor distribution in the atmosphere. Even though the wet component of the troposphere is nearer to the Earth's surface, measurements of temperature and humidity are not strong indicators of conditions on the path between the receiver and the satellite. While instruments that can provide some idea of the conditions along the line between the satellite and the receiver are somewhat more helpful in modeling the tropospheric effect, the high cost of sending water vapor radiometers and radiosondes aloft generally restricts their use to only the most high-precision GPS work. In most cases this aspect must remain in the purview of mathematical modeling, such calculations include a hydrostatic model with corrections and a horizontal gradient component. It is important to recognize that the index of tropospheric refraction decreases as height increases (see Figure 2.4).

Receiver Spacing and the Atmospheric Biases

There are other practical consequences of the atmospheric biases. As mentioned earlier, the character of the atmosphere is never homogeneous; therefore, the importance of atmospheric modeling increases as the distance between GPS receivers grows.

Consider a signal traveling from one satellite to two receivers that are close together. That signal would be subjected to very similar atmospheric effects, and, therefore, atmospheric bias modeling would be less important to the accuracy of the measurement of the relative distance between them. However, a signal traveling from the same satellite to two receivers that are far apart may pass through levels of atmosphere quite different from one another. In that case, atmospheric bias modeling would be more important. In other words, the importance of atmospheric correction increases as the *differences* in the atmosphere through which the GPS satellite signal must pass to reach the receivers increase. Such differences can generally be related to length.

MULTIPATH

Multipath is an uncorrelated error. It is a range delay symbolized by ε_{mp} in the pseudo-range equation and $\varepsilon_{m\phi}$ in the carrier phase equation. As the name implies, it is the reception of the GPS signal via multiple paths rather than from a direct line of sight (see Figure 2.5). Multipath differs from both the apparent slowing of the

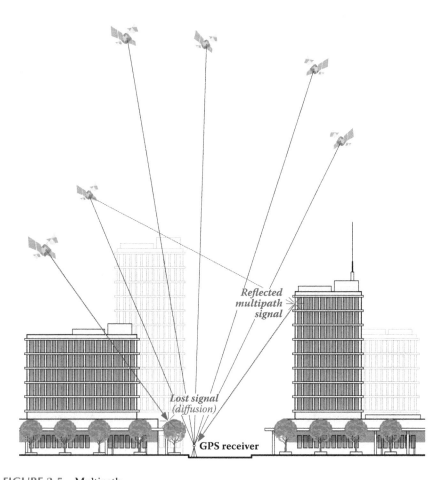

FIGURE 2.5   Multipath.

signal through the ionosphere and troposphere and the discrepancies caused by clock offsets. The range delay in multipath is the result of the reflection of the GPS signal.

Multipath occurs when part of the signal from the satellite reaches the receiver after one or more reflections or scattering from the ground, a building, or another object. These reflected signals can interfere with the signal that reaches the receiver directly from the satellite. Because carrier phase multipath is based on a fraction of the carrier wavelength and code multipath is relative to the chipping rate, the effect of multipath on pseudorange solutions is orders of magnitude larger than it is in carrier phase solutions. The effect of multipath on a carrier phase measurement can reach a quarter of a wavelength, which is about 5 cm. The effect of multipath on a pseudorange measurement can reach 1.5× the length of a chip, though it is more often a few meters. However, multipath in carrier phase is harder to mitigate than multipath in pseudoranges. Multipath can cause the correlation peak mentioned in Chapter 1 to become skewed. For example, the strategy of spacing the early, prompt, and late correlators at 1/10th of a chip rather than the standard 1 chip is not nearly as effective in mitigating carrier phase multipath as it is in pseudorange solutions.

Limiting the Effect of Multipath

The high frequency of the GPS codes tends to limit the field over which multipath can contaminate pseudorange observations. Once a receiver has achieved lock (i.e., its replica code is correlated with the incoming signal from the satellite), signals outside the expected chip length can be rejected. Generally speaking, multipath delays of less than one chip, those that are the result of a single reflection, are the most troublesome.

Fortunately, there are factors that distinguish reflected multipath signals from direct, line-of-sight signals. For example, reflected signals at the frequencies used for L1, L2, and L5 tend to be weaker and more diffuse than the directly received signals. Another difference involves the circular polarization of the GPS signal. The polarization is actually reversed when the signal is reflected. Reflected, multipath signals become *left-hand circular polarized* (LHCP), whereas the signals received directly from the GPS satellites are *right-hand circular polarized* (RHCP). RHCP means that it rotates clockwise when observed in the direction of propagation. However, while the majority of multipath signals may be LHCP, it is possible for them to arrive at the received in-phase usually through an even number of multiple reflections. These characteristics allow some multipath signals to be identified and rejected at the receiver's antenna.

Antenna Design and Multipath

GPS antenna design can play a role in minimizing the effect of multipath. Ground planes, usually a metal sheet, are used with many antennas to reduce multipath interference by eliminating signals from low elevation angles. Generally, larger ground planes, multiple wavelengths in size, have a more stabilizing influence than smaller ground planes. However, such ground planes do not provide much protection from the propagation of waves along the ground plane itself. When a GPS signal's wave front arrives at the edge of an antenna's ground plane from below, it can induce a surface wave on the top of the plane that travels horizontally (see Figure 2.6).

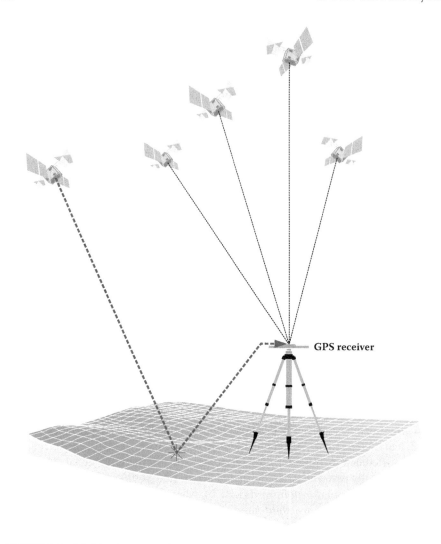

GPS receiver

FIGURE 2.6   Multipath propagation.

Another way to mitigate this problem is the use of a choke ring antenna. Choke ring antennas, based on a design first introduced by the Jet Propulsion Laboratory, can reduce antenna gain at low elevations. This design contains a series of concentric circular troughs that are a bit more than a quarter of a wavelength deep. A choke ring antenna can prevent the formation of these surface waves. However, neither ground planes nor choke rings remove the effect of reflected signals from above the antenna very effectively. There are signal processing techniques that can reduce multipath.

A widely used strategy is the 15° cutoff or *mask angle*. This technique calls for tracking satellites only after they are more than 15° above the receiver's horizon. Careful attention in placing the antenna away from reflective surfaces, such as nearby buildings, water, or vehicles, is another way to minimize the occurrence of multipath.

RECEIVER NOISE

Receiver noise is directly related to thermal noise, dynamic stress, and so on in the GPS receiver itself. Receiver noise is also an uncorrelated error source.

The effects of receiver noise on carrier phase measurements symbolized by ε_ϕ, like multipath, are small when compared to their effects on pseudorange measurements, ε_p. Generally speaking, the receiver noise error is about 1 percent of the wavelength of the signal involved. In other words, in code solutions the size of the error is related to chip width. For example, the receiver noise error in a C/A code solution can be around 3 m, which is about an order of magnitude more than it is in a P code solution, about 3 cm. Additionally, in carrier phase solutions the receiver noise error contributes millimeters to the overall error.

SOLUTIONS

There are a variety of ways to limit the effects of the biases in GPS work, but whether the techniques involve the methods of data collection or ways of processing the data, the objective is the management of errors.

SOME METHODS OF DATA COLLECTION

Static and Kinematic

GPS work is sometimes divided into three categories: positioning, navigation, and timing (PNT). Most often GPS surveying is concerned with the first of these, positioning. In general, there are two techniques used in surveying. They are kinematic and static. In static GPS surveying sessions the receivers are motionless during the observation. Because static work most often provides higher accuracy and more redundancy than kinematic work, it is usually done to establish control. The results of static GPS surveying are processed after the session is completed. In other words, the data are *post-processed GPS*.

In kinematic GPS surveying the receivers are either in periodic or continuous motion. Kinematic GPS is done when real-time, or near real-time, results are needed. When the singular objective of kinematic work is positioning, the receivers move periodically using the start and stop methodology originated by Dr. Benjamin Remondi in the 1980s. When the receivers are in continuous motion, the objective may be acquisition of the location, attitude and velocity of a moving platform (i.e., navigation), or positioning. The distinction between navigation and positioning is lessening.

Single-Point

Single-point GPS is the most familiar and ubiquitous application of the technology. It is the solution used by cell phones, GPS-enabled cameras, car navigation systems, and many more devices. Single-point positioning is also known as the *navigation solution* or *absolute positioning*. Typically, the results of this solution are in real time or near real time. It is characterized by a single receiver measuring pseudoranges to a minimum of four satellites simultaneously (see Figure 2.7). In this solution the

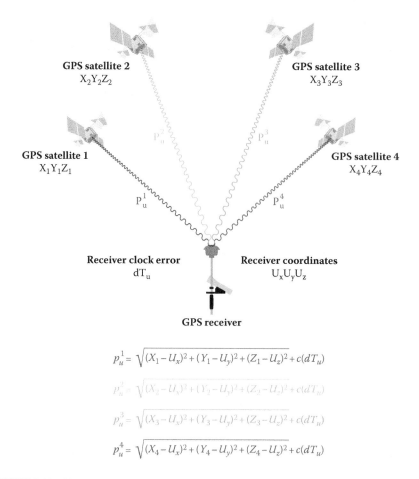

FIGURE 2.7   Single-point positioning.

receiver must also rely on the information it receives from the satellite's Navigation messages to learn the positions of the satellites, the satellite clock offset, the ionospheric correction, etc. Even if all the data in the Navigation message contained no errors, and they surely do, four unknowns remain: the position of the receiver in three Cartesian coordinates, u_x, u_y, and u_z, and the receiver's clock error dT_u. Three pseudoranges provide enough data to solve for u_x, u_y, and u_z, and the fourth pseudorange provides the information for the solution of the receiver's clock offset.

The ability to measure dT, the receiver's clock error, is one reason the moderate stability of quartz crystal clock technology is entirely adequate as a receiver oscillator. A unique solution is found here because the number of unknowns is not greater than the number of observations. The receiver tracks a minimum of four satellites simultaneously; therefore, these four equations can be solved simultaneously for every *epoch* of the observation. An epoch in GPS is a very short period of observation time and is generally just a small part of a longer measurement. However, theoretically, there is enough information in any single epoch to solve these equations.

This is the reason the trajectory of a receiver in a moving vehicle can be determined by this method. With four satellites available, resolution of a receiver's position and velocity are both available through the simultaneous solution of these four equations. Single-point positioning with its reliance on the Navigation message is in a sense the fulfillment of the original idea of GPS.

Relative Positioning

One receiver is employed in single-point positioning. A minimum of two receivers are involved in *relative positioning* or *differential GPS* positioning. The term differential GPS (DGPS) sometimes indicates the application of this technique with coded pseudorange measurements, whereas relative GPS indicates the application of this technique with carrier phase measurements. However, these definitions are by no means universal, and the use of the terms relative and differential GPS have become virtually interchangeable.

In relative positioning, one of the two receivers involved occupies a known position during the session. It is the base. The objective of the work is the determination of the position of the other, the rover, relative to the base. Both receivers observe the same constellation of satellites at the same time, and because, in typical applications, the vector between the base and the rover, known as a *baseline*, is so short compared with the 20,000 km altitude of the GPS satellites, there is extensive correlation between observations at the base and the rover. In other words, the two receivers record very similar errors and because the base's position is known, corrections can be generated there that can be used to improve the solution at the rover.

If the carrier phase observable is used in relative positioning baseline measurement accuracies of ±(1 cm + 2 ppm) are achievable. It is possible for GPS measurements of baselines to be as accurate as 1 or even 0.1 ppm. If realized that would mean that the measurement of a 9 mile baseline would approach its actual length within ±0.05 ft. (1 ppm) or ±0.005 ft. (0.1 ppm).

DIFFERENCING

Please recall that the biases in both the pseudorange and the carrier phase equations discussed at the top obscure the true geometric ranges between the receivers to the satellites that then contaminate the measurement of the baseline between the receivers. In other words, to reveal the actual vectors between two or more receivers used in relative positioning, those errors must be diminished to the degree that is possible. Fortunately, some of those embedded biases can be virtually eliminated by combining the simultaneous observables from the receivers in processes known as *differencing*. Even though the noise is increased by a factor of 2 with each differencing operation, it is typically used in commercial data processing software for both pseudorange and carrier phase measurements.

There are three types of differencing: *single difference*, *double difference*, and *triple difference*. Within the single difference category, there are the *between-receivers single difference* and the *between-satellites single difference*. Both require that all the receivers observed the same satellites at the same time.

Between-Receivers Single Difference

A between-receivers single difference involves two receivers observing a single satellite (see Figure 2.8).

A between-receivers single difference reduces the effect of biases even though it does not eliminate them. Because the two receivers are both observing the same satellite at the same time, the difference between the satellite clock bias dt at the first receiver and dt at the second receiver, Δdt, is obviously zero. Also, because the baseline is typically short compared with the 20,000 km altitude of the GPS satellites, the atmospheric biases and the orbital errors, i.e., ephemeris errors, recorded by the two receivers at each end are similar. This correlation obviously decreases as the length of the baseline increases. Generally speaking, this correlation allows centimeter-level carrier phase positioning with baselines up to 10 km or so and meter-level positioning with baselines of a few hundreds of kilometers using pseudorange observations.

The between-receivers single difference provides better position estimates for the receivers by subtracting (i.e., differencing) each receiver's observation equation from the other. For example, if one of the receivers is a base standing at a control station whose position is known, it follows that the size of the positional error of the receiver

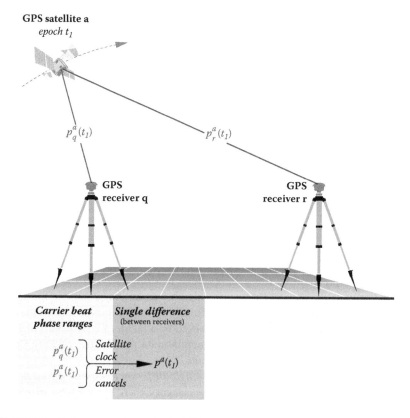

FIGURE 2.8   Between-receivers single difference.

there is knowable. Therefore, the positional error at the other end of the baseline can be estimated by finding the difference between the biases at the base and the biases at the rover. Corrections can then be generated that can reduce the three-dimensional positional error at the unknown point by reducing the level of the biases there. It is primarily this correlation and the subsequent ability to reduce the level of error that distinguishes differenced relative positioning from single-point positioning.

BETWEEN-SATELLITES SINGLE DIFFERENCE

The between-satellites single difference involves a single receiver observing two GPS satellites simultaneously and the code and/or phase measurement of one satellite are differenced, subtracted, from the other (see Figure 2.9).

The data available from the between-satellites difference allows the elimination of the receiver clock error because there is only one involved, and the atmospheric effects on the two satellite signals are again nearly identical as they come into the lone receiver, so the effects of the ionospheric and tropospheric delays are reduced. However, unlike the between-receivers single difference, the between-satellites single difference does not provide a better position estimate for the receiver involved.

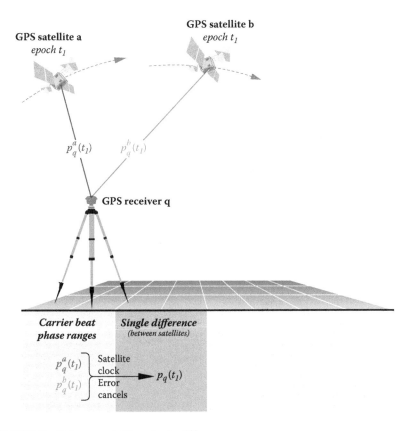

FIGURE 2.9   Between-satellites single difference.

In fact, the resulting position of the receiver is not better than would be derived from single-point positioning.

DOUBLE DIFFERENCE

When the two types of single differences are combined, the result is known as a double difference. A double difference can be said to be a between-satellites single difference of a between-receivers single difference. The improved positions from the between-receivers single difference step are not further enhanced by the combination with the between-satellites single difference. Still including the between-satellites single difference is useful because the combination virtually eliminates clock errors; both the satellite and receiver clock errors (see Figure 2.10). The removal of the receiver clock bias in the double difference makes it possible to segregate the errors

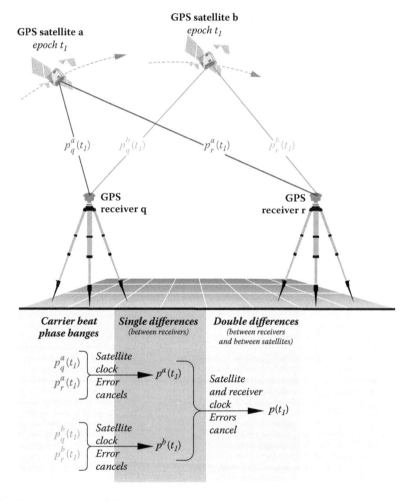

FIGURE 2.10   Double difference.

attributable to the receiver clock biases from those from other sources. This segregation improves the efficiency of the estimation of the integer cycle ambiguity in a carrier phase observation N. In other words, the reduction of all the non-integer biases makes the computation of the final accurate positions more efficient.

TRIPLE DIFFERENCE

A triple difference is the difference of two double differences over two different epochs. The triple difference has other names. It is also known as the *receiver satellite-time triple difference* and the *between-epochs difference* (see Figure 2.11). Triple differencing serves as a good pre-processing step because it can be used to detect and repair cycle slips.

A cycle slip is a discontinuity in a receiver's continuous phase lock on a satellite's signal (see Figure 2.12). A power loss, a very low signal-to-noise ratio, a failure of the receiver software, or a malfunctioning satellite oscillator can cause a cycle slip. It can also be caused by severe ionospheric conditions. Most common, however, are obstructions such as buildings and trees that are so solid they prevent the satellite signal from being tracked by the receiver. Under such circumstances, when the satellite reappears, the tracking resumes.

Coded pseudorange measurements are immune from cycle slips, but carrier phase positioning accuracy suffers if cycle slips are not detected and repaired. A cycle slip causes the critical component for successful carrier phase positioning, a resolved integer cycle ambiguity N to become instantly unknown again. In other words, lock is lost. When that happens, correct positioning requires that N be reestablished.

There are several methods of handling cycle slips. They are often controlled in post-processing rather than real time.

Repairing Cycle Slips

In post-processing, the location and their size of cycle slips must be determined; then the data set can be repaired with the application of a fixed quantity to all the subsequent phase observations. One approach is to hold the initial positions of the stations occupied by the receivers as fixed and edit the data manually. This has proven to work, but it will try the patience of Job. Another approach is to model the data on a satellite-dependent basis with continuous polynomials to find the breaks and then manually edit the data set a few cycles at a time. In fact, several methods are available to find the lost integer phase value, but they all involve testing quantities.

One of the most convenient of these methods is based on the triple difference. It can provide an automated cycle slip detection system that is not confused by clock drift, and, once least-squares convergence has been achieved, it can provide initial station positions even using the unrepaired phase combinations. They may still contain cycle slips, but the data can nevertheless be used to process approximate baseline vectors. Then the residuals of these solutions are tested, sometimes through several iterations. Proceeding from its own station solutions, the triple difference can predict how many cycles will occur over a particular time interval. Therefore, by evaluating triple-difference residuals over that particular interval, it is not only possible to determine which satellites have integer jumps but also to determine the

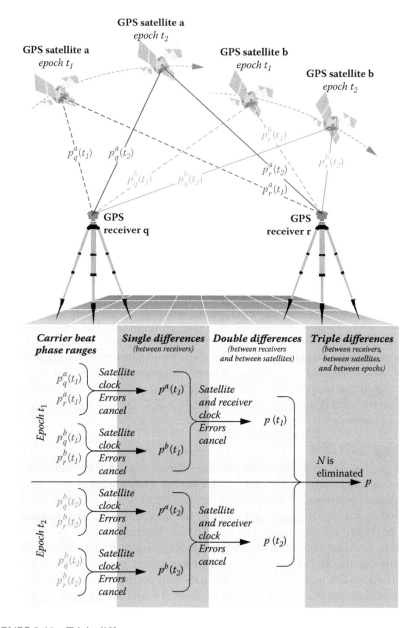

FIGURE 2.11   Triple difference.

number of cycles that have actually been lost. In a sound triple-difference solution without cycle slips, the residuals are usually limited to fractions of a cycle. Only those containing cycle slips have residuals close to one cycle or larger. Once cycle slips are discovered, their correction can be systematic.

For example, suppose the residuals of one component double difference of a triple-difference solution revealed that the residual of satellite PRN 16 minus the residual

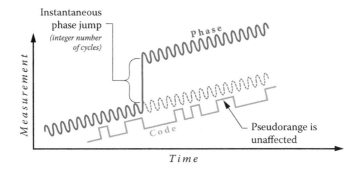

FIGURE 2.12   Cycle slip in double difference.

of satellite PRN 17 was 8.96 cycles. Further suppose that the residuals from the second component double difference showed that the residual of satellite PRN 17 minus the residual of satellite PRN 20 was 14.04 cycles. Then one might remove 9 cycles from PRN 16 and 14 cycles from PRN 20 for all the subsequent epochs of the observation. However, the process might result in a common integer error for PRNS 16, 17, and 20. Still, small jumps of a couple of cycles can be detected and fixed in the double-difference solutions.

In other words, before attempting double-difference solutions, the observations should be corrected for cycle slips identified from the triple-difference solution. Even though small jumps undiscovered in the triple-difference solution might remain in the data sets, the double-difference residuals will reveal them at the epoch where they occurred.

However, some conditions may prevent the resolution of cycle slips down to the one-cycle level. Inaccurate satellite ephemerides, noisy data, errors in the receiver's initial positions, or severe ionospheric effects all can limit the effectiveness of cycle-slip fixing. In difficult cases, a detailed inspection of the residuals might be the best way to locate the problem.

Components of the Carrier Phase Observable

From the moment a receiver locks onto a satellite to the end of the observation, the carrier phase observable can be divided into three parts. Two of them do not change during the session, and one of them does change (see Figure 2.13).

$$\phi = \alpha + \beta + N$$

where

   ϕ = total phase
   α = fractional initial cycle (phase measurement)
   β = observed cycle count
   N = carrier phase ambiguity (cycle count at lock on)

The *fractional instantaneous phase* is established at the first instant of the lock-on. When the receiver starts tracking the satellite, it is highly unlikely to acquire

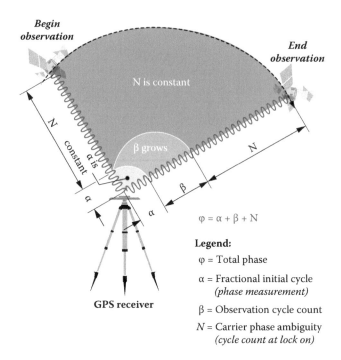

FIGURE 2.13   Carrier phase observable.

the satellite's signal precisely at the beginning of a wavelengths phase cycle. It will grab on at some fractional part of a phase, and this fractional phase will remain unchanged for the duration of the observation. It is called the fractional instantaneous phase or the phase measurement and is symbolized in the equation above by α.

The integer cycle ambiguity N represents the number of full phase cycles between the receiver and the satellite at the first instance of the receiver's lock-on. It can also be labeled the carrier phase ambiguity or the cycle count at lock-on. It does not change from the moment of the lock to the end of the observation unless that lock is lost. However, when there is a cycle slip, lock is lost, and by the time the receiver reacquires the signal, the normally constant integer ambiguity has changed. In that case the receiver loses its place in its count of the integer number of cycles, symbolized as β, and N is lost.

The value β is the count of the number of full phase cycles coming in throughout the observation. Of course, the count grows from the moment of lock on until the end of the observation. In other words, β is the receiver's record of the consecutive change in full phase cycles, 1, 2, 3, 4,…, between the receiver and the satellite as the satellite flies over. If the observation proceeds without cycle slips, the observed cycle count is the only one of three numbers that changes.

POST-PROCESSING

In many ways, post-processing is the heart of a GPS control operation. On such projects some processing should be performed on a daily basis. Blunders from operators, noisy data, and unhealthy satellites can corrupt entire sessions and left undetected, such dissolution can jeopardize the whole survey. By processing daily, these weaknesses can be discovered when they can still be eliminated and a timely amendment of the observation schedule can be done if necessary. However, even after blunders and noisy data have been removed from the observation sets, GPS measurements are still composed of fundamentally biased ranges. Therefore, GPS data-processing procedures are really a series of interconnected computerized operations designed to minimize the more difficult biases and extract the true ranges.

As has been described, the biases originate from a number of sources: imperfect clocks, atmospheric delays, cycle ambiguities in carrier phase observations, and orbital errors. If a bias has a stable, well-understood structure, it can be estimated. In other cases, multifrequency observations can be used to measure the bias directly, as in the ionospheric delay, or a model may be used to predict an effect, as in tropospheric delay. Differencing is one of the most effective strategies in eradicating biases.

Correlation of Biases

As illustrated in Figures 2.8, 2.10, and 2.12, when two or more receivers observe the same satellite constellation simultaneously, a set of correlated vectors are created between the co-observing stations. Most GPS practitioners use more than two receivers. Therefore, most GPS networks consist of many sets of correlated vectors from each individual session. The longest baselines between stations on the Earth are usually relatively short when compared with the more than 20,000 km distances from the receivers to the GPS satellites. Therefore, even when several receivers are set up on somewhat widely spaced stations, as long as the data are collected by the receivers simultaneously from the same constellation of satellites, they will record very similar errors. The resulting vectors will be correlated. It is the simultaneity of observation and the correlation of the carrier phase observables that make the extraordinary GPS accuracies possible because biases that are linearly correlated can be virtually eliminated by differencing the data sets of a session.

Organization Is Essential

One of the difficulties of post-processing is the amount of data that must be managed. For example, when even one single-frequency receiver with a 1 s sampling rate tracks 1 GPS satellite for an hour, it collects about 0.15 Mb of data. However, a more realistic scenario involves four receivers observing six satellites for 3600 epochs. There can be $4 \times 6 \times 3600$ or 86,400 carrier phase observations in such a session. In other words, to process a real-life GPS survey with many sessions and many baselines correctly, a structured approach must be implemented.

One aspect of that structure is the naming of the GPS receivers' observation files. Many manufacturers recommend a file naming format that can be symbolized by *pppp-ddd-s.yyf*. The first letters of the file name (*pppp*) indicate the point number of the station occupied. The day of the year, or Julian date, can be accommodated in the next three places (*ddd*), and the final place left of the period is the session number (*s*). The year (*yy*) and the file type (*f*) are sometimes added to the right of the period.

The first step in GPS data processing is downloading the collected data from the internal memory of the receiver or data logger into a computer. When the observation sessions have been completed for the day, the data from each receiver is transferred. Nearly all GPS systems used in surveying are computer compatible and can accommodate post-processing in the field. However, none can protect the user from a failure to back up this raw observational data onto some other form of semipermanent storage.

Receiver and data logger memory capacity is sometimes limited, and it may be necessary to clear older data to make room for new sessions. Still, it is a good policy to create the necessary space with the minimum deletion and restrict it to only the oldest files in the memory. In this way, the recent data can be retained as long as possible, and the data can provide an auxiliary backup system. However, when a receiver's memory is finally wiped of a particular session, if redundant raw data are not available, re-observation may be the only remedy.

In most GPS receivers the Navigation message, meteorological data, observables, and all other raw data are usually in a manufacturer-specific, binary form. They are also available in receiver independent exchange format (RINEX). These raw data are usually saved in several distinct files. For example, the computer operator will likely find that the phase measurements downloaded from the receiver will reside in one file and the satellite's ephemeris data in another, and so forth. Likewise, the measured pseudorange information may be found in its own dedicated file, the ionospheric information in another, and so on. The particular division of the raw data files is designed to accommodate the suite of processing software and the data management system that the manufacturer has provided its customers, so each will be somewhat unique.

CONTROL

All post-processing software suites require control. GPS static work sometimes satisfies that requirement by inclusion of National Geodetic Survey (NGS) monuments in the network design. In this approach the control stations are occupied by the surveyor building the network. There is an often used alternative. Continuously operating reference stations (CORS) already occupy many NGS control monuments and constantly collect observations; their data can be used to support carrier phase static control surveys. The most direct method is to download the CORS data files posted on the Internet. The CORS data collected during the time of the survey can be combined with those collected in the field. They can be used to post-process the baselines and derive positions for the new points.

THE FIRST POSITION

If performing in-house carrier phase post-processing the first solution is typically a simple pseudorange single-point position at each end of the baseline. These code solutions provide the approximate position of each of the two receiver's antennas. Each position establishes a search area, a three-dimensional volume of uncertainty, the size of which is defined by the accuracy of the code solution. The correct position of the receiver is contained within it somewhere. The computational time required to find it depends on the size of the search area.

Next cycle slips are addressed because the subsequent instantaneous phase jumps will defeat the upcoming double differencing. Triple differencing is not as affected by cycle slips and is insensitive to integer ambiguities so it typically precedes double differencing and is used to clean the data by detecting and repairing cycle slips. In fact, pre-processing using the triple difference weeds out and fixes cycle slips; this is one of its primary appeals.

As to detection, when a large residual appears in one of the triple-difference's component double differences, it is likely caused by a cycle slip so the satellite pairs can be sorted until the offending signal is singled-out and repaired. Triple differencing can also provide first positions for the receivers. While the receiver coordinates that result are usually more accurate than pseudorange solutions, their geometric strength is weak and not sufficiently accurate for determining short baselines. Nevertheless, they provide a starting point for the more accurate double-difference solutions that follow.

The next baseline processing steps usually involve two types of double differences that result in the float and the fixed solutions. In this first instance, there is no effort to translate the biases into integers, they are allowed to float. However, when phase measurements for even one frequency, L1, L2, or L5, are available, a sort of calculated guess at the ambiguity is the most direct route to the correct solution. Not just N but a number of unknowns, such as clock parameters and point coordinates, are estimated in this geometric approach. However, all of these estimated biases are affected by unmodeled errors and that causes the integer nature of N to be obscured. In other words, the initial estimation of N in a float solution is likely to appear as a real number rather than an integer. Where the data are sufficient, these floating real-number estimates are very close to integers—so close that they can next be rounded to their true integer values in a second adjustment of the data.

There follows a search for and identification of the integer values that minimize the residuals in the float solutions. There is then a second double-difference solution. First, the standard deviations of the adjusted integers are inspected. Those found to be significantly less than one cycle can be safely constrained to the nearest integer value. In other words, the estimations closest to integers with the minimum standard error are rounded to the nearest integer. This procedure is only pertinent to double differences because phase ambiguities are moot in triple differencing.

The number of parameters involved can be derived by multiplying one less the number of receivers by one less the number of satellites involved in the observation. For multifrequency observations the phase ambiguities for each frequency are best fixed separately. However, such a program of constraints is not typical in long baselines where the effects of the ionosphere and inaccuracies of the satellite ephemerides make the situation less determinable.

In small baselines, however, where these biases tend to closely correlate at both ends, integer fixing is almost universal in GPS processing. Once the integer ambiguities of one baseline are fixed, the way is paved for constraint of additional ambiguities in subsequent iterations. Then, step by step, more and more integers are set, until all that can be fixed have been fixed. Baselines of several thousand kilometers can be constrained in this manner.

With the integer ambiguities fixed, the GPS observations produce a series of vectors, the raw material for the final adjustment of the survey. The observed baselines represent very accurately determined relative locations between the stations they connect. However, a GPS survey is usually related to the rest by its connection with and adjustment to the national network.

LEAST-SQUARES ADJUSTMENT

There are numerous adjustment techniques, but least-squares adjustment is the most precise and most commonly used in GPS. The foundation of least-squares adjustment is the idea that the sum of the squares of all the residuals applied to the GPS vectors in their final adjustment should be held to the absolute minimum. However, minimizing this sum requires first defining those residuals approximately. Therefore, in GPS the process is based on equations where the observations are expressed as a function of unknown parameters, but parameters that are nonetheless given approximate initial values. Then by adding the squares of the terms thus formed and differentiating their sum, the derivatives can be set equal to zero. For complex work like GPS adjustments, the least-squares method has the advantage that it allows for the smallest possible changes to original estimated values.

It is important to note that the downside of least-squares adjustment is its tendency to spread the effects of even one mistake throughout the work. In other words, it can cause large residuals to show up for several measurements that are actually correct. When that happens, it can be hard to know exactly what is wrong. The adjustment may fail the *chi-square* test. That tells you there is a problem, but unfortunately, it cannot tell you where the problem is. The chi-square test is based in probability, and it can fail because there are still unmodeled biases in the measurements. Multipath, ionosphere, and troposphere biases, and so forth, may cause it to fail.

Most programs look at the residuals in light of a limit at a specific probability, and when a particular measurement goes over the limit, it gets highlighted. Trouble is, you cannot always be sure that the one that got tagged is the one that is the problem. Fortunately, least-squares adjustment does offer a high degree of comfort once all the hurdles have been cleared. If the residuals are within reason and the chi-square test is passed, it is very likely that the observations have been adjusted properly.

NETWORK ADJUSTMENT

The solution strategies of GPS adjustments themselves are best left to particular suites of software. Suffice it to say that the single baseline approach, that is, a baseline-by-baseline adjustment, has the disadvantage of ignoring the actual correlation of the observations of simultaneously occupied baselines. An alternative approach involves a network adjustment approach where the correlation between the baselines themselves can be more easily taken into account.

For the most meaningful network adjustment, the endpoints of every possible baseline should be connected to at least two other stations. Thereby, the quality of the work itself can be more realistically evaluated. For example, the most common observational mistake, the mis-measured antenna height or height of instrument (HI), is very difficult to detect when adjusting baselines sequentially, one at a time, but a network solution spots such blunders more quickly.

Most GPS post-processing adjustment begins with a minimally constrained least-squares adjustment. That means that all the observations in a network are adjusted together with only the constraints necessary to achieve a meaningful solution, e.g., the adjustment of a GPS network with the coordinates of only one station fixed. The purpose of the minimally constrained approach is to detect large mistakes, like a misidentifying one of the stations. The residuals from a minimally constrained work should come pretty close to the precision of the observations themselves. If the residuals are particularly large, there are probably mistakes; if they are really small, the network itself may not be as strong as it should be.

This minimally constrained solution is usually followed by an overconstrained solution. An overconstrained solution is a least-squares adjustment where the coordinate values of more than one selected control station are held fixed.

USING A PROCESSING SERVICE

An alternative to the just described in-house processing is the use of a processing service. There are several services available to GPS surveyors. While they differ somewhat in their requirements, they are all based on the same idea. Static GPS data collected in the field may be uploaded to a website on the Internet by the hosting organization, which will then return the final positions, often free of charge. Usually, the user is directed to an ftp site, where the results can be downloaded.

There can be advantages to the use of such a processing service. Aside from removing the necessity of having post-processing software in-house, there is the strength of the network solution available from them. In other words, rather than the data sent in by the GPS surveyor being processed against a single CORS in the vicinity of the work, it is processed against a group of the nearest CORS. There are often three in the group. Such a network-based solution improves the integrity of the final position markedly and may compensate for long baselines required by the sparseness of the CORS in the area of interest.

Among the online resources available for processing services is the National Geodetic Surveys *Online Positioning User Service* (OPUS). This service allows the user to submit RINEX files through the NGS Web page. They are processed automatically with NGS computers and software utilizing data from three CORS that may be user selected. There are others, such as the *Australian Online GPS Processing Service* (AUSPOS).

SUMMARY

Relative positioning by carrier beat phase measurement is the primary vehicle for high-accuracy GPS control surveying. Simultaneous static observations, double differencing in post-processing, and the subsequent construction of networks from GPS baselines are the hallmarks of geodetic and control work in the field. The strengths of these methods generally outweigh their weaknesses, particularly where there can be an unobstructed sky and relatively short baselines and where the length of observation sessions is not severely restricted.

Differencing is an ingenious approach to minimizing the effect of biases in both pseudorange and carrier phase measurements. It is a technique that largely overcomes the impossibility of perfect time synchronization. Double differencing is the most widely used formulation. Double differencing still contains the initial integer ambiguities, of course. The estimates of the ambiguities generated by the initial processing are usually not integers; in other words, some orbital errors and atmospheric errors remain. However, with the knowledge that the ambiguities ought to be integers, during subsequent processing, it is possible to force estimates for the ambiguities that are in fact integers. When the integers are so fixed, the results are known as a *fixed solution*, rather than a *float solution*. It is the double-differenced carrier phase–based fixed solution that makes the very high accuracy possible with GPS. However, it is important to remember that multipath, cycle slips, incorrect instrument heights, and a score of other errors whose effects can be minimized or eliminated by good practice are simply not within the purview of differencing at all. The unavoidable biases that can be managed by differencing—including clock, atmospheric, and orbital errors—can have their effects drastically reduced by the proper selection of baselines, the optimal length of the observation sessions, and several other considerations included in the design of a GPS survey. However, such decisions require an understanding of the sources of these biases and the conditions that govern their magnitudes. The adage of *garbage in, garbage out* is as true of GPS as any other surveying procedure. The management of errors cannot be relegated to mathematics alone.

EXERCISES

1. The Earth's atmosphere affects the signals from GPS satellites as they pass through. Which of these statements about the phenomenon is true?
   a. GPS signals are refracted by the ionosphere. Considering pseudoranges, the apparent length traveled by a GPS signal seems to be too short, and in the case of the carrier wave, it seems to be too long.
   b. The ionosphere is dispersive. Therefore, despite the number of charged particles the GPS signal encounters on its way from the satellite to the receiver, the time delay is directly proportional to the square of the transmission frequency. Lower frequencies are less affected by the ionosphere.
   c. The ionosphere is a region of the atmosphere where free electrons do not affect radio wave propagation owing mostly to the effects of solar ultraviolet and x-ray radiation.
   d. TEC depends on user's location, observing direction, magnetic activity, sunspot cycle, season, and time of day, among other things. In the midlatitudes the ionospheric effect is usually least between midnight and early morning and most at local noon.

2. Clocks in GPS satellites are rubidium and cesium oscillators, whereas quartz crystal oscillators provide the frequency standard in most GPS

receivers. Considering the relationship between these standards, which of the following statements is not true?

    a. Quartz clocks are not as stable as the atomic standards in the GPS satellites and are more sensitive to temperature changes, shock, and vibration.

    b. Both GPS receivers and satellites rely on their oscillators to provide a stable reference so that other frequencies of the system can be generated from or compared with them.

    c. The foundation of the oscillators in GPS receivers and satellites is the piezoelectric effect.

    d. Oscillators in the GPS satellites are also known as atomic clocks.

3. What is an advantage available using a dual-frequency GPS receiver that is not available using a single-frequency GPS receiver?

    a. A single-frequency GPS receiver cannot collect enough data to perform single-, double-, or triple-difference solutions.

    b. A dual-frequency receiver affords an opportunity to track the P code, but a single-frequency receiver cannot.

    c. A dual-frequency receiver has access to the navigation code, but a single-frequency receiver does not.

    d. Over long baselines, a dual-frequency receiver can rely on the dispersive nature of the ionosphere to model and virtually remove the ionospheric bias, whereas a single-frequency receiver cannot.

4. Which of the following statements is not correct concerning refraction of the GPS signal in the troposphere?

    a. L1 and L2 carrier waves are refracted equally.

    b. When a GPS satellite is near the horizon, its signal is most affected by the atmosphere.

    c. The density of the troposphere governs the severity of its effect on a GPS signal.

    d. The wet component of refraction in the troposphere contributes the larger portion of the range error.

5. In the between-receivers single difference across a short baseline, which of the following problems are not virtually eliminated?

    a. Satellite clock errors

    b. Atmospheric bias

    c. Integer cycle ambiguities

    d. Orbital errors

6. In the initial double difference across a short baseline, which of the following problems are not virtually eliminated?

    a. Atmospheric bias

    b. Satellite clock errors

    c. Receiver clock errors

    d. Integer cycle ambiguity

7. Which of the following statements does not correctly complete the sentence, "If cycle slips occurred in the observations…?"
   a. "…they could have been caused by intermittent power to the receiver."
   b. "…they might be detected by triple differencing."
   c. "…they are equally troublesome in pseudorange and carrier phase measurements."
   d. "…it is best if they are repaired before double differencing is done."

8. What is the correct value for the maximum time interval allowed between GPS Time and a particular GPS satellite's onboard clock?
   a. 1 ns
   b. 1 ms
   c. 1 μs
   d. 1 fs

9. Which of the following biases are not mitigated by using relative positioning GPS or DGPS?
   a. Ionospheric effect
   b. Tropospheric effect
   c. Multipath
   d. Satellite clock errors

10. Which one of the following statements is defining TEC?
    a. Total electron content in troposphere
    b. Measure of the number of free electrons in a column through the ionosphere with a cross-sectional area of 1 m^2
    c. Measurement unit for the density of the geoid
    d. Total number of the electrons in the F layer

11. Which statement concerning triple differences is most correct?
    a. Triple differences are the differences between the carrier phase observations of two receivers of the same satellite at the same epoch.
    b. Triple differences are the differences of two single differences of the same epoch that refer to two different satellites.
    c. Triple differences are the difference of two double differences of two different epochs.
    d. Triple difference does depend on the integer cycle ambiguity because this unknown is not constant in time.

12. What is the most useful aspect of triple differences in post-processing GPS observations?
    a. Triple differences can be an initial step to eliminate cycle slips.
    b. Triple differences provide the best solution for the receiver positions.
    c. Triple differences have more information than double differences.
    d. Triple differences eliminate the need to model ionospheric biases.

13. What is the most useful aspect of a double difference in post-processing GPS observations?
   a. Double differences do not need resolution of the integer ambiguity.
   b. Double differences eliminate clock errors.
   c. Double differences eliminate multipath.
   d. Double differences eliminate all atmospheric biases.

14. What would cause the integer cycle ambiguity N to change after a receiver has achieved lock-on?
   a. Incorrect HI
   b. Inclement weather
   c. Inaccurate centering over the station
   d. Cycle slip

15. What is the difference between a float and a fixed solution?
   a. Integer cycle ambiguity is resolved in a fixed solution, but not in a float solution.
   b. Float solution is processed using a single difference, unlike a fixed solution.
   c. Float solution is more accurate than a fixed solution.
   d. Float solution may have cycle slips but not a fixed solution.

16. What is the idea underlying the use of least-squares adjustment of GPS networks?
   a. Sum of the squares of all the residuals applied to the GPS vectors in their final adjustment should be held to zero.
   b. Multiplication of the GPS vectors by all the residuals in their final adjustment should be held to one.
   c. Sum of the squares of all the residuals applied to the GPS vectors in their final adjustment should be held to the absolute minimum.
   d. Least of the squares of all the residuals applied to the GPS vectors in their final adjustment should be held to zero.

17. What is usually the purpose of the initial minimally constrained least-squares adjustment in GPS work?
   a. To repair cycle slips
   b. To fix the integer cycle ambiguity
   c. To establish the correct coordinates of all the project points
   d. To find any large mistakes in the work

18. What may cause an adjustment to fail the chi-square test?
   a. Unmodeled biases in the measurements
   b. Smaller than expected residuals
   c. Measurement with too many epochs
   d. Lack of cycle slips

ANSWERS AND EXPLANATIONS

1. Answer is (d)

    Explanation: The modulations on carrier waves encountering the free electrons in the ionosphere are retarded, but the carrier wave itself is advanced. The slowing is known as the ionospheric delay. In the case of pseudoranges the apparent length of the path of the signal is stretched; it seems to be too long. However, when considering the carrier wave, the path seems to be too short. It is interesting that the absolute value of these apparent changes is almost exactly equivalent.

    The ionosphere is dispersive. Refraction in the ionosphere is a function of a wave's frequency; the higher the frequency, the less the refraction, and the shorter the delay in the case of modulations on the carrier wave. The ionospheric effect is proportional to the inverse of the frequency squared.

    The magnitude of the ionospheric effect increases with electron density. Also known as the total electron content (TEC), the electron density varies in both time and space. However, in the midlatitudes the minimum most often occurs from midnight to early morning and the maximum near local noon.

2. Answer is (c)

    Explanation: Rubidium and cesium atomic clocks are aboard the current constellation of GPS satellites. Their operation is based on the resonant transition frequency of the Rb-87 and CS-133 atoms, respectively. Alternatively, the quartz crystal oscillators in most GPS receivers utilize the piezoelectric effect. Both GPS receivers and satellites rely on their oscillators to provide a stable reference so that other frequencies of the system can be generated from or compared with them.

3. Answer is (d)

    Explanation: Sufficient data to calculate single, double, and triple differences can be available from both single- and dual-frequency receivers. The permission to track the P code is not restricted by single- or dual-frequency capability. GPS receivers have access to the navigation code whether they track L1, L2, or both.

    Using single-frequency receivers, ionospheric corrections can be computed from an ionospheric model. GPS satellites transmit coefficients for ionospheric corrections that most receivers' software use. The models assume a standard distribution of the total electron count; still, with this strategy about a quarter of the ionosphere's actual variance will be missed. However, when receivers are close together, say 30 km or less, for all practical purposes the ionospheric delay and carrier phase advance are the same for both receivers. Therefore, phase difference observations over short baselines are little affected by ionospheric bias with single-frequency receivers.

    However, that is not the case over long baselines. Dual-frequency receivers can utilize the dispersive nature of the ionosphere to overcome the effects. The resulting time delay is inversely proportional to the square of the transmission frequency. That means that L1, 1575.42 MHz, is not affected as much as L2, 1227.60 MHz. By tracking both carriers, a dual-frequency receiver has the facility of modeling and virtually removing much of the ionospheric bias.

4. Answer is (d)

   Explanation: Tropospheric effect is nondispersive for frequencies under 30 GHz. Therefore, it affects both L1 and L2 equally. Refraction in the troposphere has a dry component and a wet component. The dry component is related to the atmospheric pressure and contributes about 90 percent of the effect. It is more easily modeled than the wet component. The GPS signal that travels the shortest path through the atmosphere will be the least affected by it. Therefore, the tropospheric delay is least at the zenith and most near the horizon. GPS receivers at the ends of short baselines collect signals that pass through substantially the same atmosphere, and the tropospheric delay may not be troublesome. However, the atmosphere may be very different at the ends of long baselines.

5. Answer is (c)

   Explanation: A between-receivers single difference is the difference in the simultaneous carrier phase measurements from one GPS satellite as measured by two different receivers during a single epoch. Single differences are virtually free of satellite clock errors. The atmospheric biases and the orbital errors recorded by the two receivers in this solution are nearly identical, so they too can be virtually eliminated. However, processing does not usually end at single differences in surveying applications because the difference between the integer cycle ambiguities at each receiver and the difference between the receiver clock errors remains in the solution.

6. Answer is (d)

   Explanation: Differences of two single differences in the same epoch using two satellites are known as a double difference. The initial double difference combination is virtually free of receiver clock errors, satellite clock errors, and over short baselines, orbital and atmospheric biases. However, the integer cycle ambiguities remain.

7. Answer is (c)

   Explanation: Cycle slip is a discontinuity in a receiver's continuous phase lock on a satellites signal. When lock is lost, a cycle slip occurs. A power loss, an obstruction, a very low signal-to-noise ratio, or any other event that breaks the receiver's continuous reception of the satellite's signal causes a cycle slip. The coded pseudorange measurement is immune from this difficulty, but carrier phase measurements are not. It is best if cycle slips are removed before the double-difference solution. When a large residual appears in the component double differences of a triple difference, it is very likely caused by a cycle slip. This utility in finding and fixing cycle slips is the primary appeal of the triple difference. It can be used as a pre-processing step to weed out cycle slips and provide a first position for the receivers.

8. Answer is (b)

   Explanation: The GPS Time system is composed of the master control clock and the GPS satellite clocks and is measured from 24:00:00, January 5, 1980. However, while GPS Time itself is kept within 1 μs of UTC, excepting leap seconds, the satellite clocks can be allowed to drift up to a millisecond from GPS Time. The rates of these onboard rubidium and cesium oscillators are more stable if they are not disturbed by frequent tweaking and adjustment is kept to a minimum.

9. Answer is (c)

   Explanation: Two or more GPS receivers make relative positioning and DGPS possible. These techniques can attain high accuracy of the extensive correlation between observations taken to the same satellites at the same time from separate stations. The distance between such stations on the Earth are short compared with the 20,000 km altitude of the GPS satellites' two receivers operating simultaneously, collecting signals from the same satellites and through substantially the same atmosphere will record very similar errors of several categories. However, multipath is so dependent on the geometry of the particular location of a receiver it cannot be lessened in this way. Keeping the antenna from reflective surfaces, the use of a mask angle, or the use of a ground plane or choke ring antenna, are methods used to reduce multipath errors.

10. Answer is (b)

    Explanation: This density is often described as total electron content (TEC), a measure of the number of free electrons in a column through the ionosphere with a cross-sectional area of 1 m^2: 10^{16} is one TEC unit.

11. Answer is (c)

    Explanation: A triple difference is created by differencing two double differences at each end of the baseline. Each of the double differences involves two satellites and two receivers. A triple difference considers two double differences over two consecutive epochs. In other words, triple differences are formed by sequentially differencing double differences in time.

    Because two receivers are recording the data from the same two satellites during two consecutive epochs across a baseline, a triple difference can temporarily eliminate any concern about the integer cycle ambiguity because the cycle ambiguity is the same over the two observed epochs.

12. Answer is (a)

    Explanation: The triple difference cannot have as much information content as a double difference. Therefore, while receiver coordinates estimated from triple differences are usually more accurate than pseudorange solutions, they are less accurate than those obtained from double differences, especially fixed-ambiguity solutions. Nevertheless, the estimates that come out of triple-difference solutions refine receiver coordinates and provide a starting point for the subsequent double-difference solutions. They are also very useful in spotting and correcting cycle slips. They also provide a first estimate of the receiver's positions.

13. Answer is (b)

Explanation: Double differences have both positive and negative features. On the positive side, they make the highest GPS accuracy possible, and they remove the satellite and receiver clock errors from the observations. On the negative side, the integer cycle ambiguity, sometimes known simply as the ambiguity, cannot be ignored in the double difference. In fact, the fixed double-difference solution, usually the most accurate technique of all, requires the resolution of this ambiguity.

14. Answer is (d)

Explanation: The integer cycle ambiguity, usually symbolized by N, represents the number of full phase cycles between the receiver and the satellite at the first instance of the receiver's lock-on. N does not change from the moment lock is achieved, unless there is a cycle slip.

15. Answer is (a)

Explanation: The initial estimation of N in a float solution is likely to appear as a real number rather than an integer. However, where the data are sufficient, these floating real-number estimates are very close to integers—so close that they can be rounded to their true integer values in a second adjustment of the data. Therefore, a second double-difference solution follows. The estimation of N that is closest to an integer and has the minimum standard error is usually taken to be the most reliable and is rounded to the nearest integer. Now, with one less unknown, the process is repeated and another ambiguity can be fixed, and so on. This approach leads to the fixed solution in which N can be held to integer values. It is usually quite successful in double differences over short baselines. The resulting fixed solutions most often provide much more accurate results than were available from the initial floating estimates.

16. Answer is (c)

Explanation: The foundation of the idea of least-squares adjustment is the idea that the sum of the squares of all the residuals applied to the GPS vectors in their final adjustment should be held to the absolute minimum.

17. Answer is (d)

Explanation: Most GPS post-processing adjustment begins with a minimally constrained least-squares adjustment. That means that all the observations in a network are adjusted together with only the constraints necessary to achieve a meaningful solution; for example, the adjustment of a GPS network with the coordinates of only one station fixed. The purpose of the minimally constrained approach is to detect large mistakes, like a misidentifying one of the stations. The residuals from a minimally constrained work should come pretty close to the precision of the observations themselves. If the residuals are particularly large, there are probably mistakes; if they are really small, the network itself may not be a strong as it should be.

18. Answer is (a)

Explanation: The adjustment may fail the chi-square test. That tells you there is a problem; unfortunately, it cannot tell you where the problem is. The chi-square test is based in probability, and it can fail because there are still unmodeled biases in the measurements. Multipath, ionosphere, and troposphere biases, and so forth, may cause it to fail.

Most programs look at the residuals in light of a limit at a specific probability, and when a particular measurement goes over the limit, it gets highlighted. Trouble is you cannot always be sure that the one that got tagged is the one that is the problem. Fortunately, the least-squares method offers a high degree of comfort once all the hurdles have been cleared. If the residuals are within reason and the chi-square test is passed, it is very likely that the observations have been adjusted properly.

3 Framework

TECHNOLOGICAL FORERUNNERS

CONSOLIDATION

In the early 1970s the Department of Defense (DoD) commissioned a study to define its future positioning needs. That study found nearly 120 different types of positioning systems in place, all limited by their special and localized requirements. The study called for consolidation and NAVSTAR (navigation system with timing and ranging) Global Positioning System (GPS) was proposed. Specifications for the new system were developed to build on the strengths and avoid the weaknesses of its forerunners. Here is a brief look at some of the earlier systems and their technological contributions toward the development of GPS.

TERRESTRIAL RADIO POSITIONING

Long before the satellite era the developers of radar (radio detecting and ranging) were working out many of the concepts and terms still used in electronic positioning today. For example, the classification of the radio portion of the electromagnetic spectrum by letters, such as the L-band now used in naming the GPS carriers, was introduced during World War II to maintain military secrecy about the new technology.

Actually, the 23 cm wavelength that was originally used for search radar was given the L designation because it was long compared to the shorter 10 cm wavelengths introduced later. The shorter wavelength was called S-band, the S for short. The Germans used even shorter wavelengths of 1.5 cm. They were called K-band, for kurtz, meaning short in German. Wavelengths that were neither long nor short were given the letter C, for compromise, and P-band, for previous, was used to refer to the very first meter-length wavelengths used in radar. There is also an X-band radar used in fire-control radars and other applications.

In any case, the concept of measuring distance with electromagnetic signals (ranging in GPS) had one of its earliest practical applications in radar. Since then, there have been several incarnations of the idea.

Shoran (short-range navigation), a method of electronic ranging using pulsed 300 MHz very high frequency (VHF) signals, was designed for bomber navigation but was later adapted to more benign uses. The system depended on a signal sent by a mobile transmitter-receiver-indicator unit being returned to it by a fixed transponder. The elapsed time of the round-trip was then converted into distances. It was not long before the method was adapted for use in surveying. Using Shoran from 1949 to 1957, Canadian geodesists were able to achieve precisions as high as 1:56,000 on lines of several hundred kilometers. Shoran was used in hydrographic surveys in 1945 by the U.S. Coast and Geodetic Survey. In 1951, Shoran was used to locate islands off Alaska in the Bering Sea that were beyond positioning by visual means. Also, in the early 1950s, the U.S. Air Force created a Shoran measured trilateration net between Florida and Puerto Rico that was continued on to Trinidad and South America.

Shoran's success led to the development of Hiran (high-precision Shoran). Hiran's pulsed signal was more focused, its amplitude more precise, and its phase measurements more accurate. Hiran, also applied to geodesy, was used to make the first connection between Africa, Crete, and Rhodes in 1943. However, its most spectacular application were the arcs of triangulation joining the North American Datum (1927) with the European Datum (1950) in the early 1950s. By knitting together continental datums, Hiran surveying might be considered to be the first practical step toward positioning on a truly global scale.

Satellite Advantages

These and other radio navigation systems proved that ranges derived from accurate timing of electromagnetic radiation were viable. As useful as they were in geodesy and air navigation, they only whet the appetite for a higher platform. In 1957 the development of Sputnik, the first Earth-orbiting satellite, made that possible.

Some of the benefits of Earth-orbiting satellites were immediately apparent. It was clear that the potential coverage was virtually unlimited. The coverage of a terrestrial radio navigation system is limited by the propagation characteristics of electromagnetic radiation near the ground. To achieve long ranges, the basically spherical shape of the Earth favors low frequencies that stay close to the surface. One such terrestrial radio navigation system, *Loran-C* (long-range navigation-C), was used to determine speeds and positions of receivers up to 3000 km from fixed transmitters. Unfortunately, its frequency had to be in the low-frequency (LF) range from 90 to 110 kHz. Many nations besides the United States used Loran including Japan, Canada, and several European countries. Russia has a similar system called Chayka. In any case, Loran was phased out in the United States and Canada in 2010.

OMEGA, another low-frequency hyperbolic radio navigation system was operated from 1968 to September 30, 1997, by the U.S. Coast Guard. OMEGA was used by other countries as well. It was capable of ranges of 9000 km. Its 10 to 14 kHz frequency was so low as to be audible (the range of human hearing is about 20 Hz to 15 kHz). Such low frequencies can be profoundly affected by unpredictable ionospheric disturbances and ground conductivity, making modeling the reduced propagation velocity of a radio signal over land difficult. However, higher frequencies require line of sight.

Line of sight is no problem for Earth-orbiting satellites, of course. Signals from space overcome many low-frequency limitations, allow the use of a broader range of frequencies, and are simply more reliable. Using satellites, one could achieve virtually limitless coverage. However, development of the technology for launching transmitters with sophisticated frequency standards into orbit was not accomplished immediately.

OPTICAL SYSTEMS

Some of the earliest extraterrestrial positioning was done with optical systems. Optical tracking of satellites is a logical extension of astronomy. The astronomic determination of a position on the Earth's surface from star observations, certainly the oldest method, is actually very similar to extrapolating the position of a satellite from a photograph of it crossing the night sky. In fact, the astronomical coordinates, right ascension α and declination δ, of such a satellite image are calculated from the background of fixed stars.

Photographic images that combine reflective satellites and fixed stars are taken with *ballistic cameras* whose chopping shutters open and close very fast. The technique causes both the satellites, illuminated by sunlight or Earth-based beacons, and the fixed stars to appear on the plate as a series of dots. Comparative analysis of photographs provides data to calculate the orbit of the satellite. Photographs of the same satellite made by cameras thousands of kilometers apart can thus be used to determine the camera's positions by triangulation. The accuracy of such networks has been estimated as high as ±5 m.

Other optical systems are much more accurate. One called *SLR* (satellite laser ranging) is similar to measuring the distance to a satellite using a sophisticated EDM. A laser aimed from the Earth to satellites equipped with retro reflectors yields the range. It is instructive that two current GPS satellites carry onboard corner cube reflectors for exactly this purpose. The GPS space vehicles numbered SVN 35 (PRN 05) and SVN 36 (PRN 06) have been equipped with Laser Retro-reflector Arrays (LRA), thereby allowing ground stations to separate the effect of errors attributable to satellite clocks from errors in the satellite's ephemerides. The second of these was launched in 1994 and is still in service.

The same technique, called *LLR* (lunar laser ranging), is used to measure distances to the Moon using corner cube reflector arrays left there during manned missions. There are four available arrays. Three of them set during Apollo missions and one during the Soviet Lunokhod 2 mission. These techniques can achieve positions of centimeter precision when information is gathered from several stations. However, one drawback is that the observations must be spread over long periods, up to a month, and they, of course, depend on two-way measurement.

While some optical methods, like SLR, can achieve extraordinary accuracies, they can at the same time be subject to some chronic difficulties. Some methods require skies to be clear simultaneously at widely spaced sites. Even then, local atmospheric turbulence causes the images of the satellites to scintillate.

EXTRATERRESTRIAL RADIO POSITIONING

The earliest American extraterrestrial systems were designed to assist in satellite tracking and satellite orbit determination, not geodesy. Some of the methods used did not find their way into the GPS technology at all. Some early systems relied on the reflection of signals and transmissions from ground stations that would either bounce off the satellite or stimulate onboard transponders. However, systems that required the user to broadcast a signal from the Earth to the satellite were not favorably considered in designing the then-new GPS system. Any requirement that the user reveal his position was not attractive to the military planners responsible for developing GPS. They favored a passive system that allowed the user to simply receive the satellite's signal. So, with their two-way measurements and utilization of several frequencies to resolve the cycle ambiguity, many early extraterrestrial tracking systems were harbingers of the modern electronic distancing measuring (EDM) technology more than GPS.

Prime Minitrack

Elsewhere there were ranging techniques useful to GPS. NASA's first satellite tracking system, *Prime Minitrack*, relied on phase difference measurements of a single-frequency carrier broadcast by the satellites themselves and received by two separate ground-based antennas. This technique is called *interferometry*. Interferometry is the measurement of the difference between the phases of signals that originate at a common source but travel different paths to the receivers. The combination of such signals, collected by two separate receivers, invariably finds them out of step because one has traveled a longer distance than the other. Analysis of the signal's phase difference can yield very accurate ranges, and interferometry has become an indispensable measurement technique in several scientific fields.

VERY LONG BASELINE INTERFEROMETRY

Very long baseline interferometry (VLBI) did not originate in the field of satellite tracking or aircraft navigation but in radio astronomy. The technique was so successful it is still in use today. Radio telescopes, sometimes on different continents, tape record the microwave signals from quasars, star-like points of light billions of light-years from Earth (Figure 3.1).

These recordings are encoded with time tags controlled by hydrogen masers, the most stable of oscillators (clocks). The tapes are then brought together and played back at a central processor. Cross-correlation of the time tags reveals the difference in the instants of wave front arrivals at the two telescopes. The discovery of the time offset that maximizes the correlation of the signals recorded by the two telescopes yields the distance and direction between them within a few centimeters, over thousands of kilometers.

VLBI's potential for geodetic measurement was realized as early as 1967, but the concept of high-accuracy baseline determination using phase differencing was really proven in the late 1970s. A direct line of development leads from the VLBI work of that era by a group from the Massachusetts Institute of Technology to today's most accurate GPS ranging technique, carrier phase measurement. VLBI, along with other extraterrestrial systems like SLR, also provides valuable information on the Earth's gravitational field and rotational axis. Without those data, the high accuracy of the modern coordinate systems that are critical to the success of GPS, like the Conventional Terrestrial System, would not be possible. However, the foundation for routine satellite-based geodesy actually came even earlier and from a completely different direction. The first prototype satellite of the immediate precursor of the GPS system that was successfully launched reached orbit on June 29, 1961. Its range measurements were

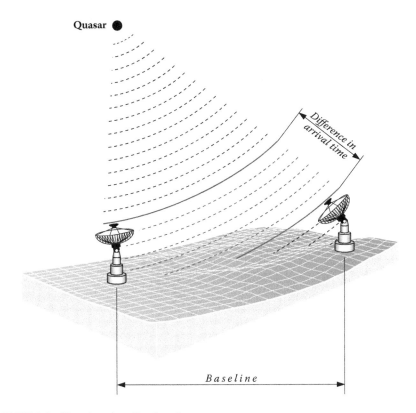

FIGURE 3.1   Very long baseline interferometry.

based on the Doppler Effect, not phase differencing, and the system came to be known as *TRANSIT*, or more formally the *Navy Navigational Satellite System*.

TRANSIT

Satellite technology and the Doppler Effect were combined in the first comprehensive Earth-orbiting satellite system dedicated to positioning. By tracking Sputnik in 1957, experimenters at Johns Hopkins University's Applied Physics Laboratory found that the Doppler shift of its signal provided enough information to determine the exact moment of its closest approach to the Earth. This discovery led to the creation of the Navy Navigational Satellite System (NNSS or NAVSAT) and the subsequent launch of six satellites specifically designed to be used for navigation of military aircraft and ships. This same system, eventually known as TRANSIT, was classified in 1964, declassified in 1967, and was widely used in civilian hydrographic and geodetic surveying for many years until it was switched off on December 31, 1996 (Figure 3.2).

The TRANSIT system had some nagging drawbacks. For example, its primary observable was based on the comparison of the nominally constant frequency generated in the receiver with the Doppler-shifted signal received from one satellite at a time. With a constellation of only six satellites, this strategy sometimes left the observer waiting up to 90 min between satellites, and at least two passes were

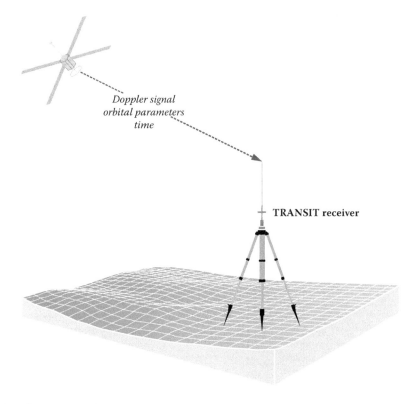

Doppler signal
orbital parameters
time

TRANSIT receiver

FIGURE 3.2   TRANSIT.

usually required for acceptable accuracy. With an orbit of only about 1100 km above the Earth the TRANSIT satellite's orbits were quite low and, therefore, unusually susceptible to atmospheric drag and gravitational variations, making the calculation of accurate orbital parameters particularly difficult.

Linking Datums

Perhaps the most significant difference between the TRANSIT system and previous extraterrestrial systems was TRANSIT's capability of linking national and international datums with relative ease. Its facility at strengthening geodetic coordinates laid the groundwork for modern geocentric datums.

System 621B and Timation

In 1963, at about the same time the Navy was using TRANSIT, the Air Force funded the development of a three-dimensional radio navigation system for aircraft. It was called System 621B. The fact that it provided a determination of the third dimension, altitude, was an improvement of some previous navigation systems. It relied on the user measuring ranges to satellites based on the time of arrival of the transmitted radio signals. With the instantaneous positions of the satellites known the user's position could be derived. The 621B program also utilized carefully designed binary codes known as PRN codes or pseudorandom noise. Even though the PRN codes appeared to be noise at first, they were actually capable of repetition and replication. This approach also allowed all of the satellites to broadcast on the same frequency. Sounds rather familiar now, doesn't it? Unfortunately, System 621B required signals from the ground to operate.

A Naval project with a name that conflated time and navigation, Timation, began in 1964. The Timation 1 satellite was launched in 1967; it was followed by Timation 2 in 1969. Both of these satellites were equipped with high-performance quartz crystal oscillators also known as *XO*. The daily error of these clocks was about 1 μs to about 300 m of ranging error. The technique they used to transmit ranging signals was called *side-tone ranging*. There was a great improvement in the clocks of Timation 3, which was launched in 1974. It was the first satellite outfitted with two rubidium clocks. Being able to have space worthy atomic frequency standards (AFS) on orbit was a big step toward accurate satellite positioning, navigation, and timing. With this development the Timation program demonstrated that a passive system using atomic clocks could facilitate worldwide navigation. The terms clock, oscillator, and frequency standard will be used interchangeably here. It was the combination of atomic clock technology, the ephemeris system from TRANSIT, and the PRN signal design from the 621B program in 1973 that eventually became GPS. There was a Department of Defence directive to the U.S. Air Force in April of 1973 that stipulated the consolidation of Timation and 621B into the navigation system called GPS. In fact, Timation 3 became part of the NAVSTAR–GPS program and was renamed Navigation Technology Satellite 1 (NTS-1). The next satellite in line, Timation 4, was known as NTS-2 (see Figure 3.3). Its onboard cesium clock had a frequency stability of 2 parts per 10^{13}. It was launched in 1977. Unfortunately, it only operated for 8 months.

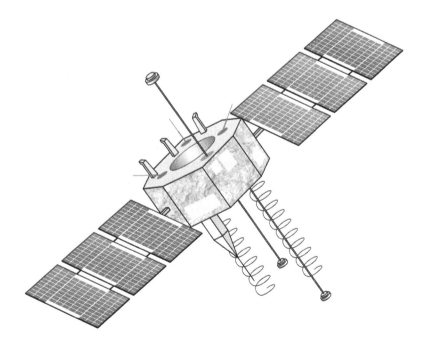

FIGURE 3.3   Timation 4 and NTS-2.

NAVSTAR

Through the decades of the 1970s and 1980s both the best and the worst aspects of the forerunner system informed GPS development. Many strategies used in TRANSIT were incorporated into GPS. For example, in the TRANSIT system the satellites broadcast their own ephemerides to the receivers and the receivers had their own frequency standards. TRANSIT had three segments: the Control Segment, including the tracking and upload facilities; the space segment, meaning the satellites themselves; and the user segment, everyone with receivers. The TRANSIT system satellites broadcast two frequencies of 400 and 150 MHz to allow compensation for the ionospheric dispersion. TRANSIT's primary observable was based on the Doppler Effect. All were used in GPS.

GPS also improved on some of the shortcomings of the previous systems. For example, the GPS satellites were placed in nearly circular orbits over 20,000 km above the Earth, where the consequences of gravity and atmospheric drag are much less severe than the lower orbits assigned to TRANSIT, Timation, and some of the other earlier systems used. GPS satellites broadcast higher-frequency signals, which reduce the ionospheric delay. The rubidium and cesium clocks pioneered in the Timation program and built into GPS satellites were a marked improvement over the quartz oscillators that were used in TRANSIT and other early satellite navigation systems. System 621B could achieve positional accuracies of approximately 16 m. TRANSIT's shortcomings restricted the practical accuracy of the system, too. It could achieve submeter work but only after long occupations on a station (at least a day) augmented by the use of a precise ephemerides for the satellites in post-processing. GPS provides much more accurate positions in a much shorter time than any of its predecessors, but these improvements were only accomplished by standing on the shoulders of the technologies that have gone before.

Requirements

The genesis of GPS was military. It grew out of the congressional mandate issued to the Departments of Defense and Transportation to consolidate the myriad of navigation systems. Its application to civilian surveying was not part of the original design. In 1973 the DoD directed the Joint Program Office (JPO) in Los Angeles, California, to establish the GPS system. Specifically, JPO was asked to create a system with high accuracy and continuous availability in real time that could provide virtually instantaneous positions to both stationary and moving receivers. The forerunners could not supply all these features. The challenge was to bring them all together in one system.

Secure, Passive, and Global

Worldwide coverage and positioning on a common coordinate grid were required of the new system—a combination that had been difficult, if not impossible, with terrestrial-based systems. It was to be a passive system, which ruled out any transmissions from the users, as had been tried with some previous satellite systems. However, the signal was to be secure and resistant to jamming, so codes in the satellite's broadcasts would need to be complex.

Expense and Frequency Allocation

The DoD also wanted the new system to be free from the sort of ambiguity problems that had plagued OMEGA and other radar systems, and DoD did not want the new system to require large expensive equipment like the optical systems. Finally, frequency allocation was a consideration. The replacement of existing systems would take time, and with so many demands on the available radio spectrum, it was going to be hard to find frequencies for GPS.

Large Capacity Signal

Not only did the specifications for GPS evolve from the experience with earlier positioning systems, so did much of the knowledge needed to satisfy them. Providing 24 hour real-time, high-accuracy navigation for moving vehicles in three dimensions was a tall order. Experience showed part of the answer was a signal that was capable of carrying a very large amount of information efficiently and that required a large bandwidth. So, the GPS signal was given a double-sided 10 MHz bandwidth, but that was still not enough, so the idea of simultaneous observation of several satellites was also incorporated into the GPS system to accommodate the requirement. That decision had far-reaching implications.

Satellite Constellation

Unlike some of its predecessors, GPS needed to have not one but at least four satellites above an observer's horizon for adequate positioning, even more if possible. Additionally, the achievement of full-time worldwide GPS coverage would require this condition be satisfied at all times, anywhere on or near the Earth.

Spread Spectrum Signal

The specification for the GPS system required all-weather performance and correction for ionospheric propagation delay. TRANSIT had shown what could be accomplished with a dual-frequency transmission from the satellites, but it had also proved that a higher frequency was needed. The GPS signal needed to be secure and resistant to both jamming and multipath. A spread spectrum, meaning spreading the frequency over a band wider than the minimum required for the information it carries, helped on all counts. This wider band also provided ample space for pseudorandom noise encoding, a fairly new development at the time. The PRN codes like those used in System 621B allowed the GPS receiver to acquire signals from several satellites simultaneously and still distinguish them from one another.

The Perfect System?

The absolute ideal navigational system, from the military's point of view, was described in the Army POS/NAV Master Plan in 1990. It should have worldwide and continuous coverage. The users should be passive. In other words, they should not be required to emit detectable electronic signals to use the system. The ideal system should be capable of being denied to an enemy, and it should be able to support an unlimited number of users. It should be resistant to countermeasures and work in real time. It should be applicable to joint and combined operations. There should be no frequency allocation problems. It should be capable of working on common grids or map datums appropriate for all users. The positional accuracy should not be degraded by changes in altitude or by the time of day or year. Operating personnel should be capable of maintaining the system. It should not be dependent on externally generated signals, and it should not have decreasing accuracy over time or the distance traveled. Finally, it should not be dependent on the identification of precise locations to initiate or update the system.

That's a pretty tall order, and GPS lives up to most, though not all, of the specifications.

GPS IN CIVILIAN SURVEYING

As mentioned, application to civilian surveying was not part of the original concept of GPS. The civilian use of GPS grew up through partnerships between various public, private, and academic organizations. Nevertheless, while the military was still testing its first receivers, GPS was already in use by civilians. Geodetic surveys were actually underway with commercially available receivers early in the 1980s.

Federal Specifications

The Federal Radionavigation Plan of 1984, a biennial document including official policies regarding radionavigation, set the predictable and repeatable accuracy of civil and commercial use of GPS at 100 m horizontally and 156 m vertically. This specification meant that the C/A code ranging for Standard Positioning Service could be defined by a horizontal circle with a radius of 100 m, 95 percent of the time. However, that same year, civilian users were already achieving results up to 6 orders of magnitude better than that limit.

Interferometry

By using interferometry, the technique that had worked so well with Prime Minitrack and VLBI, civilian users were showing that GPS surveying was capable of extraordinary results. In the summer of 1982 a research group at the Massachusetts Institute of Technology tested an early GPS receiver and achieved accuracies of 1 and 2 ppm of the station separation. Over a period of several years, extensive testing was conducted around the world that confirmed and improved on these results. In 1984, a GPS network was produced to control the construction of the Stanford Linear Accelerator. This GPS network provided accuracy at the millimeter level. In other words, by using the differentially corrected carrier phase observable instead of code ranging, private firms and researchers were going far beyond the accuracies the U.S. Government expected to be available to civilian users of GPS.

The interferometric solutions made possible by computerized processing developed with earlier extraterrestrial systems were applied to GPS by the first commercial users. The combination made the accuracy of GPS its most impressive characteristic, but it hardly solved every problem. For many years the system was restricted by the shortage of satellites as the constellation slowly grew. The necessity of having four satellites above the horizon restricted the available observation sessions to a few, sometimes inconvenient, windows of time. Another drawback of GPS for the civilian user was the cost and the limited application of both the hardware and the software. GPS was too expensive and too inconvenient for everyday use, but the accuracy of GPS surveying was already extraordinary in the beginning.

Civil Applications of GPS

Today, with a mask angle of 10°, there are some periods when 10 or more GPS satellites are above the horizon. And GPS receivers have grown from only a handful to the huge variety of receivers available today. Some push the envelope to achieve ever-higher accuracy; others offer less sophistication and lower cost. The civilian user's options are broader with GPS than any previous satellite positioning system— so broad that, as originally planned, GPS will likely replace more of its predecessors in both the military and civilian arenas. In fact, GPS has developed into a system that has more civilian applications and users than military ones.

The extraordinary range of GPS equipment and software requires the user to be familiar with an ever-expanding body of knowledge including the Global Navigation Satellite System (GNSS). GNSS includes both GPS and satellite navigation systems built by other nations and will present new options for users. However, the three segments of GPS will be presented before elaborating on GNSS.

GPS SEGMENTS

SPACE SEGMENT

Though there has been some evolution in the arrangement, the current GPS constellation under full operational capability consists of 24 satellites. However, there are more satellites than that in orbit and broadcasting at any given time, and the constellation includes several orbital spares.

As the primary satellites aged and their failure was possible, spares were launched. One reason for the arrangement was to maintain the 24 satellite constellation without interruption. It was also done to ensure that it was possible to keep four satellites, one in each of the four slots in the six orbital planes. Each of these planes is inclined to the equator by 55° (see Figure 3.4) in a symmetrical, uniform arrangement. Such a uniform design does cover the globe completely, even though the coverage is not quite as robust at high latitudes as it is at midlatitudes. The uniform design also means that multiple satellite coverage is available even if a few satellites were to fail. The satellites routinely outlast their anticipated design lives, but they are eventually worn out. However, there is always concern about the balance between cost and satellite availability, and nonuniform arrangements can be used to optimize performance.

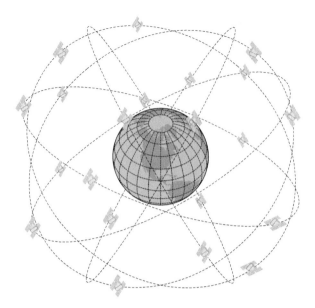

FIGURE 3.4   The GPS constellation.

GPS CONSTELLATION

ORBITAL PERIOD

NAVSTAR satellites are more than 20,000 km above the Earth in a *posigrade* orbit. A posigrade orbit is one that moves in the same direction as the Earth's rotation. Because each satellite is nearly 3× the Earth's radius above the surface, its orbital period is 12 sidereal hours.

Four-Minute Difference

When an observer actually performs a GPS survey project, one of the most noticeable aspects of a satellite's motion is that it returns to the same position in the sky about 4 min earlier each day. This apparent regression is attributable to the difference between 24 solar hours and 24 sidereal hours, otherwise known as star time. GPS satellites retrace the same orbital path twice each sidereal day, but because their observers, on Earth, measure time in solar units, the orbits do not look quite so regular to them. The satellites actually lose 3 min and 56 s with each successive solar day.

This rather esoteric fact has practical applications; for example, if the satellites are in a particularly favorable configuration for measurement, the observer may wish to take advantage of the same arrangement the following day. However, he or she would be well advised to remember the same configuration will occur about 4 min earlier on the solar time scale. Both Universal Time and GPS Time are measured in solar, not sidereal units. It is possible that the satellites will be pushed 50 km higher in the future to remove their current 4 min regression, but for now it remains. As mentioned, the GPS constellation was designed to satisfy several critical concerns.

Among them were the best possible coverage of the Earth with the fewest number of satellites, the reduction of the effects of gravitational and atmospheric drag, sufficient upload, monitoring capability, and the achievement of maximum accuracy.

Dilution of Precision

The distribution of the satellites above an observer's horizon has a direct bearing on the quality of the position derived from them. Like some of its forerunners, the accuracy of a GPS position is subject to a geometric phenomenon called *dilution of precision* (DOP). This number is somewhat similar to the strength of figure consideration in the design of a triangulation network. DOP concerns the geometric strength of the figure described by the positions of the satellites with respect to one another and the GPS receivers.

A low DOP factor is good; a high DOP factor is bad. In other words, when the satellites are in the optimal configuration for a reliable GPS position, the DOP is low; when they are not, the DOP is high (see Figure 3.5).

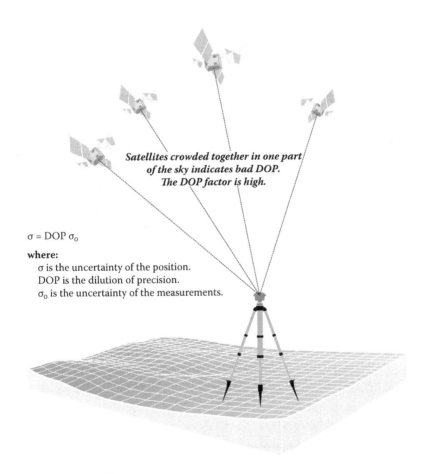

Satellites crowded together in one part of the sky indicates bad DOP. The DOP factor is high.

$\sigma = DOP\ \sigma_0$

where:
  σ is the uncertainty of the position.
  DOP is the dilution of precision.
  σ_0 is the uncertainty of the measurements.

FIGURE 3.5   Bad DOP.

Bad Dilution of Precision

Four or more satellites must be above the observer's mask angle for the simultaneous solution of the clock offset and three dimensions of the receiver's position. However, if all of those satellites are crowded together in one part of the sky, the position would be likely to have an unacceptable uncertainty and the DOP, or dilution of precision, would be high. In other words, a high DOP is a like a warning that the actual errors in a GPS position are liable to be larger than you might expect. Remember, however, it is not the errors themselves that are directly increased by the DOP factor; it is the uncertainty of the GPS position that is increased by the DOP factor. Here is an approximation of the effect:

$$\sigma = DOP \; \sigma_0$$

where
    $\sigma =$ uncertainty of the position
    $DOP =$ dilution of precision
    $\sigma_0 =$ uncertainty of the measurements (user equivalent range error)

Now because a GPS position is derived from a three-dimensional solution, there are several DOP factors used to evaluate the uncertainties in the components of a GPS position. For example, there is horizontal dilution of precision (HDOP) and vertical dilution of precision (VDOP) where the uncertainty of a solution for positioning has been isolated into its horizontal and vertical components, respectively. When both horizontal and vertical components are combined, the uncertainty of three-dimensional positions is called position dilution of precision (PDOP). There is also time dilution of precision (TDOP), which indicates the uncertainty of the clock. There is geometric dilution of precision (GDOP), which is the combination of all of the above. Finally, there is relative dilution of precision (RDOP), which includes the number of receivers, the number of satellites they can handle, the length of the observing session, as well as the geometry of the satellites' configuration.

The user equivalent range error (UERE) is the total error budget affecting a pseudo-range. It is the square root of the sum of the squares of the individual biases discussed in Chapter 2. Using a calculation like that mentioned the PDOP (position dilution of precision) factor can be used to find the positional error that will result from a particular UERE at the one sigma level (68.27 percent). For example, supposing that the PDOP factor is 1.5 and the UERE is 6 m, the positional accuracy would be 9 m at the 1 sigma level (68.27 percent). In other words, the standard deviation of the GPS position is the dilution of precision factor multiplied by the square root of the sum of the squares of the individual biases (UERE). Multiplying the 1 sigma value times 2 would provide that 95 percent level of reliability in the error estimate, which would be 18 m.

Good Dilution of Precision

As you can see in Figure 3.6, the size of the DOP factor is inversely proportional to the volume of the tetrahedron described by the satellites positions and the position of the receiver. The larger the volume of the body defined by the lines from the receiver

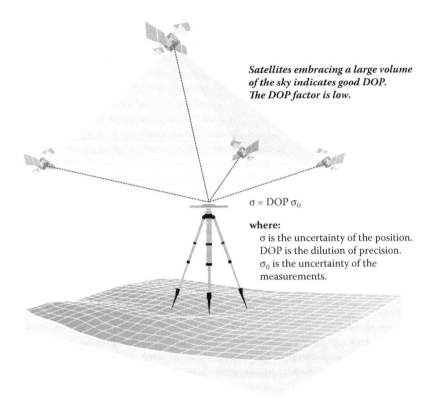

Satellites embracing a large volume
of the sky indicates good DOP.
The DOP factor is low.

$\sigma = DOP\ \sigma_0$

where:
  σ is the uncertainty of the position.
  DOP is the dilution of precision.
  σ_0 is the uncertainty of the
  measurements.

FIGURE 3.6   Good DOP.

to the satellites, the better the satellite geometry and the lower the DOP (Figure 3.6). An ideal arrangement of four satellites would be one directly above the receiver, the others 120° from one another in azimuth near the horizon. With that distribution the DOP would be nearly 1, the lowest possible value. In practice, the lowest DOPs are generally around 2.

The users of most GPS receivers can set a PDOP mask to guarantee that data will not be logged if the PDOP goes above the set value. A typical PDOP mask is 6. As the PDOP increases, the accuracy of the pseudorange positions probably deteriorate, and as PDOP decreases, the accuracy probably improves.

When a DOP factor exceeds a maximum limit in a particular location, indicating an unacceptable level of uncertainty exists over a period of time, that period is known as an *outage*. This expression of uncertainty is useful both in interpreting measured baselines and planning a GPS survey.

Satellite Positions in Mission Planning

The position of the satellites above an observer's horizon is a critical consideration in planning a GPS survey. So, most software packages provide various methods of illustrating the satellite configuration for a particular location over a specified period of time. For example, the configuration of the satellites over the entire span of the observation is important; as the satellites move, the DOP changes. Fortunately, the dilution of precision can be worked out in advance. DOP can be predicted. It depends on the orientation of the GPS satellites relative to the GPS receivers, and because most GPS software allow calculation of the satellite constellation from any given position and time, they can also provide the accompanying DOP factors.

Another commonly used plot of the satellite's tracks is constructed on a graphical representation of the half of the celestial sphere. The observer's zenith is shown in the center and the horizon on the perimeter. The program usually draws arcs by connecting the points of the instantaneous azimuths and elevations of the satellites above a specified mask angle. These arcs then represent the paths of the available satellites over the period of time and the place specified by the user.

In Figure 3.7, the plot of the polar coordinates of the available satellites with respect to time and position is just one of several tables and graphs available to help the GPS user visualize the constellation.

Figure 3.8 illustrates another useful graph that is available from many software packages. It shows the correlation between the number of satellites above a specified mask angle and the associated PDOP for a particular location during a particular span of time.

In Figure 3.8, there are four spikes of unacceptable PDOP. It might appear at first glance that these spikes are directly attributable to the drop in the number of available satellites. However, please note that while spikes 1 and 4 do indeed occur during periods of four satellite data, spikes 2 and 3 are during periods when there are seven and five satellites available, respectively. It is not the number of satellites

Point: *Kester*
Date: *Wednesday, September 29, 1993*
6 Satellites considered: *7-20-24-25-26-31*

Lat 36:50N Lon 121:45W
Threshold elevation 15 (deg)
Ephemeris: *27742652. EPH 9/22/93*
Time zone: *'Pacific Day USA'-7*

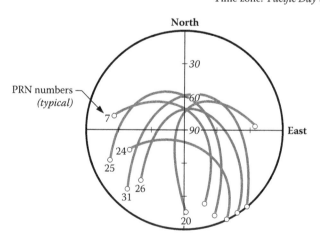

Time: *4:00 to 12:00*

FIGURE 3.7   Polar plot.

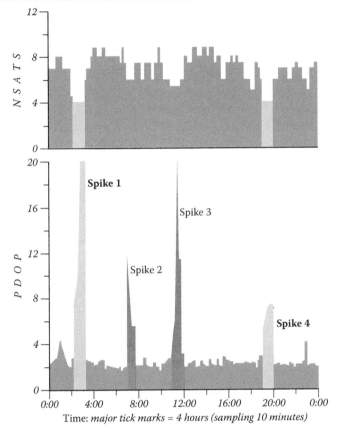

Number SVs and PDOP

Point: *Denver*   Lat: *39:47:0 N* Lon: *104:53:0 W* Almanac: *CURRENT.EPH 4/4/00*

Date: *Tuesday, November 14, 2000* Threshold Elevation: *15 (deg)*   Time Zone: *"Mountain Std USA"-7:00*

28 Satellites considered: *1,2,3,4,5,6,7,8,9,10,11,13,14,15,16,17,18,19,21,22,23,24,25,26,27,29,30,31*

Time: *major tick marks = 4 hours (sampling 10 minutes)*

FIGURE 3.8   Number of space vehicles (SVs) and PDOP.

above the horizon that determine the quality of GPS positions, one must also look at their position relative to the observer, the DOP, among other things. The variety of the tools to help the observer predict satellite visibility underlines the importance of their configuration to successful positioning.

Satellite Blocks

The 11 GPS satellites launched from Vandenberg Air Force Base between 1978 and 1985 were known as Block I satellites. The last Block I satellite was retired in late 1995. They were followed by the Block II satellites. The first of them was launched in 1989, and the last was decommissioned in 2007. During the time that the earliest of these satellites were in service, improved versions of the Block II satellites were built and launched. The first were called Block IIA satellites, and their launches began in

1990. The Block IIR satellites were next, and the first successful launch was in July 1997. In September 2005 the first of the next improved block, called Block IIR-M, was launched. Block IIF is the most recent group of GPS satellites launched, and the first of them reached orbit in May 2010. There has been steady and continuous improvement in the GPS satellite constellation from the beginning.

Satellite Names

There has always been a bit of complication in the naming of the individual GPS satellites. The first GPS satellite was launched in 1978 and was known as Navstar 1. It was also known as PRN 4 just as Navstar 2 was known as PRN 7. The Navstar number, or mission number, includes the Block name and the order of launch, for example I-1, meaning the first satellite of Block I, and the PRN number refers to the weekly segment of the P code that has been assigned to the satellite, and there are still more identifiers. Each GPS satellite has a Space Vehicle number (SVN), an Interrange Operation Number, a NASA catalog number, also known as the U.S. Space Command number, and an orbital position number. For example, the GPS satellite IIF-7 is PRN09. Its Space Vehicle number is 68. Its orbital position number is F6, and its US Space Command number is 40105. In most literature, and to the GPS receivers themselves, the PRN number is the most important.

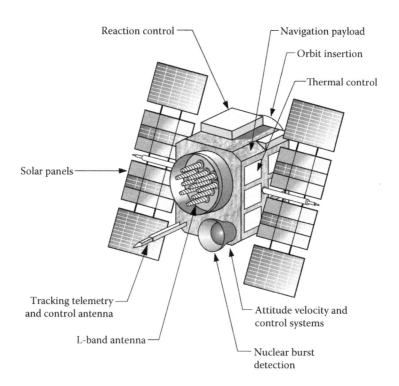

FIGURE 3.9   GPS Block II satellite.

GPS Satellites

All GPS satellites have some common characteristics (see Figure 3.9). They weigh about a ton and with solar panels extended are about 27 feet long (8.3 m). They generate about 700 watts of power. They all have three-dimensional stabilization to ensure that their solar arrays point toward the Sun and their 12 helical antennae to the Earth. GPS satellites move at a speed of about 8700 miles per hour (14,000 kmh). Even so, the satellites must pass through the shadow of the Earth from time to time, and onboard batteries provide power. All satellites are equipped with thermostatically controlled heaters and reflective insulation to maintain the optimum temperature for the oscillator's operation. Prior to launch, GPS satellites are checked at a facility in Cape Canaveral, Florida.

CONTROL SEGMENT

There are government tracking and uploading facilities distributed around the world (see Figure 3.10). These facilities not only monitor the L-band signals from the GPS satellites and update their Navigation (NAV) messages but also track the satellite's health, their maneuvers, and many other things, even battery recharging. Taken together these facilities are known as the *Control Segment*.

The Master Control Station (MCS), once located at Vandenberg Air Force Base in California, now resides at the Consolidated Space Operations Center at Schriever (formerly Falcon) Air Force Base near Colorado Springs, Colorado, and has been

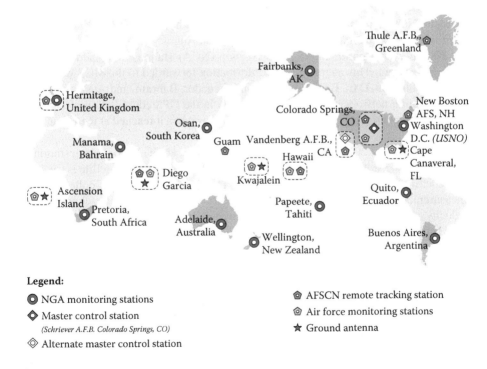

Legend:

⊙ NGA monitoring stations

◇ Master control station
   (Schriever A.F.B. Colorado Springs, CO)

◈ Alternate master control station

⬠ AFSCN remote tracking station

⬡ Air force monitoring stations

★ Ground antenna

FIGURE 3.10   Control segment.

manned by the 2nd Space Operations Squadron (2SOPS), since 1992. There is an alternate MCS at the Vandenberg Tracking Station in California.

The 2SOPS squadron controls the satellites orbits. For example, they maneuver the satellites from the highly eccentric orbits into which they are originally launched to the desired mission orbit and spacecraft orientation. They monitor the state of each satellite's onboard battery, solar, and propellant systems. They resolve satellite anomalies, activate spare satellites, and control Selective Availability (SA) and Anti-Spoofing (A/S). They dump the excess momentum from the *wheels*, the series of gyroscopic devices that stabilize each satellite. With the continuous constellation tracking data available and aided by Kalman filter estimation to manage the noise in the data, they calculate and update the parameters in the NAV message (ephemeris, almanac, and clock corrections) to keep the information within limits because the older it gets, the more its veracity deteriorates. This process is made possible by a persistent two-way communication with the constellation managed by the Control Segment that includes both monitoring and uploading accomplished through a network of ground antennas and monitoring stations.

The data that feed the MCS comes from monitoring stations. These stations track the entire GPS constellation. In the past, there were limitations. There were only six tracking stations. It was possible for a satellite to go unmonitored for up to two hours each day. It was clear that the calculation of the ephemerides and the precise orbits of the constellation could be improved with more monitoring stations in a wider geographical distribution. It was also clear that if one of the six stations went down the effectiveness of the Control Segment could be considerably hampered. These ideas, and others, led to a program of improvements known as the *Legacy Accuracy Improvement Initiative.* During this initiative from August 18 to September 7, 2005, six *National Geospatial Intelligence Agency* (NGA) stations were added to the Control Segment. This augmented the information forwarded to the MCS with data from Washington, D.C., England, Argentina, Ecuador, Bahrain, and Australia. With this 12-station network in place, every satellite in the GPS constellation was monitored almost continuously from at least two stations when it reached at least 5° above the horizon.

As of February 2015, there are 6 Air Force and the 11 NGA monitoring stations. The monitoring stations track all the satellites; in fact, every GPS satellite is tracked by at least 3 of these stations all the time. The monitoring stations collect range measurements, atmospheric information, satellite's orbital information, clock errors, velocity, right ascension, and declination and send them to the MCS. They also provide pseudorange and carrier phase data to the MCS. The MCS needs this constant flow of information. It provides the basis for the computation of the almanacs, clock corrections, ephemerides, and other components that make up the NAV message. The new stations also improve the geographical diversity of the Control Segment and that helps with the MCS isolation of errors, for example, making the distinction between the effects of the clock error from ephemeris errors. In other words, the diagnosis and solution of problems in the system are more reliable now because the MCS has redundant observations of satellite anomalies with which to work. Testing has shown that the augmented Control Segment and subsequent improved modeling has improved the accuracy of clock corrections and ephemerides in the NAV message substantially and may contribute to an increase in the accuracy of real-time GPS of 15 percent or more.

However, once the message is calculated, it needs to be sent back up to the satellites. Some of the stations have ground antennas for uploading. Four monitoring stations are collocated with such antennas. The stations at Ascension Island, Cape Canaveral, Diego Garcia, and Kwajalein upload navigation and program information to the satellites via S-band transmissions. The station at Cape Canaveral also has the capability to check satellites before launch.

The modernization of the Control Segment has been underway for some time and it continues. In 2007 the Launch/Early Orbit, Anomaly Resolution and Disposal Operations (LADO) mission PC-based ground system replaced the mainframe based Command-and-Control System. Since then, LADO has been upgraded several times. It uses Air Force Satellite Control Network (AFSCN) remote tracking stations only, not the dedicated GPS ground antennas to support the satellites from spacecraft separation through, checkout, anomaly resolution and all the way to end of life disposal. It also helps in the performance of satellite movements and the presentation of telemetry simulations to GPS payloads and subsystems. Air Force Space Command accepted the LADO capability to handle the most modern GPS satellites at the time, the Block IIF, in October 2010.

Another modernization program is known as the Next Generation Operational Control System or OCX. OCX will facilitate the full control of the new GPS signals like L5, as well as L2C and L1C and the coming GPS III program. These improvements will be discussed in Chapter 8.

Kalman Filtering

Kalman filtering is named for Rudolf Emil Kalman's linear recursive solution for least-squares filtering. It is used to smooth the effects of system and sensor noise in large data sets. In other words, a Kalman filter is a set of equations that can tease an estimate of the actual signal, meaning the signal with the minimum mean square error, from noisy sensor measurements. Kalman filtering is used to ensure the quality of some of the MCS calculations and many GPS receivers utilize Kalman filtering to estimate positions.

Kalman filtering can be illustrated by the example of an automobile speedometer (see Figure 3.11). Imagine the needle of an automobile's speedometer that is fluctuating between 64 and 72 mph as the car moves down the road. The driver might estimate the actual speed at 68 mph. Although not accepting each of the instantaneous speedometer's readings literally, the number of them is too large, he has nevertheless taken them into consideration and constructed an internal model of his velocity. If the driver further depresses the accelerator and the needle responds by moving up, his reliance on his model of the speedometer's behavior increases. Despite its vacillation, the needle has reacted as the driver thought it should. It went higher as the car accelerated. This behavior illustrates a predictable correlation between one variable, acceleration, and another, speed. Now he is more confident in his ability to predict the behavior of the speedometer. The driver is illustrating *adaptive gain*, meaning that he is fine-tuning his model as he receives new information about the measurements. As he does, a truer picture of the relationship between the readings from the

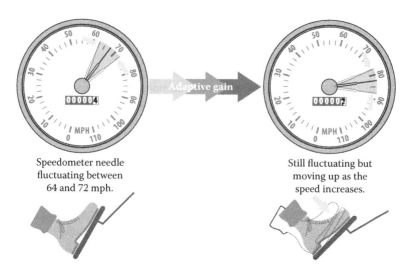

Speedometer needle
fluctuating between
64 and 72 mph.

Still fluctuating but
moving up as the
speed increases.

FIGURE 3.11   Kalman filter analogy.

speedometer and his actual speed emerges, without recording every single number as the needle jumps around. The driver's reasoning in this analogy is something like the action of a Kalman filter.

Without this ability to take the huge amounts of satellite data and condense it into a manageable number of components, GPS processors would be overwhelmed. Kalman filtering is used in the uploading process to reduce the data to the satellite clock offset and drift, six orbital parameters, three solar radiation pressure parameters, biases of the monitoring stations clock, and a model of the tropospheric effect and Earth rotational components.

User Segment

In the early years of GPS the military concentrated on testing navigation receivers. Civilians got involved much sooner than expected and took a different direction: receivers with geodetic accuracy.

Some of the first GPS receivers in commercial use were single-frequency, six-channel, and codeless instruments. Their measurements were based on interferometry. As early as the 1980s, those receivers could measure short baselines to millimeter accuracy and long baselines to 1 ppm. It is true that the equipment was cumbersome, expensive, and, without access to the NAV message, dependent on external sources for clock and ephemeris information but they were the first at work in the field and their accuracy was impressive.

During the same era a parallel trend was underway. The idea was to develop a more portable, dual-frequency, four-channel receiver that could use the NAV message. Such an instrument did not need external sources for clock and ephemeris information and could be more self-contained. Unlike the original codeless receivers that required all units on a survey brought together and their clocks synchronized twice a day, these receivers could operate independently, and while the codeless receivers needed to have satellite ephemeris information downloaded before their observations could begin, this receiver could derive its ephemeris directly from the satellite's signal. Despite these advantages, the instruments developed on this model still weighed more than 40 pounds, were very expensive, and were dependent on P code tracking.

A few years later a different kind of multichannel receiver appeared. Instead of the P code, it tracked the C/A code. Instead of using the L1 and L2 frequencies, it depended on L1 alone, and on that single frequency, it tracked the C/A code and also measured the carrier phase observable. This type of receiver established the basic design for many of the GPS receivers that surveyors use today. They are multichannel receivers, and they can recover all of the components of the L1 signal. The C/A code is used to establish the signal lock and initialize the tracking loop. Then the receiver not only reconstructs the carrier wave, it also extracts the satellite clock correction, ephemeris, and other information from the NAV message. Such receivers are capable of measuring pseudoranges, along with the carrier phase and integrated Doppler observables.

Still, as some of the earlier instruments illustrated, the dual-frequency approach does offer significant advantages. It can be used to limit the effects of ionospheric delay, it can increase the reliability of the results over long baselines, and it certainly increases the scope of GPS procedures available to a surveyor. For these reasons, a substantial number of receivers utilize both frequencies.

Receiver manufacturers are currently using several configurations in building dual- and multifrequency receivers. In order to get both carriers without knowledge of the P code, some use a combination of C/A code-correlation and codeless methods. By adding codeless technology, such receivers can avail themselves of the advantages of a dual-frequency capability while avoiding the difficulties of the P code. Anti-spoofing is the encryption of the P code on both L1 and L2. These encrypted codes are known as the Y codes, Y1 and Y2, respectively, but even though the P code has been encrypted, the carrier phase and pseudorange P code observables have been recovered successfully by many receiver manufacturers.

Dual- or multifrequency receivers are the standard for geodetic applications of GPS, and some do utilize the P code or the encrypted Y code. There are no civilian receivers that rely solely on the P code. Nearly all receivers that track the P code use the C/A code on L1. Some use the C/A code on L1 and codeless technology on L2. Some use the P code only when it is not encrypted and become codeless on L2 when AS is activated. Finally, some track all the available codes on both frequencies. In any of these configurations, the GPS surveying receivers that use the P code in combination with the C/A code and/or codeless technology are among the most expensive.

The military planned, built, and continues to maintain GPS. It is therefore no surprise that a large component of the user segment of the system is military. While surveyors and geodesists have the distinction of being among the first civilians to incorporate GPS into their practice and are sophisticated users, their number is limited. In fact, it is quite small compared with the explosion of applications the technology has now among the general public. The use of GPS in positioning, navigation, and timing (PNT) applications in precision agriculture, machine control, aviation, railroads, and the marine industry are just the beginning. With GPS incorporated into smart phones and car navigation systems it is no exaggeration that GPS technology has transformed civilian businesses and lifestyles around the world. The noncommercial uses the general public finds for GPS will undoubtedly continue to grow as the cost and size of the receivers continues to shrink.

EXERCISES

1. Which GPS satellites carry corner cube reflectors, and what is their purpose?
    a. SVN 32 and SVN 33 carry onboard corner cubes to allow photographic tracking. The purpose of the reflectors is to allow ground stations to distinguish the satellites, illuminated by Earth-based beacons, from the background of fixed stars.
    b. All GPS satellites carry corner cube reflectors. The purpose of the reflectors is to allow the users to broadcast signals to the satellites that will activate the onboard transponders.
    c. SVN 36 and SVN 37 carry onboard corner cubes to allow Satellite Laser Ranging. The purpose of the reflectors is to allow ground stations to distinguish between satellite clock errors and satellite ephemeris errors.
    d. No GPS satellites carry corner cube reflectors. Such an arrangement would require the user to broadcast a signal from the Earth to the satellite. Any requirement that the user reveal his position is not allowed by the military planners responsible for developing GPS. They have always favored a passive system that allowed the user to simply receive the satellite's signal.

2. Which of the following statements concerning the L-band designation is not true?
    a. Frequency bands used in radar were given letters to preserve military secrecy.
    b. GPS carriers L1 and L2 are named for the L-band radar designation.
    c. Frequencies broadcast by the TRANSIT satellites were within the L-band.
    d. L in L-band stands for long.

3. Which of the following is an aspect of the NAVSTAR GPS system that is an improvement on the retired TRANSIT system?
    a. Rubidium, cesium, and hydrogen maser frequency standards
    b. Satellite that broadcast two frequencies
    c. A passive system, one that does not require transmissions from the users
    d. Satellites that broadcast their own ephemerides

4. Which of the following requirements for the ideal navigational system, from the military point of view, described in the Army POS/NAV Master Plan in 1990 does GPS not currently satisfy?
   a. Users should be passive.
   b. It should be resistant to countermeasures.
   c. It should be capable of working in real time.
   d. It should not be dependent on externally generated signals.

5. Which of the following statements best explains the fact that for a stationary receiver a GPS satellite appears to return to the same position in the sky about 4 min earlier each day?
   a. Over the same period of time, vertical dilution of precision (VDOP) is frequently larger than horizontal dilution of precision (HDOP) for a stationary receiver.
   b. Apparent regression is due to the difference between the star time and solar time over a 24 hour period.
   c. Loss of time is attributable to the satellite's pass through the shadow of the Earth.
   d. Apparent regression is due to the cesium and rubidium clocks of earlier GPS satellites. They will be replaced in the Block IIR satellites by hydrogen masers. Hydrogen masers are much more stable than earlier oscillators.

6. All of the following concepts were developed in other contexts and all are now utilized in GPS. Which of them has been around the longest?
   a. Orbiting transmitters with accurate frequency standards
   b. Kalman filtering
   c. Measuring distances with electromagnetic signals
   d. Doppler shift

7. Practically speaking, which of the following was the most attractive aspect of first civilian GPS surveying in the early 1980s?
   a. Satellite availability
   b. GPS hardware
   c. Accuracy
   d. GPS software

8. Which of the following satellite identifiers is most widely used?
   a. Interrange Operation Number
   b. NASA catalog number
   c. PRN number
   d. NAVSTAR number

9. Satellites in which of the following categories are currently providing signals for positioning and navigation?
   a. Block II
   b. TRANSIT

    c.   Block IIF

    d.   Block I

10. How many stations of the Control Segment are tracking each GPS satellite at all times?

    a.   Each GPS satellite is not tracked at all times. They can go unmonitored for up to 12 hours each day.

    b.   Each GPS satellite is not tracked at all times. They can go unmonitored for up to 2 hours each day.

    c.   Each GPS satellite is tracked at all times by three stations of the Control Segment.

    d.   Each GPS satellite is tracked at all times by two stations of the Control Segment.

11. GDOP is a combination of which of the following?

    a.   PDOP, HDOP, VDOP, SDOP

    b.   PDOP, TDOP

    c.   PDOP, VDOP, TDOP, HDOP

    d.   SDOP, LDOP, MDOP, NDOP

12. Which of the following statements is correct?

    a.   The larger the volume of the body defined by the lines from the receiver to the satellites, the better the satellite geometry and the lower the DOP.

    b.   The volume of the body defined by the lines from the receivers to the satellites has no effect on DOP.

    c.   The mask angle of 30° decreases the PDOP.

    d.   Typical DOP is 0.

ANSWERS AND EXPLANATIONS

1. Answer is (c)

    Explanation: Two current GPS satellites carry onboard corner cube reflectors, SVN 36 (PRN 06) and SVN 37 (PRN 07), launched in 1994 and 1993, respectively. The purpose of the corner cube reflectors is to allow Satellite Laser Ranging (SLR) tracking. The ground stations can use this exact range information to separate the effect of errors attributable to satellite clocks from errors in the satellite's ephemerides. Remember the satellite's ephemeris can be likened to a constantly updated coordinate of the satellite's position. The SLR can nail down the difference between the satellite's broadcast ephemeris and the satellite's actual position. This allows, among other things, more proper attribution of the appropriate portion of the range error to the satellite's clock.

2. Answer is (c)

    Explanation: The original letter designations were assigned to frequency bands in radar to maintain military secrecy. The L-band was given the letter L to indicate that its wavelength was long. The frequencies within the L-band are from 3900 to 1550 MHz, approximately. Stated another way, the wavelengths within the L-band are approximately from 76 to 19 cm.

The GPS carrier frequencies L1 at 1575.42 MHz and L2 at 1227.60, with wavelengths of 19 and 24 cm respectively, and are both close to the L-band range. While one might say that the L1 frequency is not exactly within original L-band designation, the frequencies broadcast by the TRANSIT satellites certainly are not. The frequencies used in the TRANSIT system are 400 and 150 MHz. These frequencies fall much below the L-band and into the VHF range.

It is interesting to note that the old L-band designation has actually been replaced; it is now known as the D-band. However, there does not appear to be any intention to change the names of the GPS carrier frequencies.

3. Answer is (a)

Explanation: Many of the innovations used in the TRANSIT system informed decisions in creating the NAVSTAR GPS system. Both have satellites that broadcast two frequencies to allow compensation for the ionospheric dispersion. TRANSIT satellites used the frequencies of 400 and 150 MHz, while GPS uses 1575.42 and 1227.60 MHz. While the low frequencies used in the TRANSIT system were not as effective at eliminating the ionospheric delay, the idea is the same. Both systems are passive, meaning there is no transmission from the user required. Both systems use satellites that broadcast their own ephemerides to the receivers.

There are more similarities; both systems are divided into three segments: the Control Segment, including the tracking and upload facilities; the space segment, meaning the satellite themselves; and the user segment, everyone with receivers. In both the TRANSIT and GPS systems, each satellite and receiver contains its own frequency standards. However, the standards used in NAVSTAR GPS satellites are much more sophisticated than those that were used in TRANSIT. The rubidium, cesium, and hydrogen maser frequency standards used in GPS satellites far surpass the quartz oscillators that were used in the TRANSIT satellites. The TRANSIT navigational broadcasts were switched off on December 31, 1996.

4. Answer is (d)

Explanation: Not only does GPS have worldwide 24-hour coverage, it is also capable of providing positions on a huge variety of grids and datums. The system allows the user's receivers to be passive; there is no necessity for them to emit any electronic signal to use the system. GPS codes are complex, and there is more than one strategy that the Department of Defense can use to deny GPS to an enemy. At the same time, there can be a virtually unlimited number of users without overtaxing the system. However, GPS is dependent on externally generated signals for the continued health of the system. While satellites can operate for periods without uploads and orbital adjustments from the Control Segment, they certainly cannot do without them entirely.

The Control Segment's network of ground antennas and monitoring stations track the navigation signals of all the satellites. The MCS uses that information to generate updates for the satellites that it uploads through ground antennas.

5. Answer is (b)

Explanation: The difference between 24 solar hours and 24 sidereal hours, otherwise known as star time, is 3 min and 56 s, or about 4 min. GPS satellites retrace the same orbital path twice each sidereal day, but because their observers, on Earth, measure time in solar units the orbits do not look quite so regular to them and both Universal Time and GPS Time are measured in solar, not sidereal units.

6. Answer is (d)

Explanation: The concept of measuring distance with electromagnetic signals had its earliest practical applications in radar in the 1940s and during World War II. Development of the technology for launching transmitters with onboard frequency standards into orbit was available soon after the launch of Sputnik in 1957. TRANSIT 1B launched on April 13, 1960, was the first successfully launched navigation satellite.

Kalman filtering, named for R.E. Kalman's recursive solution for least-squares filtering, was developed in 1960. It is a statistical method of smoothing and condensing large amounts of data and has been used in radionavigation ever since. It is an integral part of GPS.

However, the Doppler shift was discovered in 1842, and certainly has the longest history of any of the ideas listed. The Doppler shift describes the apparent change in frequency when an observer and a source are in relative motion with respect to one another. If they are moving together, the frequency of the signal from the source appears to rise, and if they are moving apart the frequency appears to fall. The Doppler Effect came to satellite technology during the tracking of Sputnik in 1957. It occurred to observers at Johns Hopkins University's Applied Physics Laboratory that the Doppler shift of its signal could be used to find the exact moment of its closest approach to the Earth. However, the phenomenon had been described 115 years earlier by Christian Doppler using the analogy of a ship on the ocean.

7. Answer is (c)

Explanation: The interferometric solutions made possible by computerized processing developed with earlier extraterrestrial systems were applied to GPS by the first commercial users, but the software was cumbersome by today's standards. There were few satellites up in the beginning, and the necessity of having at least four satellites above the horizon restricted the available observation sessions to difficult periods of time. GPS receivers were large, unwieldy, and very expensive. Nevertheless, in the summer of 1982 a research group at the Massachusetts Institute of Technology tested an early GPS receiver and achieved accuracies of 1 and 2 ppm of the station separation. In 1984 a GPS network was produced to control the construction of the Stanford Linear Accelerator. This GPS network provided accuracy at the millimeter level. GPS was inconvenient and expensive, but the accuracy was remarkable from the outset.

8. Answer is (c)

    Explanation: The satellite that has currently been given the number Space Vehicle 32, or SV32, is also known as PRN 1. This particular satellite, which was launched on November 22, 1992, currently occupies orbital slot F-1. It has a NAVSTAR number, or mission number of IIA-16. This designation includes the Block name and the order of launch of that mission.

    This same GPS satellite also has an Interrange Operation Number and a NASA catalog number. However, the most often used identifier for this satellite is PRN 1.

9. Answer is (a)

    Explanation: The first of six TRANSIT satellites to reach orbit was launched on June 29, 1961. The constellation of satellites known as the Navy Navigational Satellite System functioned until it was switched off on December 31, 1996, and replaced by the GPS system. The first of 11 Block I GPS satellites was launched on February 22, 1978, and the last on October 9, 1985; however, no Block I satellites are operating today. The first of the Block II satellites was launched in 1989 and the last was decommissioned in 2007 after 17 years of operation. There are no Block II satellites operating today.

    However, several Block IIF GPS satellites are currently operational. The first Block IIF satellite was launched in the summer of 2010. Their design life is 12 to 15 years. Block IIF satellites have faster processors and more memory onboard. The Block IIF satellites will replace the Block IIA satellites as they age.

10. Answer is (c)

    Explanation: Today there are 6 Air Force and the 11 National Geospatial-Intelligence Agency monitoring stations; every GPS satellite is tracked by at least 3 of them at all times.

11. Answer is (c)

    Explanation: There is horizontal dilution of precision (HDOP) and vertical dilution of precision (VDOP) where the uncertainty of a solution for positioning has been isolated into its horizontal and vertical components, respectively. When both horizontal and vertical components are combined, the uncertainty of three-dimensional positions is called position dilution of precision (PDOP). There is also time dilution of precision (TDOP), which indicates the uncertainty of the clock. There is geometric dilution of precision (GDOP), which is the combination of all of the above.

12. Answer is (a)

Explanation: The larger the volume of the body defined by the lines from the receiver to the satellites, the better the satellite geometry and the lower the DOP (Figure 3.6). An ideal arrangement of four satellites would be one directly above the receiver, the others 120° from one another in azimuth near the horizon. With that distribution the DOP would be nearly 1, the lowest possible value. In practice, the lowest DOPs are generally around 2. The users of most GPS receivers can set a PDOP mask to guarantee that data will not be logged if the PDOP goes above the set value.

4 Receivers and Methods

COMMON FEATURES OF GLOBAL POSITIONING SYSTEM (GPS) RECEIVERS

A Block Diagram of a Code Correlation Receiver

Receivers for GPS Surveying

The receivers are the most important hardware in a GPS surveying operation (see Figure 4.1). Their characteristics and capabilities influence the techniques available to the user throughout the work. There are many different GPS receivers on the market. Some of them are appropriate for surveying, and they share some fundamental elements. Though no level of accuracy is ever guaranteed, with proper procedures and data handling, they are generally capable of accuracies from submeter to centimeters. Most are also capable of performing differential GPS, real-time GPS, and static GPS and are usually accompanied by processing and network adjustment software, and so on.

A GPS receiver must collect and then convert signals from GPS satellites into measurements of position, velocity, and time. There is a challenge in that the GPS signal has low power. An orbiting GPS satellite broadcasts its signal across a cone of approximately 28° of arc. From the satellite's point of view, about 11,000 miles (17,700 km) up, that cone covers a substantial portion of the whole planet. It is instructive to contrast this arrangement with a typical communication satellite that not only has much more power but also broadcasts a very directional signal. Its signals are usually collected by a large dish antenna, but the typical GPS receiver has a small, relatively nondirectional antenna. Fortunately, antennas used for GPS receivers do not have to be pointed directly at the signal source. The GPS signal also intentionally occupies a broader bandwidth than it must to carry its information. This characteristic is used to prevent jamming and mitigate multipath, but most importantly the GPS signal itself would be completely obscured by the variety of electromagnetic noise that surrounds us if it were not a spread spectrum coded signal. In fact, when a GPS signal reaches a receiver, its power is actually less than the receivers natural noise level; fortunately, the receiver can still extract the signal and achieve unambiguous satellite tracking using the correlation techniques described in Chapter 2. To do this job, the elements of a GPS receiver function cooperatively and iteratively. That means that the data stream is repeatedly refined by the several components of the device working together as it makes its way through the receiver.

Antenna

The antenna, *radio frequency* (RF) section, filtering, and intermediate frequency elements are in the front of a GPS receiver. The antenna collects the satellite's signals and converts the incoming electromagnetic waves into electric currents sensible to

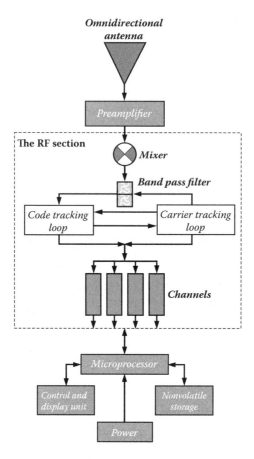

FIGURE 4.1   Block diagram of a GPS receiver.

the RF section of the receiver. Several antenna designs are possible in GPS, but the satellite's signal has such a low power density, especially after propagating through the atmosphere, that antenna efficiency is critical. Therefore, GPS antennas must have high sensitivity, also known as high gain. They can be designed to collect only the L1 frequency, L1 and L2, or all signals, including L5. In all cases they must be *right-hand circular polarized* as are the GPS signals broadcast from the satellites.

Most receivers have an antenna built in, but many can accommodate a separate tripod-mounted or range pole-mounted antenna as well. These separate antennas with their connecting coaxial cables in standard lengths are usually available from the receiver manufacturer. The cables are an important detail. The longer the cable, the more of the GPS signal is lost traveling through it.

The wavelengths of the GPS carriers are 19 cm (L1), 24 cm (L2), and 25 cm (L5), and antennas that are a quarter or half wavelength tend to be the most practical and efficient so GPS antenna elements can be as small as 4 or 5 cm. Most of the receiver manufacturers use a *microstrip* antenna. These are also known as patch antennas. The microstrip may have a patch for each frequency. Microstrip antennas are durable, compact, have a simple construction, and a low profile. The next most commonly used antenna is known as a *dipole*. A dipole antenna has a stable phase center and simple construction but needs a good ground plane. A ground plane also facilitates the use of a microstrip antenna where it not only ameliorates multipath but also tends to increase the antenna's zenith gain. A *quadrifilar* antenna is a single-frequency antenna that has two orthogonal bifilar helical loops on a common axis. Quadrifilar antennas perform better than a microstrip on crafts that pitch and roll like boats and airplanes. They are also used in many recreational handheld GPS receivers. Such antennas have a good gain pattern and do not require a ground plane, but they are not azimuthally symmetric. The least common design is the *helix* antenna. A helix is a dual-frequency antenna. It has a good gain pattern but a high profile.

Bandwidth

An antenna ought to have a bandwidth commensurate with its application; in general, the larger the bandwidth, the better the performance. However, there is a downside; increased bandwidth degrades the signal-to-noise ratio by including more interference. GPS microstrip antennas usually operate in a range from about 2 to 20 MHz, which corresponds with the null-to-null bandwidth of both new and legacy GPS signals. For example, L2C and the central lobe of the C/A code span 2.046 MHz, whereas L5 and the P(Y) code have a bandwidth of 20.46 MHz. The width of the C/A and P(Y) widths are shown in Figure 4.2.

Therefore, the antenna and front-end of a receiver designed to collect the P(Y) code on L1 and L2 would have a bandwidth of 20.46 MHz, but a system tracking the C/A code or the L2C may have a narrower bandwidth. It would need 2.046 MHz for the central lobe of the C/A code, or if it were designed to track the L1C signal,

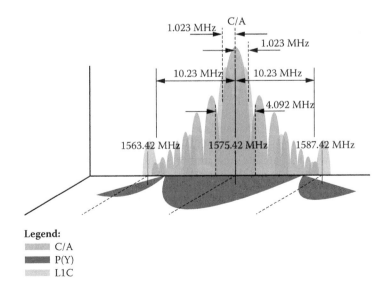

Legend:
  C/A
  P(Y)
  L1C

FIGURE 4.2   L1.

its bandwidth would need to be 4.092 MHz. A dual-frequency microstrip antenna would likely operate in a bandwidth from 10 to 20 MHz.

Nearly Hemispheric Coverage

Because a GPS antenna is designed to be omnidirectional, its gain pattern (i.e., the change in gain over a range of azimuths and elevations) ought to be nearly a full hemisphere but not perfectly hemispheric. For example, most surveying applications filter the signals from very low elevations to reduce the effects of multipath and atmospheric delays. In other words, a portion of the GPS signal may come into the antenna from below the mask angle; therefore, the antenna's gain pattern is specifically designed to reject such signals. Second, the contours of equal phase around the antenna's electronic center (i.e., the phase center) are not themselves perfectly spherical.

The gain, or gain pattern, describes the success of a GPS antenna in collecting more energy from above the mask angle and less from below the mask angle. A gain of about 3 to 5 *decibels* (dB) is typical for a GPS antenna. Decibels do not indicate the power of the antenna because the unit is dimensionless. It refers to a comparison. In this case the gain of a real GPS antenna is measured by comparing it to a theoretical *isotropic antenna*. An isotropic antenna is a hypothetical, lossless antenna that has perfectly equal capabilities in all directions. Because a decibel is a unit for the logarithmic measure of the relative power, a 3 dB increase indicates a doubling of signal strength and a 3 dB decrease indicates a halving of signal strength. This means that a typical omnidirectional GPS antenna with a gain of about 3 dB (decibels) has about 50 percent of the capability of an isotropic antenna. It is important that the GPS receiver antennas and preamplifiers be as efficient as possible because the power received from the GPS satellites is low. The minimum power received from the C/A code on L1 is about −160 dBW, and the minimum power received from the P code on L2 is even less at −166 dBW.

Antenna Orientation

In a perfect GPS antenna the phase center of the gain pattern would be exactly coincident with its actual, physical, center. If such a thing were possible, the centering of the antenna over a point on the Earth would ensure its electronic centering as well. However, that absolute certainty remains elusive for several reasons.

It is important to remember that the position at each end of a GPS baseline is the position of the phase center of the antenna at each end, not their physical centers, and the phase center is not an immovable point. The location of the phase center actually changes slightly with the satellite's signal. For example, it is different for the L2 than for L1 or L5. In addition, as the azimuth, intensity, and elevation of the received signal changes, so does the difference between the phase center and the physical center. Small azimuthal effects can also be brought on by the local environment around the antenna, but most phase center variation is attributable to changes in satellite elevation. In the end, the physical center and the phase center of an antenna may be as much as a couple of centimeters from one another. With today's patch antennas it can be as little as a few millimeters.

It is also fortunate that the errors are systematic, and to compensate for some of this offset error, most receiver manufacturers recommend users take care when making simultaneous observations on a network of points that their antennas are all oriented in the same direction. Several manufacturers even provide reference marks on the antenna so that each one may be rotated to the same azimuth, usually north, to maintain the same relative position between their physical and phase, electronic, centers. Another approach to the problem's reduction is adjusting the phase center offset out of the solution in post-processing.

Height of Instrument

The antenna's configuration also affects another measurement critical to success-ful GPS surveying—the height of instrument. The measurement of the height of the instrument in a GPS survey is normally made to some reference mark on the antenna. However, it sometimes must include an added correction to bring the total vertical distance to the antenna's phase center.

RADIO FREQUENCY (RF) SECTION

Different receiver types use different techniques to process the GPS signals but go through substantially the same steps that are contained in this section.

The preamplifier increases the signal's power, but it is important that the gain in the signal coming out of the preamplifier is considerably higher than the noise. Because signal processing is easier, if the signals arriving from the antenna are in a common frequency band, the incoming frequency is combined with a signal at a harmonic frequency. This latter, pure sinusoidal signal is the previously mentioned reference signal generated by the receiver's oscillator. The two frequencies are mul-tiplied together in a device known as a *mixer*. Two frequencies emerge: one of them is the sum of the two that went in, and the other is the difference between them.

The sum and difference frequencies then go through a *bandpass filter*, an elec-tronic filter that removes the unwanted high frequencies and selects the lower of the two. It also eliminates some of the noise from the signal. For tracking the P code this filter will have a bandwidth of about 20 MHz, but it will be around 2 MHz if the C/A code is required. In any case, the signal that results is known as the *intermediate frequency* (IF) or *beat frequency* signal.

This beat frequency is the difference between the Doppler-shifted carrier fre-quency that came from the satellite and the frequency generated by the receiver's own oscillator. In fact, to make sure that it embraces the full range of the Doppler Effect on the signals coming in from the GPS satellites, the bandwidth of the IF itself can vary from 5 to 10 kHz Doppler. That spread is typically lessened after tracking is achieved. There are usually several IF stages before copies of it are sent into separate channels, each of which extract the code and carrier information from a particular satellite.

As mentioned in Chapter 1, a replica of the C/A or P code is generated by the receiver's oscillator and now that is correlated with the IF signal. It is at this point that the pseudorange is measured. Remember the pseudorange is the time shift required to align the internally generated code with the IF signal, multiplied by the speed of light.

The receiver also generates another replica, this time a replica of the carrier. That carrier is correlated with the IF signal, and the shift in phase can be measured. The continuous phase observable, or observed cycle count, is obtained by counting the elapsed cycles since lock-on and by measuring the fractional part of the phase of the receiver generated carrier.

Channels

The antenna itself does not sort the information it gathers. The signals from several satellites enter the receiver simultaneously. However, in the *channels* of the RF section the undifferentiated signals are identified and segregated from one another.

A channel in a continuous tracking GPS receiver is not unlike a channel in a television set. It is hardware, or a combination of hardware and software, designed to separate one signal from all the others. A receiver may have 6 channels, 12 channels, or hundreds of channels. At any given moment, one frequency from one satellite can have its own dedicated channel and the channels operate in parallel. This approach allows the receiver to maintain accuracy when it is on a moving platform; it provides anti-jamming capability and shortens the time to first fix. Each channel typically operates in one of two ways, working to acquire the signal or to track it. Once the signal is acquired, it is continuously tracked unless lock is lost. If that happens, the channel goes back to acquisition mode and the process is repeated.

Multiplexing and Sequencing

While a parallel receiver has dedicated separate channels to receive the signals from each satellite that it needs for a solution, a *multiplexing* (aka muxing) *receiver* gathers some data from one satellite and then switches to another satellite and gathers more data, and so on. Such a receiver can usually perform this switching quickly enough that it appears to be tracking all of the satellites simultaneously. A multiplexing receiver must still dedicate one frequency from one satellite to one channel at a time; it just makes that time very short. It typically switches at a rapid pace, i.e., 50 Hz.

Even though multiplexing is generally less expensive, this strategy of switching channels is now little used. There are several reasons. While a parallel receiver does not necessarily offer more accurate results, parallel receivers with dedicated channels are faster; a parallel receiver has a more certain phase lock; and there is redundancy if a channel fails and they possess a superior signal-to-noise ratio (SNR). A multiplexing receiver also has a lower resistance to jamming and interference compared to continuous tracking receivers. Whether continuous or switching channels are used, a receiver must be able to discriminate between the incoming signals. They may be differentiated by their unique C/A codes on L1, their Doppler shifts, or some other method, but in the end, each signal is assigned to its own channel.

Tracking Loops

There are *code tracking loops*, the *delay lock loops*, and *carrier tracking loops*, and the *phase locking loops* in the receiver. Typically, both the code and the carrier are being tracked in phase lock. The tracking loops connected to each of the receiver's channels also work cooperatively with each other. Dual-frequency receivers have dedicated channels and tracking loops for each frequency.

Pseudoranging

In most receivers the first procedure in processing an incoming satellite signal is synchronization of the C/A code from the satellite's L1 broadcast, with a replica C/A code generated by the receiver itself, i.e., the *code phase* measurement. When there is no initial match between the satellite's code and the receiver's replica, the receiver time shifts, or *slews*, the code it is generating until the optimum correlation is found. Then a code tracking loop, the delay lock loop, keeps them aligned.

The time shift discovered in that process is a measure of the signal's travel time from the satellite to the phase center of the receiver's antenna. Multiplying this time delay by the speed of light gives a range, but it is called a pseudorange in recognition of the fact that it is contaminated by the errors and biases set out in Chapter 2.

Carrier Phase Measurement

Once the receiver acquires the C/A code, it has access to the NAV message, or the coming Civil Navigation messages (CNAV). It can read the ephemeris and the almanac information, use GPS Time, and, for those receivers that do utilize the P code, use the handover word on every subframe as a stepping stone to tracking the more precise code. However, the code's pseudoranges alone are not adequate for the majority of surveying applications. Therefore, the next step in signal processing for most receivers involves the carrier phase observable.

Just as they produce a replica of the incoming code, receivers also produce a replica of the incoming carrier wave. And the foundation of carrier phase measurement is the combination of these two frequencies. Remember the incoming signal from the satellite is subject to an ever-changing Doppler shift while the replica within the receiver is nominally constant.

Carrier Tracking Loop

The process begins after the PRN code has done its job and the code tracking loop is locked. By mixing the satellite's signal with the replica carrier, this process eliminates all the phase modulations, strips the codes from the incoming carrier, and simultaneously creates two intermediate or beat frequencies—one is the sum of the combined frequencies, and the other is the difference. The receiver selects the latter, the difference, with a bandpass filter. Then this signal is sent on to the carrier tracking loop also known as the phase locking loop, where the voltage-controlled oscillator is continuously adjusted to follow the beat frequency exactly.

Doppler Shift

As the satellite passes overhead, the range between the receiver and the satellite changes; that steady change is reflected in a smooth and continuous movement of the phase of the signal coming into the receiver. The rate of that change is reflected in the constant variation of the signal's Doppler shift, but if the receiver's oscillator frequency is matching these variations exactly, as they are happening, it will duplicate the incoming signal's Doppler shift and phase. This strategy of making measurements using the carrier beat phase observable is a matter of counting the elapsed cycles and adding the fractional phase of the receiver's own oscillator.

Doppler information has broad applications in signal processing. It can be used to discriminate between the signals from various GPS satellites to determine integer ambiguities in kinematic surveying, as a help in the detection of cycle slips, and as an additional independent observable for autonomous point positioning. Perhaps the most important application of Doppler data is the determination of the *range rate* between a receiver and a satellite. Range rate is a term used to mean the rate at which the range between a satellite and a receiver changes over a particular period of time.

Typical GPS Doppler Shift

With respect to the receiver, the satellite is always in motion even if the receiver is *static*, but the receiver may be in motion in another sense, as it is in kinematic GPS.

The ability to determine the instantaneous velocity of a moving vehicle has always been a primary application of GPS and is based on the fact that the Doppler shift frequency of a satellite's signal is nearly proportional to its range rate (see Figure 4.3).

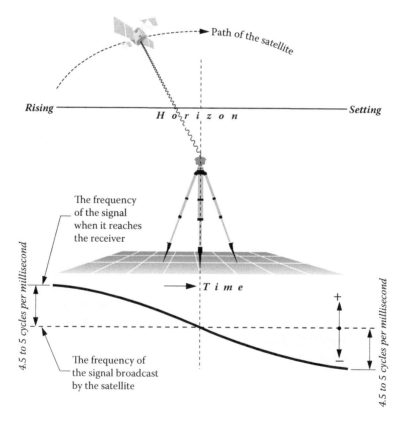

FIGURE 4.3   Typical Doppler shift.

To see how it works, let's look at a static, that is, stationary, GPS receiver. The signal received would have its maximum Doppler shift, 4.5 to 5 cycles per millisecond, when the satellite is at its maximum range, just as it is rising or setting. The Doppler shift continuously changes throughout the overhead pass. Immediately after the satellite rises, relative to a particular receiver, its Doppler shift gets smaller and smaller, until the satellite reaches its closest approach. At the instant its radial velocity with respect to the receiver is zero, the Doppler shift of the signal is zero as well, but as the satellite recedes, it grows again, negatively, until the Doppler shift once again reaches its maximum extent just as the satellite sets.

Continuously Integrated Doppler

The Doppler shift and the carrier phase are measured by first combining the received frequencies with the nominally constant reference frequency created by the receiver's oscillator. The difference between the two is the often mentioned beat frequency, an intermediate frequency, and the number of beats over a given time interval is known as the *Doppler count* for that interval. Because the beats can be counted much more precisely than their continuously changing frequency can be measured, most GPS receivers just keep track of the accumulated cycles, the Doppler count. The sum of consecutive Doppler counts from an entire satellite pass is often stored, and the data can then be treated like a sequential series of biased range differences. *Continuously integrated Doppler* is such a process. The rate of the change in the continuously integrated Doppler shift of the incoming signal is the same as that of the reconstructed carrier phase.

Integration of the Doppler frequency offset results in an accurate measurement of the advance in carrier phase between epochs, using double differences in processing the carrier phase observables removes most of the error sources other than multipath and receiver noise.

Integer Ambiguity

The solution of the integer ambiguity, the number of whole cycles on the path from satellite to receiver, would be more difficult if it was not preceded by pseudoranges or code phase measurements in most receivers. This allows the centering of the subsequent double-difference solution. In other words, a pseudorange solution provides an initial estimate of the candidates for the integer ambiguity within a smaller range than would otherwise be the case, and as more measurements become available, it can reduce them even farther.

After the code phase measurements narrows the field, there are several methods used to solve the integer ambiguity. In the geometric method the carrier phase data from multiple epochs are processed and the constantly changing satellite geometry is used to find an estimate of the actual position of the receiver. This approach is also used to show the error in the estimate by calculating how its results hold up as the geometry of the constellation changes. This strategy requires a significant amount of satellite motion to succeed and therefore, takes time to converge on a solution. In the filtering approach, independent measurements are averaged to find the estimate with the lowest noise level. Then there is the method that searches through the range of possible integer combinations and finds the one with the lowest residual. Both the search and filtering methods are *heuristic approaches*. They depend on trial and error. These methods cannot assess the correctness of a particular solution, but they can provide the probability, given certain conditions, that the answer is within given limits.

In the end, most GPS processing software uses some combination of all three ideas. All of these methods narrow the field of the integer ambiguity solution by beginning at an initial position estimate provided by code phase measurements.

Signal Squaring

There is a method that does not use the codes carried by the satellite's signal. It is called codeless tracking or *signal squaring*. It was first used in the earliest civilian GPS receivers, supplanting proposals for a TRANSIT-like Doppler solution. It makes no use of pseudoranging and relies exclusively on the carrier phase observable. Like other methods, it also depends on the creation of an intermediate or beat frequency. However, with signal squaring the beat frequency is created by multiplying the incoming carrier by itself. The result has double the frequency and half the wavelength of the original. It is squared.

There are some drawbacks to the method. For example, in the process of squaring the carrier, it is stripped of all its codes. The chips of the P code, the C/A code, and the Navigation (NAV) message normally modulated onto the carrier by 180° phase shifts are eliminated entirely. The signals broadcast by the satellites have phase shifts called *code states* that change from +1 to −1 and vice versa, but squaring the carrier converts them all to exactly 1. The result is that the codes themselves are wiped out. Therefore, this method must acquire information such as almanac data and clock corrections from other sources. Another drawback of squaring the carrier includes the deterioration of the signal-to-noise ratio because when the carrier is squared, the background noise is squared, too. Also, cycle slips occur at twice the original carrier frequency.

However, signal squaring has its upside as well. It reduces susceptibility to multipath. It has no dependence on PRN codes and is not hindered by the encryption of the P code. The technique works as well on L2 as it does on L1 or L5 and that facilitates ionospheric delay correction. Therefore, signal squaring can provide high accuracy over long baselines.

So there is a cursory look at some of the different techniques used to process the signal in the RF section. Now let's discuss the microprocessor of the receiver.

MICROPROCESSOR

The microprocessor controls the entire receiver, managing its collection of data. It controls the digital circuits that, in turn, manage the tracking and measurements, extract the ephemerides and other information from the NAV or CNAV, and mitigate multipath and noise among other things.

The GPS receivers used in surveying often send these data to the storage unit, but more and more, they are expected to process the ranging data, do datum conversion, and produce their final positions instantaneously (i.e., in *real time*). Then the receivers serve up the position through the control and display unit (CDU). There is a two-way street between the microprocessor and the CDU: each can receive information from or send information to the other.

CONTROL AND DISPLAY UNIT

A GPS receiver will often have a CDU. From handheld keyboards to soft keys around a screen to digital map displays and interfaces to other instrumentation, there are a variety of configurations. Nevertheless, they all have the same fundamental purpose, that is, facilitation of the interaction between the operator and the receiver's microprocessor. A CDU typically displays status, position data, velocity, and time. It may also be used to select different surveying methods, waypoint navigation, and/or set parameters such as epoch interval, mask angle, and antenna height. The CDU can offer a combination of help menus, prompts, datum conversions, readouts of survey results, estimated positional error, and so forth. The information available from the CDU varies from receiver to receiver, but when four or more satellites are available, they can generally be expected to display the PRN numbers of the satellites being tracked, the receiver's position in three dimensions, and velocity information. Most of them also display the dilution of precision and GPS Time.

STORAGE

Most GPS receivers today have internal data logging. The amount of storage required for a particular session depends on several things: the length of the session, the number of satellites above the horizon, the epoch interval, and so forth. For example, presuming the amount of data received from a single GPS satellite is ~100 bytes per epoch, a typical 12-channel dual-frequency receiver observing six satellites and using a 1 s epoch interval over the course of a 1 hour session would require ~2 MB of storage capacity for that session.

POWER

Battery Power

Because most receivers in the field operate on battery power, batteries and their characteristics are fundamental to GPS surveying. A variety of batteries are used, and there are various configurations. For example, some GPS units are powered by camcorder batteries, and handheld recreational GPS units often use disposable AA or even AAA batteries. However, in surveying applications, rechargeable batteries are the norm. Lithium, nickel cadmium, and nickel metal-hydride may be the most common categories, but lead acid car batteries still have an application as well.

The obvious drawbacks to lead acid batteries are size and weight, and there are a few others (e.g., corrosive acid, need to store them charged, and low cycle life). Nevertheless, lead acid batteries are especially hard to beat when high power is required. They are economical and long-lasting.

Nickel cadmium (NiCd) batteries cost more than lead acid batteries but are small and operate well at low temperatures. Their capacity does decline as the temperature drops. Like lead acid batteries, NiCd batteries are quite toxic. They self-discharge at the rate of about 10 percent per month, and even though they do require periodic full discharge, these batteries have an excellent cycle life. Nickel metal-hydride (NiMH) batteries self-discharge a bit more rapidly than NiCd batteries and have a less robust cycle life, but they are not as toxic.

Lithium-ion batteries overcome several of the limitations of the others. They have a relatively low self-discharge rate. They do not require periodic discharging and do not have memory issues as do NiCd batteries. They are light, have a good cycle life, and low toxicity. However, the others tolerate overcharging and the lithium-ion battery does not. It is best not to charge lithium-ion batteries in temperatures at or below freezing. These batteries require a protection circuit to limit current and voltage but are widely used in powering electronic devices, including GPS receivers.

About half of the available GPS carrier phase receivers have an internal power supply, and most will operate 5½ hours or longer on a fully charged 6-amp-hour battery. Most code-tracking receivers, those that do not also use the carrier phase observable, could operate for about 15 hours on the same size battery.

It is fortunate that GPS receivers operate at low power; from 9 to 36 volts DC is generally required. This allows longer observations with fewer, and lighter, batteries than might be otherwise required. It also increases the longevity of the GPS receivers themselves.

RECEIVER CATEGORIES

Receivers are generally categorized by their physical characteristics, the elements of the GPS signal they can use with advantage, and the claims about their accuracy. However, the effects of these features on a receiver's actual productivity are not always obvious. There are receivers that use only the C/A code on the L1 frequency and receivers that cross-correlate with the P code, or encrypted Y code, on L1 and L2. There are L1 carrier phase tracking receivers, dual-frequency and multi-frequency carrier phase tracking receivers, receivers that track all in view, and GPS/Global Navigation Satellite System (GNSS) receivers. The more aspects of the GPS signal a receiver can employ, the greater its flexibility but so, too, the greater its cost. It is important to understand receiver capabilities and limitations to ensure that the systematic capability of a receiver is matched to the required outcome of a project.

As shown in Figure 4.4, it is possible to divide receivers into three categories. They are recreation, mapping, and surveying. These categories can be further divided by the observables they are capable of tracking. Again, the contribution of these capabilities to the levels of possible systematic precision and accuracy are given from lower (L1 code alone) to higher (GPS + GLONASS), that is, from left to right in the illustration.

Recreation Receivers

These receivers are generally defined as L1 code receivers, which are typically not user configurable for settings such as mask angle, position dilution of precision (PDOP), the rate at which measurements are downloaded, the *logging rate*, also known as the epoch interval, and SNR. As you might expect, SNR is the ratio of the

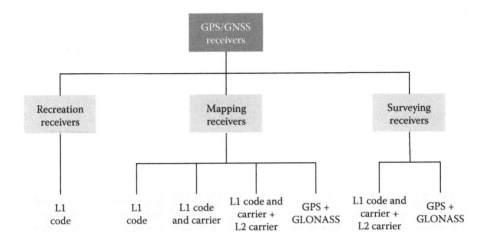

FIGURE 4.4   Receiver categories.

received signal power to the noise floor of a GPS observation. It is typical for the antenna, receiver, and CDU to be integrated into the device in these receivers.

Generally speaking, receivers that track the C/A code only provide relatively low accuracy. Most are not capable of tracking the carrier phase observable. These receivers were typically developed with basic navigation in mind. Most are designed for autonomous (stand-alone) operation to navigate, record tracks, waypoints, and routes aided by the display of onboard maps. They are sometimes categorized by the number of waypoints they can store. *Waypoint* is a term that grew out of military usage. It means the coordinate of an intermediate position a person, vehicle, or airplane must pass to reach a desired destination. With such a receiver, a user may call up a distance and direction from his present location to the next waypoint.

A single receiver operating without augmentation produces positions that are not relative to any ground control, local or national. In that context it is more appropriate to discuss the precision of the results than it is to discuss accuracy. Despite the limitations, some recreational receivers have capabilities that enhance their systematic precision and achieve a quantifiable accuracy using correction signals available from Earth-orbiting satellites such as the *Wide Area Augmentation System* (WAAS) correction. In other words, some have differential capability.

The WAAS is a U.S. *Federal Aviation Authority* (FAA) and the U.S. *Department of Transportation* system that augments GPS accuracy, availability, and integrity. The system relies on a network of ground-based reference stations at known positions that observe the GPS constellation constantly. From their data a correction message is calculated at two master stations. This message is uploaded to satellites on geostationary orbits. The satellites broadcast the message, and WAAS-enabled receivers can collect it and use the correction provided in it. For example, it provides a correction signal for precision approach aircraft navigation. Similar systems are Europe's *European Geostationary Navigation Overlay System* (EGNOS), and Japan's *Multifunction Transport Satellite* (MTSAT).

TABLE 4.1

Generalized Values for Recreational Receivers

	Autonomous Horizontal Precision	Real-Time Corrected Horizontal Network Accuracy	Post-Processed Horizontal Network Accuracy
Recreation	5–15 m	2–5 m	5–15 m

Recreational grade receivers typically do not have onboard feature data collection capabilities. They also do not usually have adequate onboard storage for recording the features (coordinates and attributes) required for a mapping project. Such capabilities are not needed for their designed applications.

When a recreation receiver is used to obtain an autonomous, or stand-alone, position, its precision may be within a range of 5–15 m as noted in Table 4.1. However, users can only expect to determine autonomous positions within 15 m of precision with this class of receiver when GPS signals of sufficient strength can be acquired under excellent satellite geometry. Under less optimal field conditions, tree cover and other obstructions, less than favorable GPS satellite geometry, etc., users can expect the precision of autonomous positions to lessen, sometimes substantially. The network accuracy had with real-time differential correction of 2–5 m is also not always achievable owing to the tendency of the WAAS being difficult to acquire particularly in the northern United States and then only with a southern sky clear of obstruction.

Local and Network Accuracy

The phrase network accuracy in Table 4.1 is used to define its difference from local accuracy. Network accuracy here concerns the uncertainty of a position relative to a datum. Local accuracy is not about a position relative to a datum, but it represents the uncertainty of a position relative to other positions nearby. In other words, local accuracy would be useful in knowing the accuracy of a line between the two positions at each end. Network accuracy would not be about the accuracy of the positions at each end of the line relative to each other but rather relative to the whole datum.

Local accuracy is also known as relative accuracy, and network accuracy is also known as absolute accuracy. The network and local accuracy values provide very different pictures. The local category represents the accuracy of a point with respect to adjacent points. The network category represents the accuracy of a point with respect to the reference system.

Local horizontal and vertical accuracies represent the averaged uncertainty in points relative to adjacent points to which they are directly connected (see Figure 4.5). Local horizontal coordinate accuracy is computed using an average error radius between the point in question and other adjacent points. Height accuracy is computed using an average of the linear vertical error between the point in question and other adjacent points.

Within a well-defined geographical area, local accuracy may be the most immediate concern. However, those tasked with constructing a control network that embraces

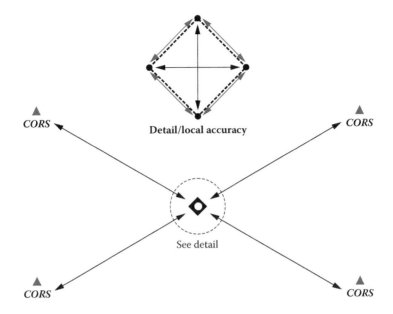

FIGURE 4.5   Local and network accuracy.

a wide geographical scope will most often need to know the positions relationship to the realization of the datum on which they are working. A point with good local accuracy may not have good network accuracy and vice versa.

Typically, network horizontal and vertical accuracies require that a point's accuracy be specified with respect to an appropriate national geodetic datum. In the United States, as a practical matter, this most often means that the work is tied to at least one of the Continuously Operating Reference Stations (CORS), which represent the most accessible realization of the National Spatial Reference System in the nation.

Mapping Receivers

These receivers are generally defined as those that allow user to configure some settings such as PDOP, SNR, elevation mask, and the logging rate. They most often have an integrated antenna and CDU in the receiver. They generally record pseudo-ranges and also can log data suitable for differential corrections either in real-time or for post-processing; many record carrier data.

Mapping receivers are often capable of storing mapped features (coordinates and attributes) and usually have adequate capacity for mapping applications. This memory is required for differential GPS, differential GPS receivers, even those that track only the C/A code. For many applications a receiver must be capable of collecting the same information as is simultaneously collected at a base station and storing it for post-processing. Receivers typically depend on proprietary post-processing software, which also includes utilities to enable GPS data to be transferred to a PC and exported in standard GIS file format(s) either over a cable or a wireless connection. Some mapping grade receivers for spatial data collection are single frequency, L1, with code only or both code and carrier.

TABLE 4.2

Generalized Values for Mapping Receivers

	Autonomous Horizontal Precision	Real-Time Corrected Horizontal Network Accuracy	Postprocessed Horizontal Network Accuracy
Mapping (L1 Code)	2–10 m	0.5–5 m	0.3–15 m
Mapping (L1 Code and Carrier)	2–10 m	0.5–5 m	0.2–1 m
Mapping (L1 Code and Carrier + L2 Carrier)	2–10 m	0.5–5 m	0.02–0.9 m
Mapping (GPS + GLONASS)	2–10 m	0.5–5 m	0.02–0.9 m

Most mapping grade receivers of all tracking configurations are WAAS (or other *satellite-based augmentation signal*) enabled and thereby offer real-time results. Such differentially corrected mapping receivers may be capable of achieving a network accuracy of ~0.5 to 5 m (Table 4.2).

Positions of submeter post-processed network accuracy can be achieved with mapping grade receivers that have the following configurations: L1 code and carrier, L1 code and carrier + L2, and L1 code and carrier + L2 (GPS and GLONASS). As noted, some mapping receivers offer tracking of the GNSS constellations including, for example, both GPS and GLONASS.

Global Navigation Satellite System

GPS is now a part of a growing international context—the GNSS. One definition of GNSS embraces any constellation of satellites providing signals from space that facilitate autonomous positioning, navigation, and timing on a global scale. Currently, there is one other system besides GPS that satisfies this definition. It is the Russian Federation's GLONASS, an acronym for Globalnaya Navigatsionnaya Sputnikovaya Sistema. There are several other satellite systems in development that will likely reach similar capability soon and will take their place within the GNSS definition. In fact, some of them are considered by many to already fall under the GNSS label. These include the GALILEO system administered by the EU, the Chinese Beidou Navigation Satellite System (BDS), and regional systems such as the Indian Regional Navigation Satellite System (IRNSS) and the Japanese Quasi-Zenith Satellite System (QZSS) are often included as well.

Surveying Receivers

Survey grade receivers are designed for the achievement of consistent network accuracy in the static or real-time mode. Positions determined by these receivers will generally provide the best accuracy of the categories listed. The components of these receivers can usually be configured in a variety of ways. The receivers are typically dual-frequency, code, and carrier receivers. Many are GNSS receivers (GPS and GLONASS). They generally provide more options for setting the observational parameters than recreational or mapping receivers. Surveying receivers are typically capable of observing the civilian code and carrier phase of all frequencies and are

TABLE 4.3

Generalized Values for Survey Receivers

	Autonomous Horizontal Precision	Real-Time Corrected Horizontal Network Accuracy	Postprocessed Horizontal Network Accuracy
Survey	2–10 m	>1 m	>0.1 m

appropriate for collecting positions on long base lines. Survey grade receivers are capable of producing network accuracies of better than 1 m with real-time differential correction and better than 0.1 m with post-processing (Table 4.3).

Most share some practical characteristics: they have multiple independent channels that track the satellites continuously, and they begin acquiring satellites' signals from a few seconds to less than a minute from the moment they are switched on. Most acquire all the satellites above their mask angle in a very few minutes, with the time usually lessened by a warm start, and most provide some sort of alert to the user that data is being recorded, and so forth. About three-quarters of them can have their sessions pre-programmed in the office before going to their field sites. Nearly all allow the user to select the logging rate, also known as epoch interval and also known as sampling interval. While a 1 s interval is often used, faster rates of 0.1 s (10 Hz) and more increments of tenths of a second are often available. This feature allows the user to stipulate the short period of time between each of the microprocessors downloads to storage. The faster the data-sampling rate, the larger the volume of data a receiver collects and the larger the amount of storage it needs. A fast rate is helpful in cycle slip detection, and that improves the receiver's performance on baselines longer than 50 km, where the detection and repair of cycle slips can be particularly difficult.

EXERCISES

1. Which of the following is not a consideration in antenna design for GPS receivers?
   a. Efficient conversion of electromagnetic waves into electric currents
   b. Directional capability
   c. Coincidence of the phase center and the physical center
   d. Reduction of the effects of multipath

2. Which of the ideas listed below is intended to limit the effect of the difference between the phase center and the physical center of GPS antennas on a baseline measurement?
   a. Ground plane antennas at each end of the baseline
   b. Choke ring antennas at each end of the baseline
   c. Rotation of the antenna's reference marks to north at each end of the baseline
   d. Use of the receiver's built-in antennas at each end of the baseline

3. Which of the following does not describe an advantage of a 12-channel parallel continuous-tracking receiver with dedicated channels over a 1-channel
multiplexing or sequencing receiver?
   a. The parallel continuous-tracking receiver has a superior signal-to-noise
      ratio
   b. The parallel continuous-tracking receiver is more accurate
   c. If a channel stops working, there is redundancy with a parallel continuous-
      tracking receiver
   d. The parallel continuous-tracking receiver has less frequent cycle slips

4. Which of the following statements concerning the intermediate frequency
(IF) in GPS signal processing is correct?
   a. In the radio frequency (RF) section of a GPS receiver, two frequencies
      go through a bandpass filter, which selects the higher of the two. This
      signal is then known as the intermediate frequency (IF)
   b. GPS signal processing usually has a single IF stage
   c. IF is a beat frequency
   d. IF is the sum of the Doppler-shifted carrier from the satellite and the
      signal generated by the receiver's own oscillator

5. Which of the following is not a drawback of signal squaring?
   a. Effect of multipath is increased
   b. Signal-to-noise ratio deteriorates
   c. Codes are stripped from the carrier
   d. Receiver must acquire information such as almanac data and clock cor-
      rections somewhere other than the NAV message

6. Which of the following is the closest to the length of a C/A code period that
is 1,023 chips and 1 ms in duaration?
   a. 300 m
   b. 300 km
   c. 66 m
   d. 20,000 m

7. What is an isotropic antenna?
   a. Hypothetical lossless antenna that has equal capabilities in all directions
   b. Durable, compact antenna with a simple construction and a low profile
   c. Antenna built with several concentric rings and designed to reduce the
      effects of multipath
   d. Antenna that has a stable phase center and simple construction but
      needs a good ground plane

8. Which of the following is still used to store data in GPS receivers?
   a. Cassette
   b. Floppy disk
   c. Tapes
   d. Internal onboard memory

9. How does the antenna bandwidth affect antenna's performance?
   a. It does not have any effect
   b. Narrow bandwidth increases antenna's performance
   c. Larger bandwidth increases antenna's performance and increases inter-ference
   d. Narrow bandwidth decreases antenna's performance

10. Which of the following describes the best absolute accuracy?
    a. Absolute accuracy and absolute precision are the same
    b. Absolute accuracy is also called local accuracy
    c. Absolute accuracy represents the location of the point with respect to the reference system
    d. Absolute accuracy is also known as relative accuracy

ANSWERS AND EXPLANATIONS

1. Answer is (b)

   Explanation: The GPS signal is actually quite weak, and it is broadcast over a very large area. The efficiency of GPS antennas is an important consideration. Limiting the effects of multipath is also a very important consideration. The perfect coincidence of the phase center with the physical center in GPS antennas has yet to be achieved, but it is much sought after and is certainly a design consideration.

   However, the GPS signal is a spread-spectrum coded signal that occupies a broader frequency bandwidth than it must to carry its information. This fact allows GPS antennas to be omnidirectional. They do not require directional orientation to properly receive the GPS signal.

2. Answer is (c)

   Explanation: Ground planes used with many GPS antennas and the choke ring antennas are designs that attack the same problem, limiting the effects of multipath. The coincidence of the phase center with the physical center in GPS antennas is not yet perfected, and because the measurements made by a GPS receiver are made to the phase center of its antenna, its orientation is paramount. The rotation of the antennas at each end of a baseline to north per an imprinted reference is a strategy to reduce the effect of the difference between the two points.

3. Answer is (b)

   Explanation: Multiplexing receivers are at somewhat of a disadvantage when compared with continuous-tracking receivers. Continuous-tracking receivers with dedicated channels have a superior signal-to-noise ratio. They are faster. There is redundancy if a channel fails. They have a more certain phase lock. There are fewer cycle slips.

   A multiplexing receiver is not necessarily less accurate than a parallel continuous-tracking receiver.

4. Answer is (c)

Explanation: The intermediate frequency (IF) is definitely a beat frequency. It is the combination of the frequency coming from the satellite and a sinusoidal signal generated by the receiver's oscillator. These frequencies are multiplied together in a mixer that produces the sum of the two and the difference. Both go through a bandpass filter, which selects the lower of the two, and it is known as the IF or beat frequency signal. This beat frequency is the difference between the Doppler-shifted carrier frequency that came from the satellite and the frequency generated by the receiver's own oscillator. There are usually several IF stages in a GPS receiver.

5. Answer is (a)

Explanation: With signal squaring, the beat or intermediate frequency is created by multiplying the incoming carrier by itself. The result has double the frequency and half the wavelength of the original. It is squared. The process of squaring the carrier strips off all the codes. The P code, C/A code, and Navigation message are not available to the receiver. With the codes wiped out, the receiver must acquire information such as almanac data and clock corrections from other sources. Also, the signal-to-noise ratio is degraded, increasing cycle slips. The effect of multipath is actually decreased, and because squaring works as well on L2 as it does on L1, dual-frequency ionospheric delay correction is possible.

6. Answer is (b)

Explanation: The C/A code *chipping rate*, the rate at which each chip is modulated onto the carrier, is 1.023 Mbps. That means, at light speed, the chip length is approximately 300 m. However, the whole C/A code period is 1023 chips, 1 ms long. That is approximately 300 km, and, of course, each satellite repeats its whole 300 km C/A code over and over.

7. Answer is (a)

Explanation: An isotropic antenna is a hypothetical, lossless antenna that has equal capabilities in all directions. This theoretical antenna is used as a basis of comparison when expressing the gain of GPS and other antennas. A gain of about 3 dB is typical for the usual omnidirectional GPS antenna. The gain, or gain pattern, describes the success of a GPS antenna in collecting more energy from above the mask angle, and less from below the mask angle. The decibel here does not indicate the power of the antenna; it refers to a comparison. In this case, 3 dB indicates that the GPS antenna has about 50 percent of the capability of an isotropic antenna.

8. Answer is (d)

Explanation: Most GPS receivers today have internal data logging. They use solid-state memory or memory cards. Most also allow the user the option of connecting to a computer and having the data downloaded directly to the hard drive. Cassette, floppy disks, and computer tapes are mostly things of the past in GPS receiver storage.

9. Answer is (c)

   Explanation: An antenna ought to have a bandwidth commensurate with its application; in general, the larger the bandwidth, the better the performance. However, there is a downside; increased bandwidth degrades the signal-to-noise ratio by including more interference.

10. Answer is (c)

   Explanation: The use of the phrase network accuracy in Table 4.1 is used to define its difference from local accuracy. Network accuracy here concerns the uncertainty of a position relative to a datum. Local accuracy is not about a position relative to a datum, but it represents the uncertainty of a position relative to other positions nearby. In other words, local accuracy would be useful in knowing the accuracy of a line between the two positions at each end. Network accuracy would not be about the accuracy of the positions at each end of the line relative to each other, but rather relative to the whole datum.

   Local accuracy is also known as relative accuracy, and network accuracy is also known as absolute accuracy. The network and local accuracy values provide very different pictures. The local category represents the accuracy of a point with respect to adjacent points. The network category represents the accuracy of a point with respect to the reference system.

5 Coordinates

A FEW PERTINENT IDEAS ABOUT GEODETIC DATUMS FOR GLOBAL POSITIONING SYSTEMS

PLANE SURVEYING

Plane surveying has traditionally relied on an imaginary, flat reference surface, or *datum*, with Cartesian axes. This rectangular system is used to describe measured positions by ordered pairs, usually expressed in northings and eastings or y and x coordinates. Even though surveyors have always known that this assumption of a flat Earth is fundamentally unrealistic, it provided, and continues to provide, an adequate arrangement for small areas. The attachment of elevations to such horizontal coordinates somewhat acknowledges the topographic irregularity of the Earth, but the whole system is always undone by its inherent inaccuracy as surveys grow large.

Development of State Plane Coordinate Systems

Designed in the 1930s, the purpose of the state plane coordinate system was to overcome some of the limitations of the horizontal plane datum when they are applied over large areas while avoiding the imposition of geodetic methods and calculations on local surveyors. Using the conic and cylindrical models of the Lambert and Mercator map projections respectively, the flat datum was curved but only in one direction. By curving the datums and limiting the area of the zones the distortion can be limited to a scale ratio of about 1 part in 10,000 without disturbing the traditional system of ordered pairs of Cartesian coordinates.

The state plane coordinate system was a step ahead at that time. To this day, it provides surveyors with a mechanism for coordination of surveying stations that approximates geodetic accuracy more closely than the commonly used methods of small-scale plane surveying. However, the state plane coordinate systems were organized in a time of generally lower accuracy and efficiency in surveying measurement. It was an understandable compromise in an age when such computation required sharp pencils, logarithmic tables, and lots of midnight oil.

The distortion of positions attributable to the transformation of geodetic coordinates into the plane grid coordinates of any one of these projections is generally small. Most Global Positioning Systems (GPS) and land surveying software packages provide routines for automatic transformation of latitude and longitude to and from these mapping projections. Similar programs are available from the *National Geodetic Survey* (NGS). Therefore, for most applications of GPS, there ought to be no technical compunction about expressing the results in grid coordinates. However, because the results are presented in plane coordinates it can be easy to lose sight of the geodetic context of the entire process that produced them. The following is an effort to provide some context to that relationship.

GPS Surveyors and Geodesy

Today, GPS has thrust surveyors into the thick of geodesy, which is no longer the exclusive realm of distant experts. Thankfully, in the age of the microcomputer, the computational drudgery can be handled with software packages. Nevertheless, it is unwise to venture into GPS believing that knowledge of the basics of geodesy is, therefore, unnecessary. It is true that GPS would be impossible without computers, but blind reliance on the data they generate eventually leads to disaster.

SOME GEODETIC COORDINATE SYSTEMS

Three-Dimensional (3-D) Cartesian Coordinates

A spatial Cartesian system with three axes lends itself to describing the terrestrial positions derived from space-based geodesy. Using three rectangular coordinates instead of two, one can unambiguously define any position on the Earth or above it for that matter. The three-dimensional Cartesian coordinates (x,y,z) derived from this system are known as Earth-Centered-Earth-Fixed (ECEF) coordinates. It is a right-handed orthogonal system that rotates with and is attached to the Earth, which is why it is called *Earth fixed*.

A three-dimensional Cartesian coordinate system is right-handed if it can be described by the following model: the extended forefinger of the right hand symbolizes the positive direction of the x axis. The middle finger of the same hand extended at right angles to the forefinger symbolizes the positive direction of the y axis. The extended thumb of the right hand, perpendicular to them both, symbolizes the positive direction of the z axis (see Figure 5.1).

However, such a system is only useful if its origin (0,0,0) and its axes (x,y,z) can be fixed to the planet with certainty, something easier said than done.

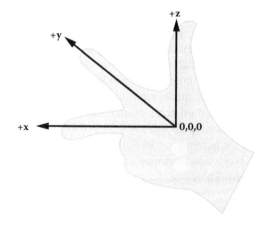

FIGURE 5.1   Three-dimensional cartesian coordinates.

The usual arrangement is known as the *Conventional Terrestrial Reference System* (CTRS), and the *Conventional Terrestrial System* (CTS). The latter name will be used here. The origin is the center of mass of the whole Earth including oceans and atmosphere, the *geocenter*. The *x* axis is a line from that geocenter through its intersection at the zero meridian or the *International Reference Meridian* (IRM) with the internationally defined conventional equator. The *y* axis is extended from the geocenter along a line perpendicular from the *x* axis in the same mean equatorial plane toward 90°E longitude. That means that the positive end of the *y* axis intersects the actual Earth in the Indian Ocean. In any case, they both rotate with the Earth around the *z* axis, a line from the geocenter through the internationally defined pole known as the *International Reference Pole* (IRP).

However, the Earth is constantly moving, of course. While one can say that the Earth has a particular axis of rotation, equator, and zero meridian for an instant, they all change slightly in the next instant. Within all this motion, how do you stabilize the origin and direction of the three axes for the long term? One way is to choose a moment in time and consider them fixed to the Earth as they are at that instant.

Polar Motion

Here is an example of that process of definition. The Earth's rotational axis wanders slightly with respect to the solid earth in a very slow oscillation called *polar motion*. The largest component of the movement relative to the Earth's crust has a 430-day cycle known as the *Chandler period*. It was named after American astronomer Seth C. Chandler, who described it in papers in the *Astronomical Journal* in 1891. Another aspect of polar motion is sometimes called *polar wander*. It is about 0.004 s of arc per year as the pole moves toward Ellesmere Island (see Figure 5.2). The actual displacement caused by the wandering generally does not exceed 12 meters. Nevertheless, the conventional terrestrial system of coordinates would be useless if its third axis was constantly wobbling. Originally, an average stable position was chosen for the position of the pole. Between 1900 and 1905, the mean position of the Earth's rotational pole was designated as the *Conventional International Origin* (CIO) and the *z* axis.

This was defined by the *Bureau International de l'Heure* (BIH). It has since been refined by the *International Earth Rotation Service* (IERS) using very long baseline interferometry (VLBI) and satellite laser ranging (SLR). It is now placed as it was midnight on New Year's Eve 1983, or January 1, 1984 (UTC). The moment is known as an *epoch* and can be written 1984.0. So we now use the axes illustrated in Figure 5.3. The name of the *z* axis has been changed to the IRP epoch 1984, but it remains within 0.005″ of the previous definition. It provides a geometrically stable and clear definition on the Earth's surface for the *z* axis.

In this 3-D, right-handed coordinate system the *x* coordinate is a distance from the *y*–*z* plane measured parallel to the *x* axis. It is always positive from the zero meridian to 90°W longitude and from the zero meridian to 90°E longitude. In the remaining 180° the *x* coordinate is negative. The *y* coordinate is a perpendicular distance from the plane of the zero meridian. It is always positive in the Eastern Hemisphere and negative in the Western Hemisphere. The *z* coordinate is a perpendicular distance from the plane of the equator. It is always positive in the Northern Hemisphere and negative in the Southern Hemisphere. For example, the position of the station CTMC

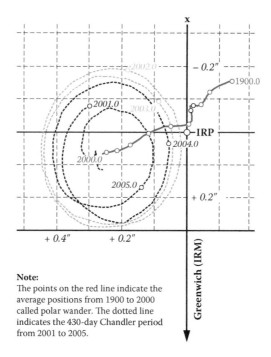

FIGURE 5.2   Polar motion.

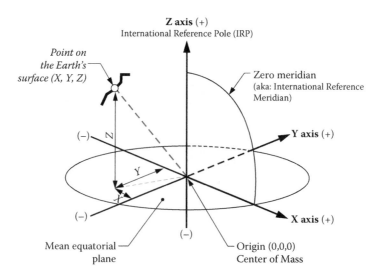

FIGURE 5.3   Three-dimensional coordinate.

TABLE 5.1

Station CTMC

CDOT GOLDEN (CTMC), COLORADO
Created on 31Aug2011 at 10:11:52

Antenna Reference Point (ARP):CDOT GOLDEN CORS ARP

PID + DM5962
NAD_83 (2011) POSITION (EPOCH 2010.0)
Transformed from IGS08 (epoch 2005.0) position in Aug 2011.

X = −1287787.768 m	latitude = 39 43 17.49315 N
Y = −4742178.364 m	longitude = 105 11 34.33560 W
Z = 4055414.614 m	ellipsoid height = 1819.836 m

expressed in 3-D Cartesian coordinates of this type expressed in meters, the native unit of the system (Table 5.1).

It is important to note that the GPS Control Segment generates the position and velocity of the satellites themselves in ECEF coordinates. It follows that most modern GPS software provides the GPS positions in ECEF as well. Further, the ends of baselines determined by GPS observation are typically given in ECEF coordinates so that the vectors themselves become the difference between those x, y, and z coordinates. The display of these differences as *DX*, *DY*, and *DZ* is a usual product of these post-processed calculations (see Figure 5.4).

Occupation Time:	00:30:30:00			
Measurement Epoch Interval (s):	15:00			
Solution Time:	Receiver/Satellite double difference			
	Iono free fixed			
Solution Acceptability:	Passed ratio test			
Baseline Slope Distance Std. Dev. (meters):	17,044.376		0.000574	
	Forward		*Backward*	
Normal Section Azimuth:	179° 04′ 38.886169″		359° 04′ 47.857511″	
	0° 07′ 00.456589″		−0° 16′ 12.548396″	
Baseline Components (meters):	dn −17,042.131 de	274.423	du	34.744

Standard Deviations:	dx −6552.297	dy −10264.496	dz −11925.530
	8.437168E-004	1.072974E-003	9.513724E-004

Aposteriori Covariance Matrix:	7.118580E-007		
	7.389287E-007	1.151273E-006	
	−6.141036E-007	−7.690377E-007	9.051094E-004

Variance Ratio Cutoff:	62.1	1.5	
Reference Variance:	0.556		
Observable Count/Rejected RMS:	Iono free phase	451/0	0.006

FIGURE 5.4   DX, DY, and DZ from GPS.

Latitude and Longitude

Despite their utility, such 3-D Cartesian coordinates are not the most common method of expressing a geodetic position. Latitude and longitude have been the coordinates of choice for centuries. The designation of these relies on the same two standard lines as 3-D Cartesian coordinates: the mean equator and the zero meridian. Unlike them, however, they require some clear representation of the terrestrial surface. In modern practice, latitude and longitude cannot be said to uniquely define a position without a clear definition of the Earth itself.

ELEMENTS OF A GEODETIC DATUM

How can latitude ϕ and longitude λ not define a unique position on the Earth? The reference lines—the mean equator and the zero meridian—are clearly defined. The units of degrees, minutes, seconds, and decimals of seconds, allow for the finest distinctions of measurement. Finally, the reference surface is the Earth itself.

The answer to the question relies, in the first instance, on the fact that there are several categories of latitude and longitude, the geographical coordinates. From the various options, astronomic, geocentric, and geodetic, the discussion here will concern geodetic latitude and longitude as they are typical. Its definition has a good deal to do with where down is.

Deflection of the Vertical

Down seems like a pretty straightforward idea. A hanging plumb bob certainly points down. Its string follows the direction of gravity. That is one version of the idea. There are others.

Imagine an optical surveying instrument set up over a point. If it is centered precisely with a plumb bob and leveled carefully, the plumb line and the line of the level telescope of the instrument are perpendicular to each other. In other words, the level line, the horizon of the instrument, is perpendicular to gravity. Using an instrument so oriented, it is possible to determine the latitude and longitude of the point. Measuring the altitude of a circumpolar star is one good method of finding the latitude of the point from which the measurement is made. The measured altitude would be relative to the horizontal level line of the instrument of course. A latitude found this way is called *astronomic latitude*.

One might expect that this astronomic latitude would be the same as the geocentric latitude of the point, but they are different. The difference is due to the fact that a plumb line coincides with the direction of gravity; it does not point to the center of the Earth where the line used to derive geocentric latitude originates.

Astronomic latitude also differs from the most widely used version of latitude, geodetic. The line from which geodetic latitude is determined is perpendicular with the surface of the ellipsoidal model of the Earth that does not match a plumb line either; more about that in the next section. In other words, there are three different versions of down and each has its own latitude. For geocentric latitude, down is along a line to the center of the Earth. For geodetic latitude, down is along a line perpendicular to the ellipsoidal model of the Earth. For astronomic latitude, down is along a line in the direction of gravity. More often than not, these are three completely different lines.

GEOCENTRIC, GEODETIC, AND ASTRONOMIC LATITUDE

Each can be extended upward, too, toward the zenith, and there are small angles between them. The angle between the vertical extension of a plumb line and the vertical extension of a line perpendicular to the ellipsoid is called the deflection of the vertical (Figure 5.5). It sounds better than the difference in down. This *deflection of the vertical* defines the actual angular difference between the astronomic latitude and longitude of a point and its geodetic latitude and longitude; latitude and

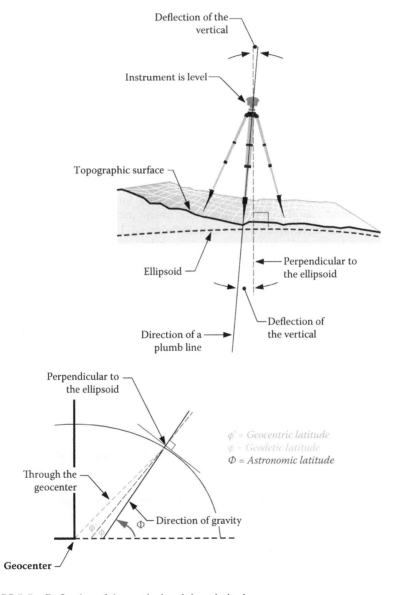

$\phi' = Geocentric\ latitude$
$\phi = Geodetic\ latitude$
$\Phi = Astronomic\ latitude$

FIGURE 5.5   Deflection of the vertical and three latitudes.

longitude because, even though the discussion has so far been limited to latitude, the deflection of the vertical usually has both a north–south and an east–west component. The deflection of the vertical also has an effect on azimuths; for example, there will be a slight difference between the azimuth of a GPS baseline and the astronomically determined azimuth of the same line.

It is interesting to note that that optical instrument set up so carefully over a point on the Earth cannot be used to measure geodetic latitude and longitude directly because they are not relative to the actual Earth but rather a model of it. Gravity does not even come into the ellipsoidal version of down. On the model of the Earth, down is a line perpendicular to the ellipsoidal surface at a particular point. On the real Earth down is the direction of gravity at the point. They are most often not the same thing, but fortunately the difference is usually very small. However, because the ellipsoidal model of the Earth is imaginary, it is quite impossible to actually set up an instrument on the ellipsoid. Yet, the measurement of latitude and longitude by astronomic observations has a very long history indeed. The most commonly used coordinates are not astronomic latitudes and longitudes but geodetic latitudes and longitudes. So conversion from astronomic latitude and longitude to geodetic latitude and longitude has a long history as well. Therefore, until the advent of GPS, geodetic latitudes and longitudes were often values ultimately derived from astronomic observations by post-observation calculation. In a sense that is still true; the change is a modern GPS receiver can display the geodetic latitude and longitude of a point to the user immediately because the calculations can be completed with incredible speed. However, a fundamental fact remains unchanged: the instruments by which latitudes and longitudes are measured are oriented to gravity, the ellipsoidal model on which geodetic latitudes and longitudes are determined is not. That is just as true for the antenna of a GPS receiver, an optical surveying instrument, a camera in an airplane taking aerial photography, or even the GPS satellite themselves.

Datums

The second part of the answer to the question posed earlier is this: if geographic coordinates are to have meaning, they must have a context, a datum.

Despite the certainty of the physical surface of the Earth, the lithosphere, it remains notoriously difficult to define in mathematical terms. The dilemma is illustrated by the ancient struggle to represent its curved surface on flat maps. There have been a whole variety of map projections developed over the centuries that rely on mathematical relationships between positions on the Earth's surface and points on the map. Each projection serves a particular application well, but none of them can represent the Earth without distortion. For example, no modern surveyor would presume to promise a client a high-precision control network with data scaled from a map. As the technology of measurement has improved, the pressure for greater exactness in the definition of the Earth's shape has increased. Even with electronic tools that widen the scope and increase the precision of the data, perfection is nowhere in sight.

Development of the Ellipsoidal Model

Despite the fact that local topography is obvious to an observer standing on the Earth, efforts to grasp the more general nature of the planet's shape and size have been occupying scientists for at least 2300 years. There have, of course, been long intervening periods of unmitigated nonsense on the subject. Ever since 200 B.C. when Eratosthenes almost calculated the planet's circumference correctly, geodesy has been getting ever closer to expressing the actual shape of the Earth in numerical terms. A leap forward occurred with Newton's thesis that the Earth was an ellipsoid rather than a sphere in the first edition of his *Principia* in 1687.

Newton's idea that the actual shape of the Earth was slightly ellipsoidal was not entirely independent. There had already been some other suggestive observations. For example, 15 years earlier, astronomer Jean Richer had found that to maintain the accuracy of the 1-s clock he used in his observations in Cayenne, French Guiana, he had to shorten its pendulum significantly. The clock's pendulum, regulated in Paris, tended to swing more slowly as it approached the equator. Newton reasoned that the phenomenon was attributable to a lessening of the force of gravity. On the basis of his own theoretical work, he explained the weaker gravity by the proposition, "the earth is higher under the equator than at the poles, and that by an excess of about 17 miles" (*Philosophiae naturalis principia mathematica*, Book III, Proposition XX).

Although Newton's model of the planet bulging along the equator and flattened at the poles was supported by some of his contemporaries, notably Huygens, the inventor of Richer's clock, it was attacked by others. The director of the Paris Observatory, Jean Dominique Cassini, for example, took exception to Newton's concept. Even though the elder Cassini had himself observed the flattening of the poles of Jupiter in 1666, neither he nor his equally learned son Jacques were prepared to accept the same idea when it came to the shape of the Earth. It appeared they had some empirical evidence on their side.

For geometric verification of the Earth model, scientists had employed arc measurements at various latitudes since the early 1500s. Establishing the latitude of their beginning and ending points astronomically, they measured a cardinal line to discover the length of 1° of longitude along a meridian arc. Early attempts assumed a spherical Earth, and the results were used to estimate its radius by simple multiplication. In fact, one of the most accurate of the measurements of this type, begun in 1669 by the French Abbé J. Picard, was actually used by Newton in formulating his own law of gravitation. However, Cassini noted that close analysis of Picard's arc measurement, and others, seemed to show the length of 1° of longitude actually decreased as it proceeded northward. He concluded that the Earth was not flattened as proposed by Newton but was rather elongated at the poles.

The argument was not resolved until two expeditions between about 1733 and 1744 were completed. They were sponsored by the Paris Académie Royale des Sciences and produced irrefutable proof. One group that included Clairaut and Maupertuis was sent to measure a meridian arc near the Arctic Circle, 66°20′ Nφ, in Lapland.

Another expedition with Bouguer and Godin to what is now Ecuador measured an arc near the equator, 01°31′ Sφ. Newton's conjecture was proved correct, and the contradictory evidence of Picard's arc was charged to errors in the latter's measurement of the astronomic latitudes.

The ellipsoidal model (see Figure 5.6), bulging at the equator and flattened at the poles, has been used ever since as a representation of the general shape of the Earth's surface. It is called an oblate spheroid. In fact, several reference ellipsoids

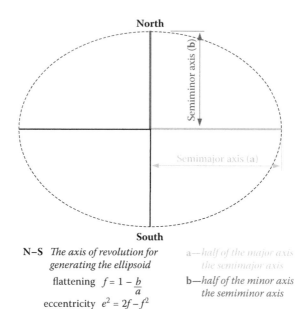

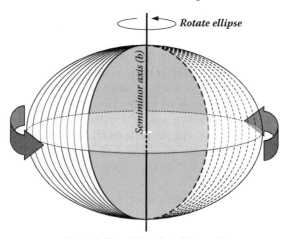

Biaxial ellipsoid model of the earth

FIGURE 5.6   An ellipsoid.

have been established for various regions of the planet. They are precisely defined by their semimajor axis and flattening. The relationship between these parameters is expressed in the formula:

$$f = \frac{a-b}{c}$$

where
  f = flattening
  a = semimajor axis
  b = semiminor axis

BIAXIAL ELLIPSOIDAL MODEL OF THE EARTH

Role of an Ellipsoid in a Datum

The semimajor axis and flattening can be used to completely define an ellipsoid of revolution. The ellipsoid is revolved around the minor axis. However, in the traditional approach, six additional elements are required if that ellipsoid is to be used as a geodetic datum: three to specify its center and three more to clearly indicate its orientation around that center. The Clarke 1866 spheroid is one of many reference ellipsoids. Its shape is completely defined by a semimajor axis, a, of 6378.2064 km and a flattening f of 1/294.9786982. It is the reference ellipsoid of the datum known to surveyors as the North American Datum of 1927 (NAD27), but it is not the datum itself.

For the Clarke 1866 spheroid to become NAD27, the ellipsoid of reference had to be attached at a point and specifically oriented to the actual surface of the Earth. However, even this ellipsoid, which fits North America well, could not conform to that surface perfectly. Therefore, the initial point was chosen near the center of the anticipated geodetic network to best distribute the inevitable distortion between the ellipsoid and the actual surface of the Earth as the network extended beyond the initial point. The attachment was established at Meades Ranch, Kansas, 39°13′26.686″ Nϕ, 98°32′30.506″ Wλ and *geoidal height* was considered to be zero. Those coordinates were not sufficient, however. The establishment of directions from this initial point was required to complete the orientation. The azimuth from Meades Ranch to station Waldo was fixed at 75°28′09.64″ and the deflection of the vertical set at zero.

Once the initial point and directions were fixed, the whole orientation of NAD27 was established, including the center of the reference ellipsoid. Its center was imagined to reside somewhere around the center of mass of the Earth. However, the two points were certainly not coincident nor were they intended to be. In short, NAD27 does not use a geocentric ellipsoid.

REGIONAL ELLIPSOIDS

Measurement Technology and Datum Selection

In the period before space-based geodesy was tenable, a regional datum was not unusual. The Australian Geodetic Datum 1966, the European Datum 1950, and the South American Datum 1969, among others, were also designed as nongeo-centric systems. Achievement of the minimum distortion over a particular region was the primary consideration in choosing their ellipsoids, not the relationship of their centers to the center of mass of the Earth (see Figure 5.7). For example, in the Conventional Terrestrial System (CTS) the 3-D Cartesian coordinates of the center of the Clarke 1866 spheroid as it was used for NAD27 are about $X = -4$ m, $Y = +166$ m, and $Z = +183$ m.

This approach to the design of datums was bolstered by the fact that the vast majority of geodetic measurements they would be expected to support were of the classical variety. That is, the work was done with theodolites, towers, and tapes. They were Earth bound. Even after the advent of electronic distance measurement, the general approach involved the determination of horizontal coordinates by measur-ing from point to point on the Earth's surface and adding heights, otherwise known as *elevations*, through a separate leveling operation. As long as this methodological separation existed between the horizontal and vertical coordinates of a station, the difference between the ellipsoid and the true Earth's surface was not an overriding concern. Such circumstances did not require a geocentric datum.

However, as the sophistication of satellite geodesy increased, the need for a truly global, geocentric datum became obvious. The horizontal and vertical information were no longer separate. Because satellites orbit around the center of mass of the Earth, a position derived from space-based geodesy can be visualized as a vector originating from that point.

So, today, not only are the horizontal and vertical components of a position derived from precisely the same vector, the choice of the coordinate system used to express them is actually a matter of convenience. The position vector can be transformed

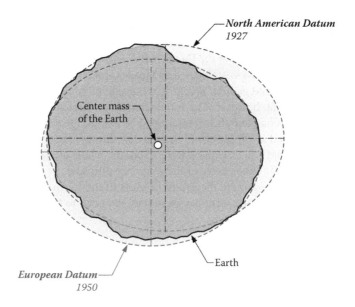

FIGURE 5.7   Nongeocentric datums.

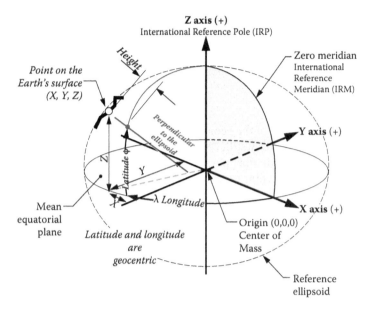

Z axis (+)
International Reference Pole (IRP)

Height

Point on the Earth's surface (X, Y, Z)

Zero meridian
International
Reference
Meridian (IRM)

Perpendicular to the ellipsoid

Y axis (+)

Z

Latitude of

Y

λ *Longitude*

X axis (+)

Mean equatorial plane

Origin (0,0,0)
Center of Mass

Latitude and longitude are geocentric

Reference ellipsoid

FIGURE 5.8   A few fundamentals.

into the 3-D Cartesian ECEF system, the traditional latitude, longitude, and height, or virtually any other well-defined coordinate system. However, because the orbital motion and the subsequent position vector derived from satellite geodesy are themselves Earth centered, it follows that the most straightforward representations of that data are Earth centered as well (see Figure 5.8).

POSITION DERIVED FROM GPS

Development of a Geocentric Model

Satellites have not only provided the impetus for a geocentric datum, they have also supplied the means to achieve it. In fact, the orbital perturbations of man-made near-Earth satellites have probably brought more refinements to the understanding of the shape of the Earth in a shorter span of time than was ever before possible. For example, the analysis of the precession of Sputnik 2 in the late 1950s showed researchers that the Earth's semiminor axis was actually 85 m shorter than had been previously thought. In 1958, while studying the tracking data from the orbit of Vanguard I, Ann Bailey of the Goddard Spaceflight Center discovered that the planet is shaped a bit like a pear. There is a slight protuberance at the North Pole, a little depression at the South Pole, and a small bulge just south of the equator.

These formations and others have been discovered through the observation of small distortions in satellites' otherwise elliptical orbits, little bumps in their road, so to speak. The deviations are caused by the action of Earth's gravity on the satellites as they travel through space. Just as Richer's clock reacted to the lessening of gravity at the equator and thereby revealed one of the largest features of the Earth's shape to Newton, small perturbations in the orbits of satellites, also responding to gravity, reveal details of Earth's shape to today's scientists. The common aspect of these examples is the direct relationship between direction and magnitude of gravity and the planet's form. In fact, the surface that best fits the Earth's gravity field has been given a name. It is called the geoid.

GEOID

An often-used description of the geoidal surface involves idealized oceans. Imagine the oceans of the world utterly still, completely free of currents, tides, friction, variations in temperature and all other physical forces, except gravity. Reacting to gravity alone, these unattainable calm waters would coincide with the figure known as the geoid (see Figure 5.9). Admitted by small frictionless channels or tubes and allowed to migrate across the land, the water would then, theoretically, define the same geoidal surface across the continents, too.

Of course, the 70 percent of the Earth covered by oceans is not so cooperative nor is there any such system of channels and tubes. In addition, the physical forces eliminated from the model cannot be avoided in reality. These unavoidable forces actually cause Mean Sea Level to deviate from the geoid. This is one of the reasons that Mean Sea Level and the surface of the geoid are not the same, and it is a fact frequently mentioned to emphasize the inconsistency of the original definition of the geoid as it was offered by J.B. Listing in 1872. Listing thought of the geoidal surface as equivalent to Mean Sea Level. Even though his idea does not stand up to scrutiny today, it can still be instructive.

Equipotential Surface

Gravity is not consistent across the topographic surface of the Earth. At every point it has a magnitude and a direction. In other words, anywhere on the Earth, gravity can be described by a mathematical vector. Along the solid earth, such vectors do not

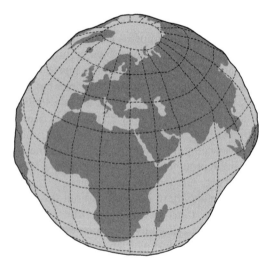

FIGURE 5.9 Exaggerated representation of the geoid.

have all the same direction or magnitude, but one can imagine a surface of constant gravity potential. Such an *equipotential* surface would be level in the true sense. It would coincide with the top of the hypothetical water in the previous example. Mean Sea Level does not define such a figure; nevertheless, the geoidal surface is not just a product of imagination. For example, the vertical axis of any properly leveled surveying instrument and the string of any stable plumb bob are perpendicular to the geoid. Just as pendulum clocks and Earth-orbiting satellites, they clearly show that the geoid is a reality.

Geoidal Undulation

The geoid does not precisely follow Mean Sea Level nor does it exactly correspond with the topography of the dry land. It is irregular like the terrestrial surface. It is bumpy. Uneven distribution of the mass of the planet makes it maddeningly so, because if the solid earth had no internal anomalies of density, the geoid would be smooth and almost exactly ellipsoidal. In that case, the reference ellipsoid could fit the geoid to near perfection and the lives of geodesists would be much simpler. However, like the Earth itself, the geoid defies such mathematical consistency and departs from true ellipsoidal form by as much as 100 m in places (see Figure 5.10).

MODERN GEOCENTRIC DATUM

Three distinct figures are involved in a geodetic datum for latitude, longitude, and height: the geoid, the reference ellipsoid, and the topographic surface. Owing in large measure to the ascendancy of satellite geodesy, it has become highly desirable that they share a common center.

While the level surface of the geoid provides a solid foundation for the definitions of heights and the topographic surface of the Earth is necessarily where measurements are made, neither can serve as the reference surface for geodetic positions.

From the continents to the floors of the oceans, the solid Earth's actual surface is too irregular to be represented by a simple mathematical statement. The geoid, which is sometimes under, and sometimes above, the surface of the Earth, has an overall shape that also defies any concise geometrical definition. The ellipsoid not only has the same general shape as the Earth, but, unlike the other two figures, can be described simply and completely in mathematical terms.

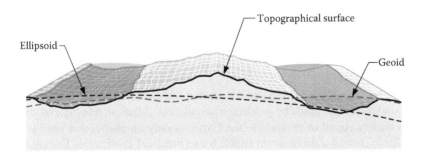

FIGURE 5.10   Three surfaces.

Therefore, a global geocentric system has been developed based on the ellipsoid adopted by the *International Union of Geodesy and Geophysics* (IUGG) in 1979. It is called the *Geodetic Reference System 1980* (GRS80). Its semimajor axis a is 6,378.137 km and is probably within a few meters of the Earth's actual equatorial radius. Its flattening f is 1/298.25722 and likely deviates only slightly from the true value, a considerable improvement over Newton's calculation of a flattening ratio of 1/230, but then he did not have orbital data from near-Earth satellites to check his work.

World Geodetic System 1984 (WGS84)

However, GRS80 is not the reference ellipsoid for the GPS system that is the WGS84 ellipsoid. The WGS84 ellipsoid is actually very similar to the GRS80 ellipsoid. The difference between them is chiefly found in the flattening. The WGS84 ellipsoid is the foundation of the coordinate system, known as the *World Geodetic System 1984* (WGS84). This datum has been used by the U.S. Military since January 21, 1987. However, there have been six incarnations of WGS84 since then.

While WGS84 has always been the basis for the GPS Navigation message computations, the particular version of the datum has changed. As of this writing the latest version of WGS84 is *WGS84* (G1762). The number following the letter G is the number of the GPS week during which the coordinates first were used in the *National Geospatial Intelligence Agency* (NGA) precise ephemeris estimations. Therefore, coordinates provided today by GPS receivers are based in WGS84 (G1762), which is the sixth update to the realization of the WGS84 Reference Frame. The original WGS 84 was based on observations from more than 1900 Doppler stations. It was revised to become WGS84 (G730) to incorporate GPS observations. That realization was implemented in GPS by the *operational control segment* (OCS) on June 29, 1994. More GPS-based realizations of WGS84 followed, WGS84 (G873) on January 29, 1997, and WGS84 (G1150) was implemented on January 20, 2002, and WGS84 (G1674) on February 8, 2012. Today the epoch of WGS84 is (G1762).

However, most available GPS software can transform those coordinates to a number of other datums as well. The one that is probably of greatest interest to surveyors in the United States today is the *North American Datum 1983* (NAD83). Originally, the difference between WGS84 as originally rolled out in 1987, and NAD83 as first introduced in 1986 coordinates was so small that transformation was unnecessary. That is no longer the case when it comes to NAD83 (2011) and WGS84 (G1674); the difference can be up to 1 or 2 m.

North American Datum 1983

NAD27

The Clarke 1866 ellipsoid was the foundation of NAD27, and the blocks that built that foundation were made by geodetic triangulation. After all, an ellipsoid, even one with a clearly stated orientation to the Earth, is only an abstraction until physical, identifiable control stations are available for its practical application. During the tenure of NAD27, control positions were tied together by tens of thousands of miles of triangulation and some traverses. Its measurements grew into chains of figures from

Canada to Mexico and coast to coast, with their vertices perpetuated by bronze disks set in stone, concrete, and other permanent media (Figure 5.11).

These tri-stations, also known as brass caps, and their attached coordinates have provided a framework for all types of surveying and mapping projects for many years. They have served to locate international, state, and county boundaries. They have provided geodetic control for the planning of national and local projects, development of natural resources, national defense, and land management. They have enabled surveys to be fitted together, provided checks, and assisted in the perpetuation of their marks. They have supported scientific inquiry, including crustal monitoring studies and other geophysical research. However, even as application of the nationwide control network grew, the revelations of local distortions in NAD27 were reaching unacceptable levels.

Judged by the standards of newer measurement technologies, the quality of some of the observations used in the datum were too low. That, and its lack of an

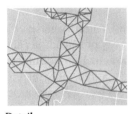

Detail
Idaho, Utah, Colorado, Wyoming,
Nevada, and Arizona

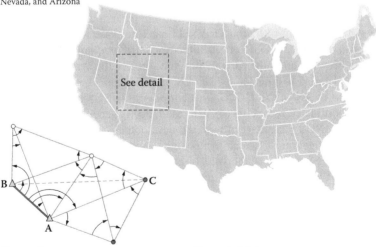

Known data:
Length of baseline AB.
Latitude and longitude of points A and B.
Azimuth of line AB.

Measured data:
Angles to the new control points.

Computed data:
Latitude and longitude of point C, and other new points.
Length and azimuth of line AC.
Length and azimuth of all other lines.

FIGURE 5.11   Triangulation.

internationally viable geocentric ellipsoid, finally drove its positions into obsolescence. The monuments remain, but it was clear early on that NAD27 had some difficulties. There were problems from too few baselines, Laplace azimuths, and other deficiencies. By the early 1970s the NAD27 coordinates of the national geodetic control network were no longer adequate.

Development of the North American Datum 1983 (NAD83)

While a committee of the National Academy of Sciences advocated the need for a new adjustment in its 1971 report, work on the new datum, NAD83, did not really begin until after July 1, 1974. Leading the charge was an old agency with a new name. Called the *U.S. Coast & Geodetic Survey* in 1878, and then the *Coast and Geodetic Survey* (C&GS) from 1899, the agency is now known as the *National Geodetic Survey* (NGS). It is within the *National Oceanic and Atmospheric Administration* (NOAA). The first ancestor of today's NGS was established back in 1807 and was known as the *Survey of the Coast*. Its current authority is contained in U.S. Code, Title 33, USC 883a.

NAD83 includes not only the United States but also Central America, Canada, Greenland, and Mexico. The NGS and the Geodetic Survey of Canada set about the task of attaching and orienting the GRS80 ellipsoid to the actual surface of the Earth, as it was defined by the best positions available at the time. It took more than 10 years to readjust and redefine the horizontal coordinate system of North America into what is now NAD83. More than 1.7 million weighted observations derived from classical surveying techniques throughout the Western Hemisphere were involved in the least squares adjustment. They were supplemented by approximately 30,000 electronic distance measured (EDM) baselines, 5000 astronomic azimuths, and more than 650 Doppler stations positioned by the TRANSIT satellite system. Over 100 VLBI vectors were also included, but GPS, in its infancy, contributed only five points.

GPS was growing up in the early 1980s, and some of the agencies involved in its development decided to join forces. NOAA, the *National Aeronautics and Space Administration* (NASA), the *U.S. Geological Survey* (USGS), and the Department of Defense coordinated their efforts. As a result, each agency was assigned specific responsibilities. NGS was charged with the development of specifications for GPS operations, investigation of related technologies, and the use of GPS for modeling crustal motion. It was also authorized to conduct its subsequent geodetic control surveys with GPS. So, despite an initial sparseness of GPS data in the creation of NAD83, the stage was set for a systematic infusion of its positions as the datum matured. The work was officially completed on July 31, 1986.

International Terrestrial Reference System

As NAD83 has aged, there has been constant improvement in geodesy. When NAD83 was created, it was intended to be geocentric. It is now known that the center of the reference ellipsoid of NAD83, GRS80, is about 2.24 m from the true geocenter (Figure 5.12).

However, there is a reference system that is geocentric. It is known as the International Terrestrial Reference System (ITRS). The reference frame derived from it is the *International Terrestrial Reference Frame* (ITRF). Its origin is at the center

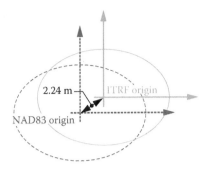

FIGURE 5.12   International Terrestrial Reference System (ITRS).

of mass of the whole Earth including the oceans and atmosphere. The unit of length is the meter. The orientation of its axes was established as consistent with that of the IERS's predecessor, Bureau International de l'Heure (BIH) at the beginning of 1984.

Today, the ITRF is maintained by the International Earth Rotation Service (IERS), which monitors Earth Orientation Parameters (EOP) for the scientific community through a global network of observing stations. This is done with GPS, VLBI, lunar laser ranging (LLR), satellite laser ranging (SLR), the Doppler Orbitography and Radiopositioning Integrated by Satellite (DORIS), and the positions of the observing stations are now considered to be accurate to the centimeter level. Recognizing that the several hundred control stations worldwide for which it publishes yearly coordinates are actually in motion due to the shifting of approximately 20 tectonic plates worldwide, the IERS also provides velocities for them. The International Terrestrial Reference Frame is actually a series of realizations. The first was in 1988. In other words, it is revised and published on a regular basis.

ITRF, WGS84, and NAD83

The North American Datum of 1983 (NAD83) is used everywhere in North America except Mexico. The latest realization of the datum as of this writing is *NAD83 (2011)*. This realization in the coterminous United States and Alaska is available through the National *Continuously Operating Reference Stations* (CORS). There are nearly 2000 National CORS and Cooperative CORS sites with more than 200 organizations participating in the network, and this number continuously grows with the addition of several new stations each month.

In the past we did not have to be concerned with the shift between NAD83 (1986) and WGS84 as introduced in 1987 because the discrepancy easily fell within our overall error budget. NAD83 and WGS84 originally differed by only a centimeter or two. That is no longer true. In their new definitions, NAD83 (2011) and WGS84 (G1762) differ up to 1 or 2 m within the continental United States. However, ITRF08 and WGS84 (G1762) are virtually identical (Table 5.2). NGS has developed a program called Horizontal Time Dependent Positioning (HTDP) to transform ITRF positions of a given epoch to NAD83 (2011) and vice versa. This program allows the movement of positions from one date to another, transformation from one reference frame to another, and supports the recent realizations of the NAD 83, ITRS, and WGS84.

TABLE 5.2

WGS84, NAD83, and ITRF

Year	Realization (Epoch)	For All Practical Purposes Equivalent to
1987	WGS1984 (ORIG)	NAD83 (1986)
1994	WGS84 (G730)	ITRF91/92
1997	WGS84 (G873)	ITRF94/96
2002	WGS84 (G1150)	ITRF00
2012	WGS (G1674)	ITRF08
2013	WGS (G1762)	Compares with ITRF08 within 1 cm root mean square (RMS) overall

It is important to note that the current ITRF and WGS84 systems are global and their realizations take into account the fact that the Earth is in constant motion because of the shifting of tectonic plates around the world. However, NAD83 is fixed to one plate, the North American plate, and moves with it. Consequently, NAD83, in the continental United States, moves approximately 10 to 20 mm/y in relation to the ITRF, WGS84, and also in relation with IGS08, a new reference frame that is virtually coincident with ITRF08 that has been in use by the International GNSS Service and NGS since 2011.

Management of NAD83

Because geodetic accuracy with GPS depends on relative positioning, surveyors continue to rely on NGS stations to control their work just as they have for generations. Today, it is not unusual for surveyors to find that some NGS stations have published coordinates in NAD83 and others, perhaps needed to control the same project, only have positions in NAD27. In such a situation, it is often desirable to transform the NAD27 positions into coordinates of the newer datum. Unfortunately, there is no single-step mathematical approach that can do it accurately.

The distortions between the original NAD27 positions are part of the difficulty. The older coordinates were sometimes in error as much as 1 part in 15,000. Problems stemming from the deflection of the vertical, lack of correction for geoidal undulations, low-quality measurements, and other sources contributed to inaccuracies in some NAD27 coordinates that cannot be corrected by simply transforming them into another datum.

Transformations from NAD27 to NAD83

Nevertheless, various approximate methods are used to transform NAD27 coordinates into NAD83 values. For example, the computation of a constant local translation is sometimes attempted using stations with coordinates in both systems as a guide. Another technique is the calculation of two translations, one rotation and one scale parameter, for particular locations based on the latitudes and longitudes of three or more common stations. Perhaps the best results derive from polynomial expressions developed for coordinate differences, expressed in Cartesian $(\Delta x, \Delta y, \Delta z)$ or ellipsoidal coordinates $(\Delta\phi, \Delta\lambda, \Delta h)$, using a 3-D Helmert transformation. However, besides requiring seven parameters (three shift, one scale, and three rotation components), this approach is at its best when ellipsoidal heights are available for all the points involved. Where adequate information is available, software packages such as the NGS program NADCON can provide coordinates.

Even if a local transformation is modeled with these techniques, the resulting NAD27 positions might still be plagued with relatively low accuracy. The NAD83 adjustment of the national network is based on nearly 10× the number of observations that supported the NAD27 system. This larger quantity of data, combined with the generally higher quality of the measurements at the foundation of NAD83, can have some rather unexpected results. For example, when NAD27 coordinates are transformed into the new system, the shift of individual stations may be quite different from what the regional trend indicates. In short, when using control from both NAD83 and NAD27 simultaneously on the same project, surveyors have come to expect difficulty.

In fact, the only truly reliable method of transformation is not to rely on coordinates at all, but to return to the original observations themselves. It is important to remember, for example, that geodetic latitude and longitude, as other coordinates, are specifically referenced to a given datum and are not derived from some sort of absolute framework. However, the original measurements, incorporated into a properly designed least squares adjustment, can provide most satisfactory results.

Densification and Improvement of NAD83

Inadequacies of NAD27 and even NAD83 positions in some regions are growing pains of a fundamentally changed relationship. In the past, relatively few engineers and surveyors were employed in geodetic work. Perhaps the greatest importance of the data from the various geodetic surveys was that they furnished precise points of reference to which the multitude of surveys of lower precision could then be tied. This arrangement was clearly illustrated by the design of state plane coordinates systems, devised to make the national control network accessible to surveyors without geodetic capability.

However, the situation has changed. The gulf between the precision of local surveys and national geodetic work is virtually closed by GPS, and that has changed the relationship between local surveyors in private practice and geodesists. For example, the significance of state plane coordinates as a bridge between the two groups has been drastically reduced. Today's surveyor has relatively easy and direct access to the geodetic coordinate systems themselves through GPS. In fact, the 1 to 2 ppm probable error in networks of relative GPS-derived positions frequently exceeds the accuracy of the NAD83 positions intended to control them.

High-Accuracy Reference Networks

Other significant work along this line was accomplished in the state-by-state super-net programs. The creation of *High-Accuracy Reference Networks* (HARN) was cooperative ventures between NGS and the states, and often include other organizations as well. The campaign was originally known as *High-Precision Geodetic Networks* (HPGN).

A station spacing of not more than about 62 miles and not less than about 16 miles was the objective in these statewide networks. The accuracy was intended to be 1 part per million, or better between stations. In other words, with heavy reliance on GPS observations, these networks were intended to provide extremely accurate, vehicle-accessible, regularly spaced control point monuments with good overhead visibility. These stations were intended to provide control superior to the vectors derived from the day-to-day GPS observations that are tied to them. In that way the HARN points provide the user with a means to avoid any need to warp vectors to fit inferior control. That used to sometime happen in the early days of GPS. To further ensure such coherence in the HARN, when the GPS measurements were complete, they were submitted to NGS for inclusion in a statewide readjustment of the existing NGRS covered by the state. Coordinate shifts of 0.3 to 1.0 m from NAD83 values were typical in these readjustments, which were concluded in 1998.

The most important aspect of HARN positions was the accuracy of their final positions. Entirely new orders of accuracy were developed for GPS relative position-ing techniques by the *Federal Geodetic Control Committee*. Its 1989 provisional standards and specifications for GPS work include orders AA, A, and B, which are now defined as having minimum geometric accuracies of 3 mm ± 0.01 ppm, 5 mm ± 0.1 ppm, and 8 mm ± 1 ppm, respectively, at the 95 percent, or 2σ, confidence level. The publication of up-to-date geodetic data, always one of the most important functions of NGS, is even more crucial today. Today the Federal Geodetic Control Subcommittee is within the Federal Geographic Data Committee and has published accuracy standards for geodetic networks in part 2 of the Geospatial Positioning Standards (FGDC-007-1998).

The original NAD83 adjustment is indicated with a suffix including the year 1986 in parentheses, that is, NAD83 (1986). However, when a newer realization is avail-able, the year in the parentheses will be the year of the adjustment. The most recent realization is NAD83 (2011).

Continuously Operating Reference Stations

From 1998 to 2004, NGS introduced another series of observations in each state designed to tie the network to the CORS. This work resulted in the *Federal Base Network* (FBN), which is a nationwide network of monumented stations. These spa-tial reference positions are among the most precise available and are particularly dense in crustal motion areas. In general, these points are spaced at approximately 100 km apart. The accuracies intended are 1 cm latitudes and longitudes, 2 cm ellip-soidal heights, and 3 cm orthometric heights. These stations are few compared to the much more numerous *Cooperative Base Network* (CBN). This is a high-accuracy network of monumented control stations spaced at 25 to 50 km apart throughout the United States and its territories. The CBN was created and is maintained by state and private organizations with the help of NGS.

In about 1992 the NGS began establishing a network of CORS throughout the country. The original idea was to provide positioning for navigational and marine needs. There were about 50 CORS in 1996. Their positional accuracies are 3 cm horizontal and 5 cm vertical. They also must meet NOAA geodetic standards for installation, operation, and data distribution. Today there are nearly 2000 CORS online.

The CORS in the NGS network are mostly to provide support for carrier phase observations. Information is available for post-processing on the Internet.

STATE PLANE COORDINATES

NAD83 POSITIONS AND PLANE COORDINATES

NGS published data for stations also include state plane coordinates in the appropriate zone. As before, the easting and northing are accompanied by the mapping angle and grid azimuths, but a scale factor is also included for easy conversions. *Universal Transverse Mercator* (UTM) coordinates are among the new elements offered by NGS in the published information for NAD83 stations.

These plane coordinates, both state plane and UTM, are far from an anachronism. The UTM projection has been adopted by the IUGG, the same organization that reached the international agreement to use GRS80 as the reference ellipsoid for the modern geocentric datum. The U.S. NASA and other military and civilian organizations worldwide also use UTM coordinates for various mapping needs. UTM coordinates are often useful to those planning work that embraces large areas. In the United States, state plane systems based on the transverse Mercator projection, an oblique Mercator projection, and the Lambert conic map projection grid every state, Puerto Rico, and the U.S. Virgin Islands into their own plane rectangular coordinate system. GPS surveys performed for local projects and mapping are frequently reported in the plane coordinates of one of these systems.

State plane coordinates rely on an imaginary flat reference surface with Cartesian axes. They describe measured positions by ordered pairs, expressed in northings and eastings, or y and x coordinates. Despite the fact that the assumption of a flat Earth is fundamentally wrong, calculation of areas, angles, and lengths using latitude and longitude can be complicated, so plane coordinates persist. Therefore, the projection of points from the Earth's surface onto a reference ellipsoid and finally onto flat maps is still viable.

In fact, many agencies of government, particularly those that administer state, county, and municipal databases, prefer coordinates in their particular *State Plane Coordinate System* (SPCS). The systems are, as the name implies, state specific. In many states the system is officially sanctioned by legislation. Generally speaking, such legislation allows surveyors to use state plane coordinates to legally describe property corners.

MAP PROJECTION

State Plane Coordinate Systems are built on *map projections*. Map projection means representing a portion of the actual Earth on a plane. Done for hundreds of years to create paper maps, it continues, but map projection today is most often really a mathematical procedure done in a computer. Nevertheless, even in an electronic world, it cannot be done without distortion.

The problem is often illustrated by trying to flatten part of an orange peel. The orange peel stands in for the surface of the Earth. A small part, say a square a quarter of an inch on the side, can be pushed flat without much noticeable deformation, but when the portion gets larger, problems appear. Suppose a third of the orange peel is involved, as the center is pushed down, the edges tear and stretch, or both, and if the peel gets even bigger, the tearing gets more severe. So if a map is drawn on the orange before it is peeled, the map gets distorted in unpredictable ways when it is flattened. Therefore, it is difficult to relate a point on one torn piece with a point on another in any meaningful way.

These are the problems that a map projection needs to solve to be useful. The first problem is the surface of an ellipsoid, like the orange peel, is *nondevelopable*. In other words, flattening it inevitably leads to distortion. So, a useful map projection ought to start with a surface that is *developable*, a surface that may be flattened without all that unpredictable deformation. It happens that a paper cone or cylinder both illustrate this idea nicely. They are illustrations only, models for thinking about the issues involved.

If a right circular cone is cut up and one of its elements is perpendicular from the base to its apex, the cone can then be made completely flat without trouble. The same may be said of a cylinder cut up perpendicular from base to base (see Figure 5.13).

Alternatively, one could use the simplest case, a surface that is already developed. A flat piece of paper is an example. If the center of a flat plane is brought tangent

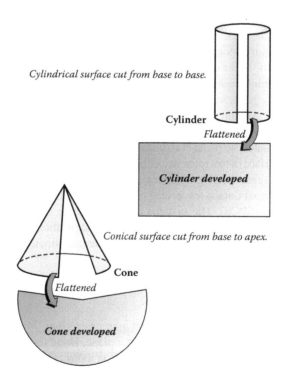

Cylindrical surface cut from base to base.

Cylinder

Flattened

Cylinder developed

Conical surface cut from base to apex.

Cone

Flattened

Cone developed

FIGURE 5.13   Development of a cylinder and a cone.

to the Earth, a portion of the planet can be mapped on it; that is, it can be projected directly onto the flat plane. In fact, this is the typical method for establishing an independent *local coordinate system*.

These simple Cartesian systems are convenient and satisfy the needs of small projects. The method of projection, onto a simple flat plane, is based on the idea that a small section of the Earth, as with a small section of the orange mentioned previously, conforms so nearly to a plane that distortion on such a system is negligible.

Subsequently, local tangent planes have been long used by land surveyors. Such systems demand little if any manipulation of the field observations and the approach has merit as long as the extent of the work is small, but the larger each of the planes grows, the more untenable it becomes. As the area being mapped grows, the reduction of survey observations becomes more complicated because it must take account of the actual shape of the Earth. This usually involves the ellipsoid, the geoid, and geographical coordinates, latitude and longitude. At that point, surveyors and engineers rely on map projections to mitigate the situation and limit the now troublesome distortion. However, a well-designed map projection can offer the convenience of working in plane Cartesian coordinates and still keep the inevitable distortion at manageable levels at the same time.

DISTORTION

Design of such a projection must accommodate some awkward facts. For example, while it would be possible to imagine mapping a considerable portion of the Earth using a large number of small individual planes, like facets of a gem, it is seldom done because when these planes are brought together they cannot be edge-matched accurately (see Figure 5.14).

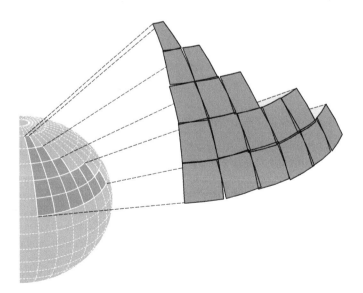

FIGURE 5.14   Local coordinate systems do not edge match.

They cannot be joined properly along their borders. The problem is unavoidable because the planes, tangent at their centers, inevitably depart more and more from the reference ellipsoid at their edges, and the greater the distance between the ellipsoidal surface and the surface of the map on which it is represented, the greater the distortion on the resulting flat map. This is true of all methods of map projection. Therefore, one is faced with the daunting task of joining together a mosaic of individual maps along their edges where the accuracy of the representation is at its worst, and even if one could overcome the problem by making the distortion, however large, the same on two adjoining maps, another difficulty would remain. Typically, each of these planes has a unique coordinate system. The orientation of the axes, the scale, and the rotation of each one of these individual local systems will not be the same as those elements of its neighbor's coordinate system. Subsequently, there are gaps and overlaps between adjacent maps, and their attendant coordinate systems, because there is no common reference system.

So the idea of a self-consistent, local map projection based on small, flat planes tangent to the Earth, or the reference ellipsoid, is convenient, but only for small projects that have no need to be related to adjoining work. As long as there is no need to venture outside the bounds of a particular local system, it can be entirely adequate, but, generally speaking, if a significant area needs coverage, another strategy is needed.

Decreasing Distortion

Decreasing distortion is a constant and elusive goal in map projection. It can be done in several ways. Most involve reducing the distance between the map projection surface and the ellipsoidal surface. One way this is done is to move the mapping surface from tangency with the ellipsoid and make it actually cut through it. This strategy produces what is known as a secant projection. A secant projection is one way to shrink the distance between the map projection surface and the ellipsoid, thereby the area where distortion is in an acceptable range on the map can be effectively increased (see Figure 5.15).

The distortion can be reduced even further when one of those developable surfaces mentioned earlier is added to the idea of a secant map projection plane (see Figure 5.16).

Secant and Cylindrical Projections

Both cones and cylinders have an advantage over a flat map projection plane. They are curved in one direction and can be designed to follow the curvature of the area to be mapped in that direction. Also, if a large portion of the ellipsoid is to be mapped, several cones or several cylinders may be used together in the same system to further limit distortion. In that case, each cone or cylinder defines a *zone* in a larger coverage. This is the approach used in State Plane Coordinate Systems.

As mentioned, when a conic or a cylindrical map projection surface is made secant, it intersects the ellipsoid, and the map is brought close to its surface. For example, the conic and cylindrical projections shown in Figure 5.16 cut through the ellipsoid. The map is projected both inward and outward onto it, and two lines of exact scale, standard lines, are created along the small circles where the cone and

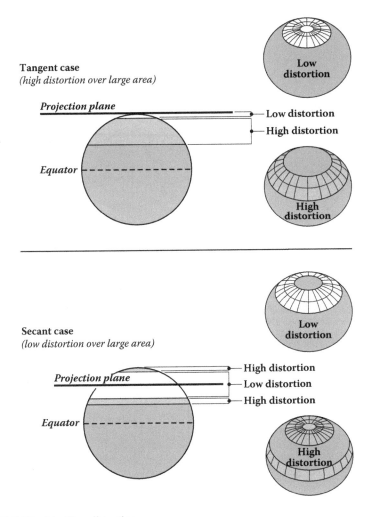

FIGURE 5.15   Limiting distortion.

the cylinder intersect the ellipsoid. They are called *small circles* because they do not describe a plane that goes through the center of the Earth as do the previously mentioned great circles.

Where the ellipsoid and the map projection surface touch, in this case, intersect, there is no distortion. However, between the standard lines the map is under the ellipsoid and outside of them the map is above it. That means that between the standard lines a distance from one point to another is actually longer on the ellipsoid than it is shown on the map and outside the standard lines a distance on the ellipsoid is shorter than it is on the map. Any length that is measured along a standard line is the same on the ellipsoid and on the map, which is why another name for standard parallels is *lines of exact scale.*

Ultimately, the goal is very straightforward relating each position on one surface, the reference ellipsoid, to a corresponding position on another surface as faithfully

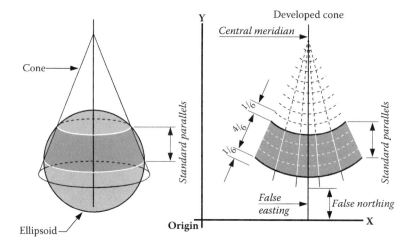

Lambert Conformal Conic projection

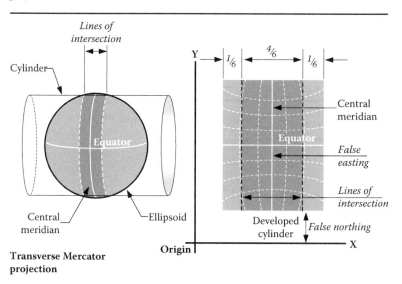

Transverse Mercator projection

FIGURE 5.16 Two projections.

as possible and then flattening that second surface to accommodate Cartesian coordinates. In fact, the whole procedure is in the service of moving from geographic to Cartesian coordinates and back again. These days the complexities of the mathematics are handled with computers. Of course, that was not always the case.

Origin of State Plane Coordinates

In 1932, two engineers in North Carolina's highway department, O.B. Bester and George F. Syme, appealed to the then Coast & Geodetic Survey (C&GS, now NGS) for help. They had found that the stretching and compression inevitable in the representation of the curved Earth on a plane was so severe over long route surveys that they could not check into the C&GS geodetic control stations across a state within reasonable limits. The engineers suggested that a plane coordinate grid system be developed that was mathematically related to the reference ellipsoid but could be utilized using plane trigonometry.

Dr. Oscar Adams of the Division of Geodesy, assisted by Charles Claire, designed the first State Plane Coordinate System to mediate the problem. It was based on a map projection called the *Lambert Conformal Conic Projection*. Adams realized that it was possible to use this map projection and allow one of the four elements of area, shape, scale, or direction to remain virtually unchanged from its actual value on the Earth, but not all four. On a perfect map projection, all distances, directions, and areas could be conserved. They would be the same on the ellipsoid and on the map. Unfortunately, it is not possible to satisfy all of these specifications simultaneously, at least not completely. There are inevitable choices. It must be decided which characteristic will be shown the most correctly, but it will be done at the expense of the others, and there is no universal best decision. Still a solution that gives the most satisfactory results for a particular mapping problem is always available.

Adams chose the Lambert Conformal Conic Projection for the North Carolina system. On the Lambert Conformal Conic Projection parallels of latitude are arcs of concentric circles and meridians of longitude are equally spaced straight radial lines, and the meridians and parallels intersect at right angles. The axis of the cone is imagined to be a prolongation of the polar axis. The parallels are not equally spaced because the scale varies as you move north and south along a meridian of longitude. Adams decided to use this map projection in which shape is preserved based on a developable cone.

Map projections in which shape is preserved are known as *conformal* or *orthomorphic*. Orthomorphic means right shape. In a conformal projection the angles between intersecting lines and curves retain their original form on the map. In other words, between short lines, meaning lines under about 10 miles, a 45° angle on the ellipsoid is a 45° angle on the map. It also means that the scale is the same in all directions from a point; in fact, it is this characteristic that preserves the angles. These aspects were certainly a boon for the North Carolina Highway engineers and benefits that all State Plane Coordinate System users have enjoyed since. On long lines, angles on the ellipsoid are not exactly the same on the map projection. Nevertheless, the change is small and systematic.

Actually, all three of the projections that were used in the designs of the original State Plane Coordinate Systems were conformal. Each system was based on the North American Datum 1927 (NAD27). Along with the Oblique Mercator projection, which was used on the panhandle of Alaska, the two primary projections were the Lambert Conic Conformal Projection and the Transverse Mercator projection. For North Carolina, and other states that are longest east-west, the Lambert Conic projection works best. State Plane Coordinate Systems in states that are longest north-south were built on the Transverse Mercator projection. There are exceptions to this general rule. For example, California uses the Lambert Conic projection even though the state could be covered with fewer Transverse Mercator zones. The Lambert Conic projection is a bit simpler to use, which may account for the choice.

The Transverse Mercator projection is based on a cylindrical mapping surface much like that illustrated in Figure 5.16. However, the axis of the cylinder is rotated so that it is perpendicular with the polar axis of the ellipsoid. Unlike the Lambert Conic projection, the Transverse Mercator represents meridians of longitude as curves rather than straight lines on the developed grid. The Transverse Mercator projection is not the same thing as the *Universal Transverse Mercator System* (UTM). UTM was originally a military system that covers the entire Earth and differs significantly from the Transverse Mercator System used in State Plane Coordinates System.

In using these projections as the foundation of the State Plane Coordinate Systems, Adams wanted to have the advantage of conformality and also cover each state with as few zones as possible. A zone in this context is a belt across the state that has one Cartesian coordinate grid with one origin and is projected onto one mapping surface. One strategy that played a significant role in achieving that end was Adams's use of *secant* projections in both the Lambert and Transverse Mercator systems.

STATE PLANE COORDINATE SYSTEM MAP PROJECTIONS

Using a single secant cone in the Lambert projection and limiting the extent of a zone, or belt, across a state to about 158 miles, approximately 254 km, Adams limited the distortion of the length of lines. Not only were angles preserved on the final product, but there were also no radical differences between the length of a measured line on the Earth's surface and the length of the same line on the map projection. In other words, the scale of the distortion was pretty small.

He placed four-sixths of the map projection plane between the standard lines, one-sixth outside at each extremity. The distortion was held to 1 part in 10,000. A maximum distortion in the lengths of lines of 1 part in 10,000 means that the difference between the length of a 2-mile (3.22 km) line on the ellipsoid and its representation on the map would only be about 1 foot (0.31 m) at the most.

State Plane Coordinates were created to be the basis of a method that approximates geodetic accuracy more closely than the then commonly used methods of small-scale plane surveying. Today surveying methods can easily achieve accuracies beyond 1 part in 100,000 and better, but the State Plane Coordinate Systems were designed in a time of generally lower accuracy and efficiency in surveying measurement. Today computers easily handle the lengthy and complicated mathematics of geodesy, but the first State Plane Coordinate System was created when such computation required sharp pencils and logarithmic tables. In fact, the original State Plane Coordinate System was so successful in North Carolina that similar systems were devised for all the states in the Union within a year or so. The system was successful because, among other things, it overcame some of the limitations of mapping on a horizontal plane while avoiding the imposition of strict geodetic methods and calculations. It managed to keep the distortion of the scale ratio under 1 part in 10,000 and preserved conformality. It did not disturb the familiar system of ordered pairs of Cartesian coordinates, and it covered each state with as few zones as possible whose boundaries were constructed to follow county lines. County lines were generally used so that those relying on state plane coordinates could work in one zone throughout a jurisdiction.

SPCS27 TO SPCS83

In several instances the boundaries of State Plane Coordinate Zones today, *SPCS83*, the State Plane Coordinate System based on NAD83 and its reference ellipsoid GRS80, differ from the original zone boundaries. The foundation of the original State Plane Coordinate System, SPCS27 was NAD27 and its reference ellipsoid Clarke 1866. As mentioned earlier, NAD27 geographical coordinates, latitudes and longitudes, differ significantly from those in NAD83. In fact, conversion from geographic coordinates, latitude and longitude, to grid coordinates, *y* and *x*, and back is one of the three fundamental conversions in the State Plane Coordinate System. It is important because the whole objective of the SPCS is to allow the user to work in plane coordinates but still have the option of expressing any of the points under consideration in either latitude and longitude or state plane coordinates without significant loss of accuracy. Therefore, when geodetic control was migrated from NAD27 to NAD83, the State Plane Coordinate System had to go along.

When the migration was undertaken in the 1970s, it presented an opportunity for an overhaul of the system. Many options were considered, but in the end, just a few changes were made (Figure 5.17). One of the reasons for the conservative approach was the fact that 37 states had passed legislation supporting the use of state plane coordinates. Nevertheless, some zones got new numbers, and some of the zones changed.

The zones are numbered in the SPCS83 system known as *FIPS*. FIPS stands for *Federal Information Processing Standard*, and each SPCS83 zone has been given a FIPS number. These days the zones are often known as *FIPS zones*. SPCS27 zones did not have these FIPS numbers. As mentioned above, the original goal was to keep each zone small enough to ensure that the scale distortion was 1 part in 10,000 or less, but when the SPCS83 was designed that scale was not maintained in some states.

Changes in Zones

In five states, some SPCS27 zones were eliminated altogether, and the areas they had covered consolidated into one zone or added to adjoining zones. In three of those states the result was one single large zone. Those states are South Carolina, Montana, and Nebraska. In SPCS27, South Carolina and Nebraska had two zones, in SPCS83 they have just one, FIPS zone 3900 and FIPS zone 2600, respectively. Montana previously had three zones. It now has one, FIPS zone 2500. Therefore, because the area covered by these single zones has become so large, they are not limited by the 1 part in 10,000 standard. California eliminated zone 7 and added that area to FIPS zone 0405, formerly zone 5. Two zones previously covered Puerto Rico and the U.S. Virgin Islands. They now have one. It is FIPS zone 5200. In Michigan, three Transverse Mercator zones were entirely eliminated.

In both the Transverse Mercator and the Lambert projections the positions of the axes are similar in all SPCS zones. As you can see in Figure 5.16, each zone has a central meridian. These central meridians are true meridians of longitude near the geometric center of the zone. Please note that the central meridian is not the *y* axis. If it were the *y* axis, negative coordinates would result. To avoid them, the actual *y* axis is moved far to the west of the zone itself. In the old SPCS27 arrangement the *y* axis was 2,000,000 feet west from the central meridian in the Lambert Conic

FIGURE 5.17   State Plane Coordinate Systems 1983.

projection and 500,000 feet in the Transverse Mercator projection. In the SPCS83 design those constants have been changed. The most common values are 600,000 m for the Lambert Conic and 200,000 m for the Transverse Mercator. However, there is a good deal of variation in these numbers from state to state and zone to zone. In all cases, however, the y axis is still far to the west of the zone and there are no negative state plane coordinates. No negative coordinates, because the x axis, also known as the baseline, is far to the south of the zone. Where the x axis and y axis intersect is the origin of the zone, and that is always south and west of the zone itself. This configuration of the axes ensures that all state plane coordinates occur in the first quadrant and are, therefore, always positive.

It is important to note that the fundamental unit for SPCS27 is the U.S. survey foot and for SPCS it is the meter. The conversion from meters to U.S. survey feet is correctly accomplished by multiplying the measurement in meters by the fraction 3937/1200.

STATE PLANE COORDINATES SCALE AND DISTANCE

Geodetic Lengths to Grid Lengths

This brings us to the scale factor, also known as the K factor and the projection factor. It was this factor that the original design of the State Plane Coordinate System sought to limit to 1 part in 10,000. As implied by that effort, scale factors are ratios that can be used as multipliers to convert ellipsoidal lengths, also known as geodetic distances, to lengths on the map projection surface, also known as grid distances, and vice versa. In other words, the geodetic length of a line, on the ellipsoid, multiplied by the appropriate scale factor will give you the grid length of that line on the map, and the grid length multiplied by the inverse of that same scale factor would bring you back to the geodetic length again.

The projection used most on states that are longest from east to west is the Lambert Conic. In this projection the scale factor for east–west lines is constant. In other words, the scale factor is the same all along the line. One way to think about this is to recall that the distance between the ellipsoid and the map projection surface does not change east to west in that projection. However, along a north–south line, the scale factor is constantly changing on the Lambert Conic, and it is no surprise then to see that the distance between the ellipsoid and the map projection surface is always changing north to south line in that projection. Looking at the Transverse Mercator projection, the projection used most on states longest north to south, the situation is exactly reversed. In that case, the scale factor is the same all along a north–south line, and changes constantly along an east–west line.

Both the Transverse Mercator and the Lambert Conic used a secant projection surface and originally restricted the width to 158 miles, or about 254 km. These were two strategies used to limit scale factors when the State Plane Coordinate Systems were designed. Where that was not optimum, the width was sometimes made smaller, which means the distortion was lessened. As the belt of the ellipsoid projected onto the map narrows, the distortion gets smaller. For example, Connecticut is less than 80 miles, or nearly 129 km. wide north to south. It has only one zone. Along its northern and southern boundaries, outside of the standard parallels, the scale factor is 1 part in 40,000, a fourfold improvement over 1 part in 10,000. In the middle of the state the scale factor is 1 part in 79,000, nearly an eightfold increase. The scale factor was allowed to get a little bit smaller than 1 part in 10,000 in Texas. By doing that the state was covered completely with five zones. Covering the states with as few zones as possible was among the guiding principles in 1933. Another principle was having zone boundaries follow county lines. Still it requires ten zones and all three projections to cover Alaska.

In Figure 5.18 a typical State Plane Coordinate zone is represented by a grid plane of projection cutting through the ellipsoid of reference. Between the intersections of the standard lines, the grid is under the ellipsoid. There, a distance from one point to another is longer on the ellipsoid than on the grid. This means that right in the middle of a SPCS zone the scale factor is at its minimum. In the middle a typical minimum SPCS scale factor is not less than 0.9999, though there are exceptions. Outside the intersections, the grid is above the ellipsoid where a distance from one point to another is shorter on the ellipsoid than it is on the grid. There, at the edge of

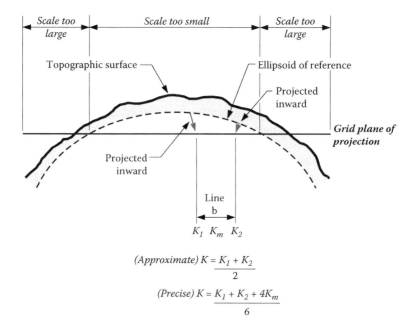

FIGURE 5.18   Scale factor.

the zone a maximum typical SPCS scale factor is generally not more than 1.0001, but again there are now exceptions.

When SPCS 27 was current, scale factors were interpolated from tables published for each state. In the tables for states in which the Lambert Conic projection was used scale factors change north–south with the changes in latitude. In the tables for states in which the Transverse Mercator projection was used, scale factors change east–west with the changes in x coordinate. Today scale factors are not interpolated from tables for SPCS83. For both the Transverse Mercator and the Lambert Conic projections they are calculated directly from equations.

There are also several software applications that can be used to automatically calculate scale factors for particular stations. They can be used to convert latitudes and longitudes to state plane coordinates. Given the latitude and longitude of the stations under consideration part of the available output from these programs is typically the scale factors for those stations.

To illustrate the use of these factors, consider line b to have a length on the ellipsoid of 130,210.44 feet, a bit over 24 miles (38.62 km). That would be its geodetic distance. Suppose that the scale factor for that line was 0.9999536, then the grid distance along line b would be:

$$\text{Geodetic Distance} \times \text{Scale Factor} = \text{Grid Distance}$$
$$130,210.44\,\text{ft.} \times 0.999953617 = 130,204.40\,\text{ft.}$$

The difference between the longer geodetic distance and the shorter grid distance here is a little more than 6 feet (1.83 m). That is actually better than 1 part in 20,000; please recall that the 1 part in 10,000 ratio was originally considered the maximum. Distortion lessens and the scale factor approaches 1 as a line nears a standard parallel.

Please also recall that on the Lambert projection an east–west line, i.e., a line that follows a parallel of latitude, has the same scale factor at both ends and throughout. However, a line that bears in any other direction will have a different scale factor at each end. A north–south line will have a great difference in the scale factor at its north end compared with the scale factor of its south end. In this vein, here is an approximate formula.

$$K = \frac{K_1 + K_2}{2}$$

where K is the scale factor for a line, K_1 is the scale factor at one end of the line, and K_2 is the scale factor at the other end of the line. Scale factor varies with the latitude in the Lambert projection. For example, suppose the point at the north end of the 24-mile (38.62 km) line is called Stormy and has a geographic coordinate of

37°46′00.7225″
103°46′35.3195″

and at the south end the point is known as Seven with a geographic coordinate of

37°30′43.5867″
104°05′26.5420″

The scale factor for point Seven is 0.99996113, and the scale factor for point Stormy is 0.99994609. It happens that point Seven is further south and closer to the standard parallel than is point Stormy, and it, therefore, follows that the scale factor at Seven is closer to 1 (Figure 5.19). It would be exactly 1 if it were on the standard parallel, which is why the standard parallels are called lines of exact scale. The typical scale factor for the line is the average of the scale factors at the two end points:

$$K = \frac{K_1 + K_2}{2}$$
$$0.99995361 = \frac{0.99996113 + 0.99994609}{2}$$

Deriving the scale factor at each end and averaging them is the usual method for calculating the scale factor of a line. The average of the two is sometimes called K_m.

However, that is not the whole story when it comes to reducing distance to the State Plane Coordinate grid. Measurement of lines must always be done on the

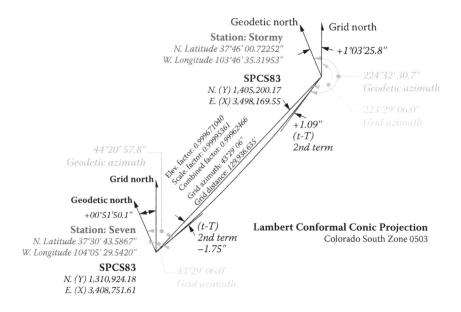

FIGURE 5.19　Stormy-Seven.

topographic surface of the Earth, and not on the ellipsoid. Therefore, the first step in deriving a grid distance must be moving a measured line from the Earth to the ellipsoid. In other words, converting a distance measured on the topographic surface to a geodetic distance on the reference ellipsoid. This is done with another ratio that is also used as a multiplier. Originally, this factor had a rather unfortunate name. It used to be known as the *sea level factor* in SPCS27. It was given that name because as you may recall that when NAD27 was established using the Clarke 1866 reference ellipsoid, the distance between the ellipsoid and the geoid was declared to be zero at Meades Ranch in Kansas. That meant that in the middle of the country the *sea level* surface, the geoid, and the ellipsoid were coincident by definition. Because the Clarke 1866 ellipsoid fit the United States quite well, the separation between the two surfaces, the ellipsoid and geoid, only grew to about 12 m anywhere in the country. With such a small distance between them, many practitioners at the time took the point of view that, for all practical purposes, the ellipsoid and the geoid were in the same place. That place was called sea level. Hence reducing a distance measure on the surface of the Earth to the ellipsoid was said to be reducing it to sea level.

Today that idea and that name for the factor are misleading because, of course, the GRS80 ellipsoid on which NAD83 is based is certainly not the same as Mean Sea Level. Now the separation between the geoid and ellipsoid can grow as large as 53 m, and technology by which lines are measured has improved dramatically. Therefore, in SPCS83 the factor for reducing a measured distance to the ellipsoid is known as the *ellipsoid factor*. In any case, both the old and the new name can be covered under the name the *elevation factor*.

Regardless of the name applied to the factor, it is a ratio. The ratio is the relationship between an approximation of the Earth's radius and that same approximation

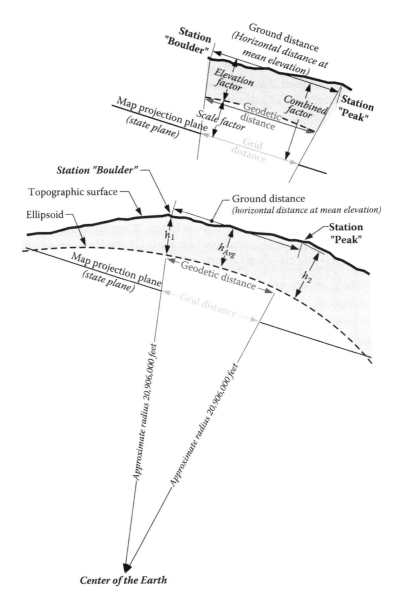

FIGURE 5.20   Example of distances.

with the mean ellipsoidal height of the measured line added to it. For example, consider station Boulder and station Peak illustrated in Figure 5.20.

Boulder
    39°59′29.1299″
    105°15′39.6758″

Peak
    40°01′19.1582″
    105°30′55.1283″

The distance between these two stations is 72,126.21 feet. This distance is sometimes called the *ground distance*, or the *horizontal distance at mean elevation*. In other words, it is not the slope distance but rather the distance between them corrected to an averaged horizontal plane, as is common practice. For practical purposes, this is the distance between the two stations on the topographic surface of the Earth. On the way to finding the grid distance Boulder to Peak, there is the interim step, calculating the geodetic distance between them, i.e., the distance on the ellipsoid. We need the elevation factor, and here is how it is determined.

The ellipsoidal height of Boulder, h_1, is 5437 feet. The ellipsoidal height of Peak, h_2, is 9099 feet. The approximate radius of the Earth, traditionally used in this work, is 20,906,000 feet. The elevation factor is calculated:

$$\text{Elevation Factor} = \frac{R}{R + h_{avg}}$$

$$\text{Elevation Factor} = \frac{20,906,000 \text{ ft.}}{20,906,000 \text{ ft.} + 7268 \text{ ft.}}$$

$$\text{Elevation Factor} = \frac{20,906,000 \text{ ft.}}{20,913,268 \text{ ft.}}$$

$$\text{Elevation Factor} = 0.99965247$$

This factor then is the ratio used to move the ground distance down to the ellipsoid, down to the geodetic distance.

Ground Distance Boulder to Peak = 72,126.21 ft.
Geodetic Distance = Ground Distance × Elevation Factor
Geodetic Distance = 72,126.21 × 0.99965247
Geodetic Distance = 72,101.14 ft.

It is possible to refine the calculation of the elevation factor by using an average of the actual radial distances from the center of the ellipsoid to the end points of the line rather than the approximate 20,906,000 feet. In the area of stations Boulder and Peak the average ellipsoidal radius is actually a bit longer, but it is worth noting that within the continental United States, such variation will not cause a calculated geodetic distance to differ significantly. However, it is worthwhile to take care to use the ellipsoidal heights of the stations when calculating the elevation factor rather than the orthometric heights.

In calculating the elevation factor in SPCS27, no real distinction is made between ellipsoid height and orthometric height. However, in SPCS83 the averages of the ellipsoidal heights at each end of the line are used. If the ellipsoid height is not directly available, it can be calculated from

$$h = H + N$$

where
  h = ellipsoid height
  H = orthometric height
  N = geoid height

Converting a geodetic distance to a grid distance is done with an averaged scale factor:

$$K = \frac{K_1 + K_2}{2}$$

In this instance the scale factor at Boulder is 0.99996703 and at Peak it is 0.99996477.

$$0.99996590 = \frac{0.99996703 + 0.99996477}{2}$$

Using the scale factor, it is possible to reduce the geodetic distance 72,101.14 ft. to a grid distance:

$$\text{Geodetic Distance} \times \text{Scale Factor} = \text{Grid Distance}$$
$$72,101.14\,\text{ft.} * 0.99996590 = 72,098.68\,\text{ft.}$$

There are two steps: (1) from ground distance to geodetic distance using the elevation factor and (2) from geodetic distance to grid distance using the scale factor that can be combined into one. Multiplying the elevation factor and the scale factor produces a single ratio that is usually known as the *combined factor* or the *grid factor*. Using this grid factor, the measured line is converted from a ground distance to a grid distance in one jump. Here is how it works. In the example above, the elevation factor for the line from Boulder to Peak is 0.99965247 and the scale factor is 0.99996590:

$$\text{Grid Factor} = \text{Scale Factor} \times \text{Elevation Factor}$$
$$0.99961838 = 0.99996590 * 0.99965247$$

Then using the grid factor, the ground distance is converted to a grid distance.

$$\text{Grid Distance} = \text{Grid Factor} \times \text{Ground Distance}$$
$$72,098.68\,\text{ft} \times 0.99961838 = 72,126.21\,\text{ft}.$$

Also, the grid factor can be used to go the other way. If the grid distance is divided by the grid factor, it is converted to a ground distance.

$$\text{Ground Distance} = \text{Grid Distance/Grid Factor}$$
$$72,126.21\,\text{ft.} = \frac{72,098.68\,\text{ft.}}{0.99961838}$$

UNIVERSAL TRANSVERSE MERCATOR COORDINATES

A plane coordinate system that is convenient for GIS work over large areas is the Universal Transverse Mercator (UTM) system. UTM with the Universal Polar Stereographic system covers the world in one consistent system. It is 4× less accurate than typical State Plane Coordinate Systems with a scale factor that typically reaches 0.9996. Yet the ease of using UTM and its worldwide coverage makes it very attractive for work that would otherwise have to cross many different SPCS zones. For example, nearly all *National Geospatial-Intelligence Agency* (NGA) topographic maps, *U.S. Geological Survey* (USGS) quad sheets, and many aeronautical charts show the UTM grid lines.

It is often said that UTM is a military system created by the U.S. Army, but several nations, and the North Atlantic Treaty Organization (NATO), played roles in its creation after World War II. At that time the goal was to design a consistent coordinate system that could promote cooperation between the military organizations of several nations. Before the introduction of UTM, allies found that their differing systems hindered the synchronization of military operations.

Conferences were held on the subject from 1945 to 1951 with representatives from Belgium, Portugal, France, and Britain, and the outlines of the present UTM system were developed. By 1951 the U.S. Army introduced a system that was very similar to that currently used.

The UTM projection divides the world into 60 zones that begin at longitude 180°, the International Date Line. Zone 1 is from 180° to the 174°W longitude. The coterminous United States are within UTM zones 10 to 19.

Here is a convenient way to find the zone number for a particular longitude. Consider west longitude negative and east longitude positive, add 180° and divide by 6. Any answer greater than an integer is rounded to the next highest integer, and you have the zone. For example, Denver, Colorado, is near 105°W longitude, −105°.

$$-105° + 180° = 75°$$

$$75°/6 = 12.50$$

Round up to 13

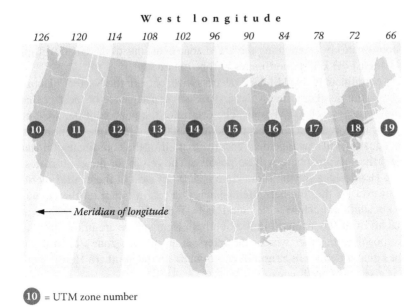

FIGURE 5.21   UTM zones in coterminous United States.

Therefore, Denver, Colorado, is in UTM Zone 13 as shown in Figure 5.21.

All UTM zones have a width of 6° of longitude. From north to south the zones extend from 84°N latitude to 80°S latitude. Originally, the northern limit was at 80°N latitude and the southern 80°S latitude. On the south the latitude is a small circle that conveniently traverses the ocean well south of Africa, Australia, and South America. However, 80°N latitude was found to exclude parts of Russia and Greenland and was extended to 84°N latitude.

UNIVERSAL TRANSVERSE MERCATOR ZONES OF THE WORLD

These zones nearly cover the Earth, except the polar regions, which are covered by two azimuthal polar zones called the Universal Polar Stereographic (UPS) projection. The foundation of the 60 UTM zones is a secant Transverse Mercator projection very similar to those used in some State Plane Coordinate Systems (Figure 5.21).

The central meridian of the zones is exactly in the middle. For example, in Zone 1 from 180° W to the 174°W longitude the central meridian is 177°W longitude, so each zone extends 3 degrees east and west from its central meridian.

The UTM secant projection gives approximately 180 km between the lines of exact scale where the cylinder intersects the ellipsoid. The scale factor grows from 0.9996 along the central meridian of a UTM zone to 1.00000 at 180 km to the east and west. Recall that SPCS zones are usually more limited in width, ~158 miles (or about 254 km), and, therefore, have a smaller range of scale factors than do the UTM zones. In state plane coordinates, the scale factor is usually no more than 1 part in 10,000. In UTM coordinates it can be as large as 1 part in 2500.

The reference ellipsoids for UTM coordinates vary among five different figures, but in the United States it is the Clarke 1866 ellipsoid. However, one can obtain 1983 UTM coordinates by referencing the UTM zone constants to the GRS80 ellipsoid of NAD83, then 1983 UTM coordinates will be obtained.

Every coordinate in a UTM zone occurs twice, once in the Northern Hemisphere and once in the Southern Hemisphere. This is a consequence of the fact that there are two origins in each UTM zone. The origin for the portion of the zone north of the equator is moved 500 km west of the intersection of the zone's central meridian and the equator. This arrangement ensures that all of the coordinates for that zone in the Northern Hemisphere will be positive. The origin for the coordinates in the Southern Hemisphere for the same zone is 500 km west of the central meridian as well. However, in the Southern Hemisphere the origin is not on the equator, it is 10,000 km south of it, close to the South Pole. This orientation of the origin guarantees that all of the coordinates in the Southern Hemisphere are in the first quadrant and are positive. In other words, the intersection of each zone's central meridian with the equator defines its origin of coordinates. In the Southern Hemisphere, each origin is given the coordinates

$$\text{easting} = X_0 = 500{,}000 \text{ m, and northing} = Y_0 = 10{,}000{,}000 \text{ m}$$

In the Northern Hemisphere, the values are

$$\text{easting} = X_0 = 500{,}000 \text{ m, and northing} = Y_0 = 0 \text{ m, at the origin}$$

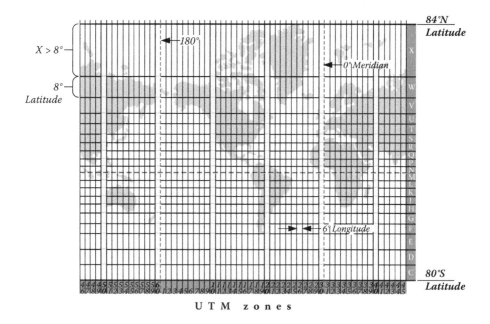

FIGURE 5.22   UTM zones around the world.

In fact, in the official version of the UTM system, there are actually more divisions in each UTM zone than the north–south demarcation at the equator. As shown in Figure 5.22, each zone is divided into 20 subzones. Each of the subzones covers 8° of latitude and is lettered from C on the south to X on the north. Actually, subzone X is a bit longer than 8°; remember the extension of the system from 80°N latitude to 84°N latitude. That all went into subzone X. It is also interesting that I and O are not included. They resemble one and zero too closely. In any case these subzones are little used outside of military applications of the system.

The developed UTM grid is defined in meters. Each zone is projected onto the cylinder that is oriented in the same way as that used in the Transverse Mercator SPCS, and the radius of the cylinder is chosen to keep the scale errors within acceptable limits (see Figure 5.23).

A word or two about the polar zones that round out the UTM system. The UPS are azimuthal stereographic projections. The projection has two zones. The North

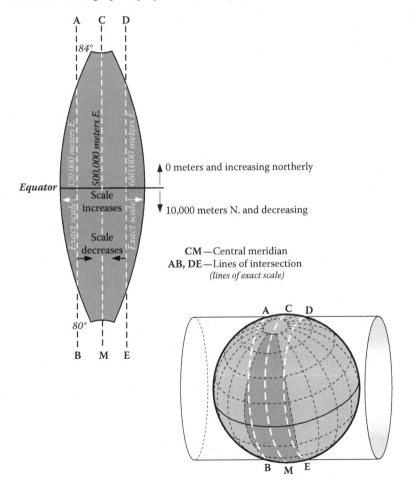

FIGURE 5.23   UTM zone.

zone covers latitudes 84°N to 90°N. The South zone covers latitudes 80°S to 90°S. The scale factor is 0.994, and the false easting and northing are both 2000 km. Their foundation is the International Ellipsoid, and its units are meters, just as in the UTM system in general.

HEIGHTS

ELLIPSOIDAL HEIGHTS

A point on the Earth's surface is not completely defined by its latitude and longitude. In such a context, there is, of course, a third element, that of height. Surveyors have traditionally referred to this component of a position as its elevation. One classical method of determining elevations is spirit leveling. A level, correctly oriented at a point on the surface of the Earth, defines a line parallel to the geoid at that point. Therefore, the elevations determined by level circuits are orthometric; that is, they are defined by their vertical distance above the geoid as it would be measured along a plumb line. However, orthometric elevations are not directly available from the geocentric position vectors derived from GPS measurements.

Modern geodetic datums rely on the surfaces of geocentric ellipsoids to approximate the surface of the Earth, but the actual surface of the Earth does not coincide with these nice smooth surfaces, even though that is where the points represented by the coordinate pairs lay. Abstract points may be on the ellipsoid, but the physical features those coordinates intend to represent are on the Earth. Though the intention is for the Earth and the ellipsoid to have the same center, the surfaces of the two figures are certainly not in the same place. There is a distance between them.

The distance represented by a coordinate pair on the reference ellipsoid to the point on the surface of the Earth is measured along a line perpendicular to the ellipsoid. This distance is known by more than one name. It is called the *ellipsoidal height*, and it is also called the *geodetic height* and is usually symbolized by *h*.

In Figure 5.24 the ellipsoidal height of a station is illustrated. The concept of an ellipsoidal height is straightforward. A reference ellipsoid may be above or below the surface of the Earth at a particular place. If the ellipsoid's surface is below the surface of the Earth, at the point the ellipsoidal height has a positive sign, if the ellipsoid's surface is above the surface of the Earth, at the point the ellipsoidal height has a negative sign. However, it is important to remember that the measurement of an ellipsoidal height is along a line perpendicular to the ellipsoid, not along a plumb line. Most often they are not the same, and because a reference ellipsoid is a geometric imagining, it is quite impossible to actually set up an instrument on it. That makes it tough to measure ellipsoidal height using surveying instruments. In other words, ellipsoidal height is not what most people think of as an elevation. Said another way, an ellipsoidal height is not measured in the direction of gravity. It is not measured in the conventional sense of down or up.

Nevertheless, the ellipsoidal height of a point is readily determined using a GPS receiver. GPS can be used to discover the distance from the geocenter of the Earth to any point on the Earth, or above it for that matter. In other words, it has the

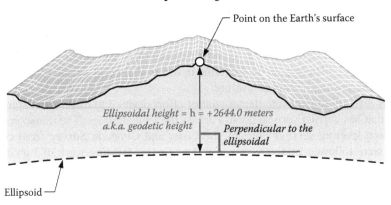

FIGURE 5.24   Ellipsoidal height.

capability of determining three-dimensional coordinates of a point in a short time. It can provide latitude and longitude, and if the system has the parameters of the reference ellipsoid in its software, it can calculate the ellipsoidal height. The relationship between points can be further expressed in the ECEF coordinates, X, Y, and Z, or in a *Local Geodetic Horizon System* of north, east, and up. Actually, in a manner of speaking, ellipsoidal heights are new, at least in common usage, since they could not be easily determined until GPS became a practical tool in the 1980s.

However, ellipsoidal heights are not all the same, because reference ellipsoids or sometimes just their origins can differ. For example, an ellipsoidal height expressed in ITRF would be based on an ellipsoid with exactly the same shape as the NAD83 ellipsoid, GRS80. Nevertheless the heights would be different because the origin has a different relationship with the Earth's surface (see Figure 5.25).

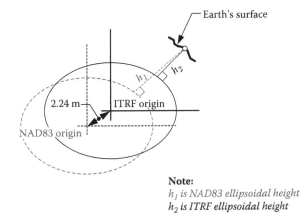

Note:
h_1 *is NAD83 ellipsoidal height*
h_2 *is ITRF ellipsoidal height*

FIGURE 5.25   A shift.

ORTHOMETRIC HEIGHTS

Spirit Leveling

Long before ellipsoidal heights were so conveniently available, knowing the elevation of a point was critical to the complete definition of a position. In fact, there are more than 200 different vertical datums in use in the world today. They were, and still are, determined by spirit leveling.

It is difficult to overstate the amount of effort devoted to differential spirit level work that has carried vertical control across the United States. The transcontinental precision leveling surveys done by the Coast and Geodetic Survey from coast to coast were followed by thousands of miles of spirit leveling work of varying precision. When the 39th parallel survey reached the west coast in 1907, there were approximately 19,700 miles (31,789 km) of geodetic leveling in the national network. That was more than doubled 22 years later in 1929 to approximately 46,700 miles (75,159 km). As the quantity of leveling information grew so did the errors and inconsistencies. The foundation of the work was ultimately intended to be *Mean Sea Level* (MSL) as measured by *tide station gauges*. Inevitably, this growth in leveling information and benchmarks made a new general adjustment of the network necessary to bring the resulting elevations closer to their true values relative to MSL.

There had already been four previous general adjustments to the vertical network across the United States by 1929. They were done in 1900, 1903, 1907, and 1912. The adjustment in 1900 was based on elevations held to MSL as determined at five tide stations. The adjustments in 1907 and 1912 left the eastern half of the United States fixed as adjusted in 1903. In 1927, there was a special adjustment of the leveling network. This adjustment was not fixed to MSL at all tide stations, and after it was completed, it became apparent that the MSL surface as defined by tidal observations had a tendency to slope upward to the north along both the Pacific and Atlantic coasts, with the Pacific being higher than the Atlantic.

In the adjustment that established the Sea Level Datum of 1929, the determinations of MSL at 26 tide stations, 21 in the United States, and 5 in Canada, were held fixed. Sea level was the intended foundation of these adjustments, and it might make sense to say a few words about the forces that shape it.

EVOLUTION OF A VERTICAL DATUM

Sea Level

Both the Sun and the Moon exert tidal forces on the Earth, but the Moon's force is greater. The Sun's tidal force is about half of that exerted on the Earth by the Moon. The Moon makes a complete elliptical orbit around the Earth every 27.3 days. There is a gravitational force between the Moon and the Earth. Each pulls on the other. At any particular moment, the gravitational pull is greatest on the portion of the Earth that happens to be closest to the Moon. That produces a bulge in the waters on the Earth in response to the tidal force. On the side of the Earth opposite the bulge, centrifugal force exceeds the gravitational force of the Earth, and water in this area is forced out away from the surface of the Earth, creating another bulge. The two bulges are not stationary; they move across the surface of the Earth. They move

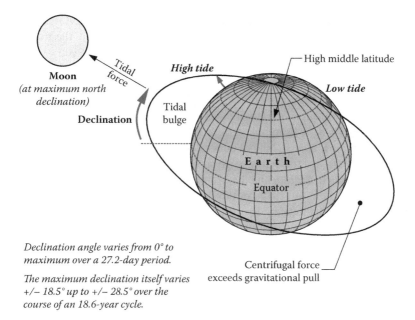

Declination angle varies from 0° to
maximum over a 27.2-day period.

The maximum declination itself varies
+/− 18.5° up to +/− 28.5° over the
course of an 18.6-year cycle.

FIGURE 5.26   Tides.

because not only is the Moon moving slowly relative to the Earth as it proceeds
along its orbit, but more importantly, the Earth is rotating in relation to the Moon.
The Earth's rotation is relatively rapid in comparison with the Moon's movement.
Therefore, a coastal area in the high middle latitudes may find itself with a high tide
early in the day when it is close to the Moon and a low tide in the middle of the day
when it has rotated away from it. This cycle will begin again with another high tide a
bit more than 24 hours after the first high tide. It's a bit more than 24 hours because
from the moment the Moon reaches a particular meridian to the next time it is there
is actually about 24 hours and 50 minutes, a period called a lunar day.

This sort of tide with one high water and one low water in a lunar day is known
as a *diurnal tide*. This characteristic tide would be most likely to occur in the middle
latitudes to the high latitudes when the Moon is near its maximum declination as you
can see from Figure 5.26.

DIURNAL TIDE

Declination of a celestial body is similar to the latitude of a point on the Earth. It is
an angle measured at the center of the Earth from the plane of the equator, positive
to the north and negative to the south, to the subject, which is in this case the Moon.
The Moon's declination varies from its minimum of 0° at the equator to its maximum over a 27.2-day period, and that maximum declination oscillates, too. It goes
from +/−18.5° up to +/−28.5° over the course of an 18.6-year cycle.

Another factor that contributes to the behavior of tides is the elliptical nature
of the Moon's orbit around the Earth. When the Moon is closest to the Earth, i.e.,

its *perigee*, the gravitational force between the Earth and the Moon is 20% greater than usual. At *apogee*, when the Moon is farthest from the Earth, the force is 20% less than usual. The variations in the force have exactly the affect you would expect on the tides, making them higher and lower than usual. It is about 27.5 days from perigee to perigee.

To summarize, the Moon's orbital period is 27.3 days. It also takes 27.2 days for the Moon to move from its maximum declinations back to 0° directly over the equator. There are 27.5 days from one perigee to the next. You can see that these cycles are almost the same, almost, but not quite. They are just different enough that it takes from 18 to 19 years for the Moon to go through all the possible combinations of its cycles with respect to the Sun and the Moon. Therefore, if you want to be certain that you have recorded the full range of tidal variation at a place, you must observe and record the tides at that location for 19 years.

This 19 year period, sometimes called the *Metonic* cycle, is the foundation of the definition of Mean Sea Level (MSL). MSL can be defined as the arithmetic mean of hourly heights of the sea at a primary-control tide station observed over a period of 19 years. The mean in MSL refers to the average of these observations over time at one place. It is important to note that it does not refer to an average calculation made from measurements at several different places. Therefore, when the Sea Level Datum of 1929 (SLD29) was fixed to MSL at 26 tide stations that meant it was made to fit 26 different and distinct Local MSL. In other words, it was warped to coincide with 26 different elevations.

The topography of the sea changes from place to place and that means, for example, MSL in Florida is not the same as MSL in California. The fact is MSL varies, and the water's temperature, salinity, currents, density, wind, and other physical forces all cause changes in the sea surface's topography. For example, the Atlantic Ocean north of the Gulf Stream's strong current is around 1 m lower than it is farther south, and the more dense water of the Atlantic is generally about 40 cm lower than the Pacific. At the Panama Canal the actual difference is about 20 cm from the east end to the west end.

A Different Approach

After it was formally established, thousands of miles of leveling were added to the SLD29. The Canadian network also contributed data to SLD29, but Canada did not ultimately use what eventually came to be known as the *National Geodetic Vertical Datum of 1929* (NGVD29). The name was changed May 10, 1973, because in the end the final result did not really coincide with MSL. It became apparent that the precise leveling done to produce the fundamental data had great internal consistency, but when the network was warped to fit so many tidal station determinations of MSL that consistency suffered.

By the time the name was changed to NGVD29 in 1973, there were more than 400,000 miles (643,700 km) of new leveling work included. There were distortions in the network. Original benchmarks had been disturbed, destroyed, or lost. The NGS thought it time to consider a new adjustment. This time there was a different approach. Instead of fixing the adjustment to tidal stations, the new adjustment would be minimally constrained. That means that it would be fixed to only one station, not 26. That station turned out to be Father Point/Rimouski, an *International Great Lakes Datum of 1985* (IGLD85) station near the mouth of the St. Lawrence River and on its southern bank. In other words, for all practical purposes the new adjustment of the huge network was not intended to be a sea level datum at all. It was a change in thinking that was eminently practical.

While is it relatively straightforward to determine MSL in coastal areas, carrying that reference reliably to the middle of a continent is quite another matter. Therefore, the new datum would not be subject to the variations in sea surface topography. It was unimportant whether the new adjustment's zero elevation and MSL were the same thing or not.

Zero Point

Precise leveling proceeded from the zero reference established at Pointe-au-Père, Quebec, in 1953. The resulting benchmark elevations were originally published in September 1961. The result of this effort was *International Great Lakes Datum 1955*. After nearly 30 years, the work was revised. The revision effort began in 1976, and the result was IGLD 1985. It was motivated by several developments including deterioration of the zero reference point gauge location and improved surveying methods. One of the major reasons for the revision was the movement of previously established benchmarks due to *isostatic rebound*. This effect is literally the Earth's crust rising slowly, rebounding, from the removal of the weight and subsurface fluids caused by the retreat of the glaciers from the last ice age.

The choice of the tide gauge at Pointe-au-Père, Quebec, as the zero reference for IGLD was logical in 1955. It was reliable. It had already been connected to the network with precise leveling. It was at the outlet of the Great Lakes. By 1984 the wharf at Pointe-au-Père had deteriorated and the gauge was moved, and it was subsequently moved about 3 miles (about 4 1/2 km) to Rimouski, Quebec, and precise levels were run between the two. It was there that the zero reference for IGLD 1985 and what became a new adjustment called *North American Vertical Datum 1988* (NAVD88) was established.

The re-adjustment, known as NAVD88, was begun in the 1970s. It addressed the elevations of benchmarks all across the nation. The effort also included field work. Destroyed and disturbed benchmarks were replaced and over 50,000 miles (over 80,000 km) of leveling were actually redone before NAVD88 was ready in June 1991. The differences between elevations of benchmarks determined in NGVD29 compared with the elevations of the same benchmarks in NAVD88 vary from approximately −1.3 feet (approximately −1/2 m) in the east to approximately +4.9 feet (approximately +1 1/2 m) in the west in the 48 coterminous states of the United States. The larger differences tend to be on the coasts, as one would expect since NGVD29 was forced to fit MSL at many tidal stations and NAVD88 was held to just one.

Geoid

Any object in the Earth's gravitational field has *potential energy* derived from being pulled toward the Earth. Quantifying this potential energy is one way to talk about height, because the amount of potential energy an object derives from the force of gravity is related to its height. There are an infinite number where the potential of gravity is always the same. They are known as *equipotential surfaces*.

MSL itself is not an equipotential surface at all, of course. Forces other than gravity affect it, forces such as temperature, salinity, currents, wind, and so forth. The geoid, however, is defined by gravity alone. The geoid is the particular equipotential surface arranged to fit MSL as well as possible, in a least squares sense (see Figure 5.27).

So while there is a relationship between MSL and the geoid, they are not the same. They could be the same if the oceans of the world could be utterly still, completely free of currents, tides, friction, variations in temperature, and all other physical forces, except gravity.

Reacting to gravity alone, these unattainable calm waters would coincide with the geoid. If the water was then directed by small frictionless channels or tubes and allowed to migrate across the land, the water would then, theoretically, define the same geoidal surface across the continents, too. Of course, the 70 percent of the Earth covered by oceans is not so cooperative, and the physical forces cannot really be eliminated. These unavoidable forces actually cause MSL to deviate up to 1 m, even 2 m, from the geoid.

Because the geoid is completely defined by gravity, it is not smooth and continuous. It is lumpy because gravity is not consistent across the surface of the Earth. At every point, gravity has a magnitude and a direction. Anywhere on the Earth, a vector can describe gravity, but these vectors do not all have the same direction or magnitude. Some parts of the Earth are denser than others. Where the Earth is denser, there is more gravity, and the fact that the Earth is not a sphere also affects gravity. It follows then that defining the geoid precisely involves actually measuring the direction and magnitude of gravity at many places.

The geoid undulates with the uneven distribution of the mass of the Earth. It has all the irregularity that the attendant variation in gravity implies. In fact, the separation between the lumpy surface of the geoid and the smooth GRS80 ellipsoid worldwide varies from about +85 m west of Ireland to about −106 m, the latter in the area south of India near Ceylon.

In the coterminous United States, sometimes abbreviated CONUS, the distances between the geoid and the GRS80 ellipsoid, known as *geoid heights*, are less. They vary from about −8 m to about −53 m.

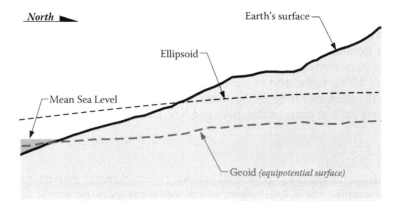

FIGURE 5.27   Ellipsoid–geoid–Mean Sea Level.

A geoid height is the distance measured along a line perpendicular to the ellipsoid of reference to the geoid. Also, as you can see, these geoid heights are negative. They are usually symbolized by N. If the geoid is above the ellipsoid, N is positive; if the geoid is below the ellipsoid, N is negative. It is negative here because the geoid is underneath the ellipsoid throughout the coterminous United States. In Alaska it is the other way around; the ellipsoid is underneath the geoid and N is positive.

Recall that an ellipsoid height is symbolized by h. The ellipsoid height is also measured along a line perpendicular to the ellipsoid of reference but to a point on the surface of the Earth. However, an orthometric height, symbolized by H, is measured along a plumb line from the geoid to a point on the surface of the Earth. In either case, by using

$$H = h - N$$

one can convert an ellipsoidal height h, derived say from a GPS observation, into an orthometric height H by knowing the extent of geoid-ellipsoid separation, also known as the geoidal height N at that point.

The ellipsoid height of a particular point is actually smaller than the orthometric height throughout the coterminous United States.

The formula $H = h - N$ does not account for the fact that the plumb line along which an orthometric height is measured is curved as you see in Figure 5.28. Curved because it is perpendicular with each and every equipotential surface through which it passes. Because equipotential surfaces are not parallel with each other, the plumb line must curve to maintain perpendicularity with them. This deviation of a plumb line from the perpendicular to the ellipsoid reaches about 1 minute of arc in only the most extreme cases. Therefore, any height difference that is caused by the curvature is negligible. It would take a height of over 6 miles (over 9.6 km) for the curvature to amount to even 1 mm of difference in height.

Geoid Models

Major improvements have been made over the past quarter century or so in mapping the geoid on both national and global scales. Because there are large complex variations in the geoid related to both the density and relief of the Earth, geoid models and interpolation software have been developed to support the conversion of GPS elevations to orthometric elevations. For example, in early 1991, NGS presented a program known as GEOID90. This program allowed a user to find N, the geoidal height, in meters for any NAD83 latitude and longitude in the United States.

The GEOID90 model was computed at the end of 1990, using over a million gravity observations. It was followed by the GEOID93 model. It was computed at the beginning of 1993 using more than 5× the number of gravity values used to create GEOID90. Both provided a grid of geoid height values in a 3 minutes of latitude by 3 minutes of longitude grid with an accuracy of about 10 cm. Next the GEOID96 model resulted in a gravimetric geoid height grid in a 2 minutes of latitude by 2 minutes of longitude grid.

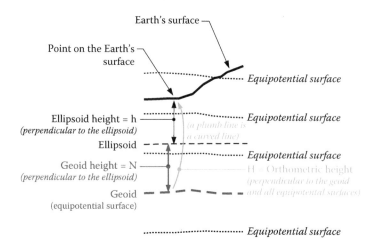

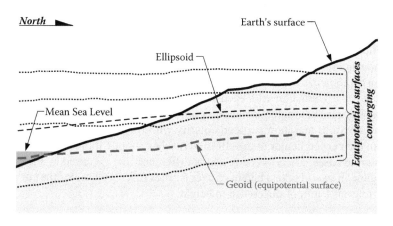

FIGURE 5.28 Height conversion.

GEOID99 covered the coterminous United States, and it includes U.S. Virgin Islands, Puerto Rico, Hawaii, and Alaska.

GEOID03 superseded the previous models for the continental United States. It was followed by GEOID09 and the current geoid model in place is known as GEOID12A.

EXERCISES

1. What is the datum used for the GPS Navigation message?
   a. NAD83 (CORS96)
   b. NAD27
   c. GRS80
   d. WGS84 (G1762)

2. What technology contributed the least number of positions to the original least squares adjustment of NAD83?
   a. EDM baselines
   b. TRANSIT Doppler positions
   c. Conventional optical surveying
   d. GPS

3. In the State Plane Coordinate Systems in the United States, which mapping projection listed below is not used?
   a. Transverse Mercator projection
   b. Oblique Mercator projection
   c. Lambert Conformal Conic projection
   d. Universal Transverse Mercator projection

4. What information is necessary to convert an ellipsoidal height to an orthometric height?
   a. Geoid height
   b. State Plane Coordinate
   c. Semimajor axis of the ellipsoid
   d. GPS Time

5. Which statement about the geoid is correct?
   a. Geoid's surface is always perpendicular to gravity.
   b. Geoid's surface is the same as Mean Sea Level.
   c. Geoid's surface is always parallel with an ellipsoid.
   d. Geoid's surface is the same as the topographic surface.

6. What acronym is used to describe the cooperative ventures that took place between NGS and the states to provide extremely accurate, vehicle-accessible, regularly spaced control points with good overhead visibility for GPS?
   a. ITRS
   b. HARN
   c. CORS
   d. NAVD

7. Which UTM zones cover the coterminous United States?
   a. Zones 10 North to 18 North
   b. Zones 1 North to 12 North
   c. Zones 6 North to 30 North
   d. Zones 20 North to 30 North

8. Which of the following organizations currently maintains the International Terrestrial Reference System (ITRS)?
   a. NGS
   b. IERS
   c. BIH
   d. C&GS

9. The combined factor is used in SPCS conversion. How is the combined factor calculated?
   a. Scale factor is multiplied by the elevation factor.
   b. Scale factor is divided by the grid factor.
   c. Grid factor added to the elevation factor and the sum is divided by 2.
   d. Scale factor is multiplied by the grid factor.

10. Which of the following statements is correct?
    a. Geodetic and astronomic latitudes are identical.
    b. Geocentric and geodetic latitudes are the same.
    c. Geocentric, astronomic, and geodetic latitudes are not the same and have different values.
    d. All three latitudes are the same.

11. Deflection of the vertical is what?
    a. Angle between the gravity vector and plumb line
    b. Angle between the plumb line and the line perpendicular to the ellipsoid
    c. The deviation of the declination of the instrument from the plumb line
    d. The distance between the gravity vector and plumb line

12. Orthomorphic map projection preserves what?
    a. Azimuth and distance
    b. Angles and distances
    c. Areas and shapes
    d. Directions and areas

ANSWERS AND EXPLANATIONS

1. Answer is (c)

   Explanation: With very slight changes, GRS80 became WGS84, which is the reference ellipsoid for the coordinate system, known as the World Geodetic System 1984 (WGS84). This datum has been used by the U.S. Military since January 21, 1987, as the basis for the GPS Navigation message computations. Therefore, coordinates provided directly by GPS receivers are based in WGS84. The newest incarnation of WGS84 is WGS84 (G1762). It was implemented by GPS Operational Control Segment in 2013.

2. Answer is (d)

   Explanation: It took more than 10 years to readjust and redefine the horizontal coordinate system of North America into what is now NAD83. More than 1.7 million positions derived from classical surveying techniques throughout the Western Hemisphere were involved in the least squares adjustment. They were supplemented by approximately 30,000 EDM measured baselines 5000 astronomic azimuths, and 650 Doppler stations positioned by the TRANSIT satellite system. Over 100 Very Long Baseline

Interferometry (VLBI) vectors were also included, but GPS, in its infancy, contributed only five points.

3. Answer is (d)

Explanation: In the United States, state plane systems are based on the transverse Mercator projection, an oblique Mercator projection, and the Lambert conic map projection grid. Every state, Puerto Rico, and the U.S. Virgin Islands have their own plane rectangular coordinate system.

4. Answer is (a)

Explanation: The geoid undulates with the uneven distribution of the mass of the Earth and has all the irregularity that implies. In fact, the separation between the bumpy surface of the geoid and the smooth GRS80 ellipsoid varies from 0 up to 100 m. Therefore, the only way a surveyor can convert an ellipsoidal height from a GPS observation on a particular station into a useable orthometric elevation is to know the extent of geoid-ellipsoid separation, also known as the geoid height, at that point.

Toward that end, major improvements have been made over the past quarter century or so in mapping the geoid on both national and global scales. This work has gone a long way toward the accurate determination of the geoid-ellipsoid separation, or geoid height, known as N. The formula for transforming ellipsoidal heights h into orthometric elevations H is $H = h - N$.

5. Answer is (a)

Explanation: The geoid is a representation of the Earth's gravity field. It is an equipotential surface that is everywhere perpendicular to the direction of gravity. In other words, it is perpendicular to a plumb line at every point.

Mean Sea Level is the average height of the surface of the sea for all stages of the tide. It was and sometimes still is used as a reference for elevations. However, it is not the same as the geoid. Mean Sea Level departs from the surface of the geoid; these displacements are known as the sea surface topography. Neither is the ellipsoid, a smooth mathematically defined surface, always parallel to the bumpy geoid. Finally, the geoid is certainly not coincident with the topographic surface of the Earth.

6. Answer is (b)

Explanation: The creation of High-Accuracy Reference Networks (HARN) was a cooperative venture between NGS and the states and often included other organizations as well. With heavy reliance on GPS observations, these networks are intended to provide extremely accurate, vehicle-accessible, regularly spaced control points with good overhead visibility. To ensure coherence, when the GPS measurements are complete, they were submitted to NGS for inclusion in a statewide readjustment of the existing

NGRS covered by the state. Coordinate shifts of 0.3 to 1.0 m from NAD83 values have been typical in these readjustments.

7. Answer is (a)

   Explanation: The UTM projection divides the world into 60 zones that begin at λ 180°, each with a width of 6° of longitude, extending from 84° Nφ and 80° Sφ. Its coverage is completed by the addition of two polar zones. The coterminous United States are within UTM zones 10 to 19.

   The UTM grid is defined in meters. Each zone is projected onto a cylinder that is oriented in the same way as that used in the transverse Mercator state plane coordinates described above. The radius of the cylinder is chosen to keep the scale errors within acceptable limits. Coordinates of points from the reference ellipsoid within a particular zone are projected onto the UTM grid.

   The intersection of each zone's central meridian with the equator defines its origin of coordinates. In the Southern Hemisphere, each origin is given the coordinates: easting = X_0 = 500,000 m and northing = Y_0 = 10,000,000 m to ensure that all points have positive coordinates. In the Northern Hemisphere, the values are easting = X_0 = 500,000 m and northing = Y_0 = 0 m at the origin. The scale factor grows from 0.9996 along the central meridian of a UTM zone to 1.00000 at 180,000 m to the east and west.

8. Answer is (b)

   Explanation: The best geocentric reference frame currently available is the International Terrestrial Reference Frame (ITRF). Its origin is at the center of mass of the whole Earth including the oceans and atmosphere. The unit of length is the meter. The orientation of its axes was established as consistent with that of the IERS's predecessor, Bureau International de l'Heure (BIH), at the beginning of 1984.

   Today, the ITRF is maintained by the International Earth Rotation Service (IERS), which monitors Earth Orientation Parameters (EOP) for the scientific community through a global network of observing stations. This is done with GPS, Very Long Baseline Interferometry (VLBI), Lunar Laser Ranging (LLR), satellite laser ranging (SLR), Doppler Orbitography and Radiopositioning Integrated by Satellite (DORIS), and the positions of the observing stations are now considered to be accurate to the centimeter level.

   The ITRF is actually a series of realizations. In other words, it is revised and published on a regular basis. Today NAD83 can be realizably defined in terms of a best-fit transformation from ITRF96.

9. Answer is (a)

   Explanation: The grid factor changes with the ellipsoidal height of the line. It also changes with its location in relation to the standard lines of its SPCS zone. The grid factor is derived by multiplying the scale factor by the elevation factor. The product is nearly 1 and is known as either the grid factor or the combination factor. There is a different combination factor for every line in the correct application of SPCS.

10. Answer is (c)

Explanation: One might expect that this astronomic latitude would be the same as the geocentric latitude of the point, but they are different. The difference is due to the fact that a plumb line coincides with the direction of gravity; it does not point to the center of the Earth where the line used to derive geocentric latitude originates.

11. Answer is (b)

Explanation: The angle between the vertical extension of a plumb line and the vertical extension of a line perpendicular to the ellipsoid is called the deflection of the vertical. It sounds better than the difference in down. This *deflection of the vertical* defines the actual angular difference between the astronomic latitude and longitude of a point and its geodetic latitude and longitude; latitude and longitude because, even though the discussion has so far been limited to latitude, the deflection of the vertical usually has both a north–south and an east–west component. The deflection of the vertical also has an effect on azimuths; for example, there will be a slight difference between the azimuth of a GPS baseline and the astronomically determined azimuth of the same line.

12. Answer is (b)

Explanation: Map projections in which shape is preserved are known as *conformal* or *orthomorphic*. Orthomorphic means right shape. In a conformal projection the angles between intersecting lines and curves retain their original form on the map. In other words, between short lines, meaning lines under about 10 miles (about 16 km), a 45° angle on the ellipsoid is a 45° angle on the map. It also means that the scale is the same in all directions from a point; in fact, it is this characteristic that preserves the angles.

6 Static Global Positioning System Surveying

Static Global Positioning System (GPS) surveying has been used on control surveys from a local to statewide extent and will probably continue to be the preferred technique in that category. If a static GPS control survey is carefully planned, it usually progresses smoothly. The technology has virtually conquered two stumbling blocks that have defeated the plans of conventional surveyors for generations. Inclement weather does not disrupt GPS observations, and a lack of intervisibility between stations is of no concern whatsoever, at least in post-processed GPS. Still, GPS is far from so independent of conditions in the sky and on the ground that the process of designing a survey can now be reduced to points-per-day formulas, as some would like. Even with falling costs, the initial investment in GPS remains large by most surveyors' standards. However, there is seldom anything more expensive in a GPS project than a surprise.

Static GPS was the first method of GPS surveying used in the field. Relative static positioning involves several stationary receivers simultaneously collecting data from at least four satellites during observation sessions that usually last from 30 min to 2 hours. A typical application of this method would be the determination of vectors, or baselines, between several static receivers to accuracies from 1 to 0.1 ppm over tens of kilometers. There are few absolute requirements for relative static positioning. The requisites include more than one receiver, four or more satellites, and a mostly unobstructed sky above the stations to be occupied. However, as in most of surveying, the rest of the elements of the system are dependent on several other considerations, and this implies planning.

PLANNING

A FEW WORDS ABOUT ACCURACY

When planning a GPS or Global Navigation Satellite System (GNSS) survey, one of the most important parameters is the accuracy specification. A clear accuracy goal avoids ambiguity both during and after the work is done. First, it is important to remember that there is a difference between precision and accuracy.

One aspect of precision can be visualized as the tightness of the clustering of measurements; the closer the grouping, the more precise the measurement. Accuracy, however, requires one more element.

It has to have a *truth set*. For example, the *truth* in Figure 6.1 for a, b, and c is the center of the target; without that, accuracy is indefinable. In other words, accuracy is not determined by measurement alone. There must also be a standard value or values

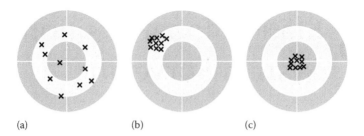

(a)   (b)   (c)

FIGURE 6.1   Precision and accuracy 1. (a) Accurate but not precise, (b) precise but not accurate, and (c) accurate and precise.

involved. It is through the comparison of the measurements with such standard values that the outcome of the work can be found to be sufficiently near the ideal or true value, or not.

For example, in Figure 6.2a it may seem at first that the average of the measurements in the GPS-A group are more accurate than the average of those in GPS-B because the GPS-A group is more precise. However, when the true position is introduced in Figure 6.2b, it is revealed that the GPS-B group's average is the more accurate of the two, because accuracy and precision are not the same.

When it comes to accuracy, there are other important details, too. Local accuracy and network accuracy are not the same. As mentioned in Chapter 4, local accuracy, also known as relative accuracy, represents the uncertainty in the positions relative to the other adjacent points to which they are directly connected. Network accuracy, also known as absolute accuracy, requires that a position's accuracy be specified with respect to an appropriate truth set such as a national geodetic datum. Differentially corrected GPS survey procedures that are tied to continuously operating reference stations (CORS), which represent the National Spatial Reference System (NSRS) of the United States, provide information from which network accuracy can be derived.

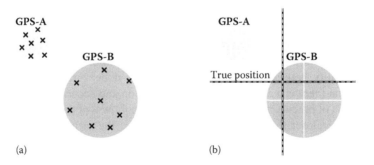

(a)   (b)

FIGURE 6.2   Precision and accuracy 2. (a) GPS-A is more precise than GPS-B but (b) after averaging GPS-B is more accurate than GPS-A.

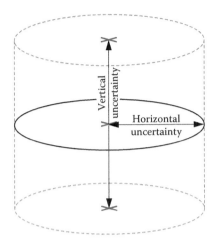

FIGURE 6.3   Horizontal and vertical accuracy.

However, autonomous GPS positioning, i.e., a single receiver without augmentation, is not operating relative to any control, local or national. In that context it is more appropriate to discuss the precision of the results than it is to discuss accuracy.

It is typical for uncertainty in horizontal accuracies to be expressed in a number that is radial. The uncertainties in vertical accuracies are given similarly, but they are linear, not radial. In both cases the limits are always plus or minus (±) (see Figure 6.3).

In other words, the reporting standard in the horizontal component is the radius of a circle of uncertainty, such that the true location of the point falls within that circle at some level of reliability, i.e., 95 percent of the time. Also, the reporting standard in the vertical component is a linear uncertainty value, such that the true location of the point falls within ± of that linear uncertainty to some degree of reliability. In GPS positioning it is reasonable to expect that the vertical accuracy will be about 1/3 that of horizontal accuracy. In other words, if the absolute horizontal accuracy of a GPS position is ±1 m, then the estimate of the absolute vertical accuracy of the same GPS position would be approximately ±3 m.

Here is bit more on horizontal accuracy. Figure 6.4 shows a spread of positions around a center of the range. As the radius of the error circle grows larger, the certainty that the center of the range is the true position increases (it never reaches 100 percent).

STANDARDS OF ACCURACY

The Federal Geodetic Control Committee (FGCC) wrote accuracy standards for GPS relative positioning techniques. These were preceded by older standards of first, second, and third order that then became subsumed under the group C in the newer scheme. Until the last decades of the twentieth century, the cost of achieving first-order accuracy was considered beyond the reach of most conventional surveyors. Besides, surveyors often said that such results were far in excess of their needs anyway. The burden of the equipment, techniques, and planning that is required to reach

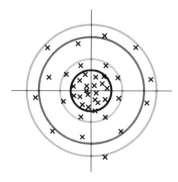

Circular error probable (CEP) = 50% = 0.5 m error circle radius = ±0.5 m
1-Sigma (1σ) = 68% = 0.6 m error circle radius = ±0.6 m
E90 = 90% = 0.9 m error circle radius = ±0.9 m
E95 = 95% = 1.0 m error circle radius = ±1.0 m

FIGURE 6.4   Horizontal point accuracy.

its 2σ relative error ratio of 1 part in 100,000 of the old first order was something most surveyors were happy to leave to government agencies. However, the FGCC's standards of B, A, and AA are respectively 10, 100, and 1000× more accurate than first order. With the advent of GPS the attainment of these accuracies did not require corresponding 10-, 100-, and 1000-fold increases in equipment, training, personnel, or effort. They were now well within the reach of private GPS surveyors both economically and technically. These accuracy standards are now superseded.

In 1998 the FGCC under the Federal Geographic Data committee published the Geospatial Positioning Accuracy Standards Part 2: Standards for Geodetic Networks (FGDC 1998). These standards, shown in Table 6.1, supplant the earlier standards of 1984 and 1989.

New Design Criteria

These upgrades in accuracy standards not only accommodate control by static GPS, but they also have cast survey design into a new light for many surveyors. Nevertheless, it is not correct to say that every job suddenly requires the highest achievable accuracy, nor is it correct to say that every GPS survey now demands an elaborate design. In some situations, a crew of two or even one surveyor on-site may carry a GPS survey from start to finish with no more planning than minute-to-minute decisions can provide even though the basis and the content of those decisions may be quite different from those made in a conventional survey.

In areas that are not heavily treed and generally free of overhead obstructions, sufficient accuracy may be possible without a prior design of any significance. While it is certainly unlikely that a survey of photocontrol or work on a cleared construction site would present overhead obstruction problems comparable with a static GPS control survey in the Rocky Mountains, even such open work may demand preliminary attention. For example, just the location of appropriate vertical and horizontal

TABLE 6.1
Accuracy Standards

Horizontal, Ellipsoid Height, and Orthometric Height

Accuracy Classification	95% Confidence (Less than or Equal to)
1-Millimeter	0.001 meters
2-Millimeter	0.002 meters
5-Millimeter	0.005 meters
1-Centimeter	0.010 meters
2-Centimeter	0.020 meters
5-Centimeter	0.050 meters
1-Decimeter	0.100 meters
2-Decimeter	0.200 meters
5-Decimeter	0.500 meters
1-Meter	1.000 meters
2-Meter	2.000 meters
5-Meter	5.000 meters
10-Meter	10.000 meters

Source: Federal Geographic Data Committee FGDC-STD-007.2-1998 Draft Geospatial Positioning Accuaracy Standards Part 2: Standard for Geodetic Networks.

control stations or obtaining permits for access across privately owned property or government installations can be critical to the success of the work.

Lay of the Land

An initial visit to the site of the survey is not always possible. Today, online mapping browsers are making virtual site evaluation possible as well. Topography as it affects the line of sight between stations is of no concern on a static GPS project, but its influence on transportation from station to station is a primary consideration. Perhaps some areas are only accessible by helicopter or other special vehicle. Initial inquiries can be made. Roads may be excellent in one area of the project and poor in another. The general density of vegetation, buildings, or fences may open general questions of overhead obstruction or multipath. The pattern of land ownership relative to the location of project points may raise or lower the level of concern about obtaining permission to cross property.

Maps

Maps, both digital and hard copy, are particularly valuable resources for preparing a static GPS survey design. Local government and private sources can sometimes provide appropriate mapping, or it may be available online.

NATIONAL GEODETIC SURVEY (NGS) CONTROL

NGS CONTROL DATA SHEETS

Monuments that are the physical manifestation of the NSRS and can be occupied with survey equipment are known as *passive* marks. They can provide reliable control when properly utilized. That utilization should be informed by an understanding of the data sheet that accompanies each station and is easily available online.

There is a good deal of information about the passive survey monuments on each individual sheet (see Figure 6.5). In addition to the latitude and longitude, the published data include the State Plane Coordinates in the appropriate zones. The first line of each data sheet includes the retrieval date. Then the station's category is indicated. There are several, and among them are Continuously Operating Reference Station, Federal Base Network Control Station, and Cooperative Base Network Control Station. This is followed by the station's designation, which is its name, and its Permanent Identifier (PID). Either of these may be used to search for the station in the NGS database. The PID is also found all along the left side of each data sheet record and is always two uppercase letters followed by four numbers. The state, county, country, and U.S. Geological Survey (USGS) 7.5 minute quad name follows. Even though the station is located in the area covered by the quad sheet, it may not actually appear in the map. Under the heading, "Current Survey Control," you will find the latitude and longitude of the station in NAD83, which is fixed to the North American plate, currently, in NAD83 (2011), and its height in NAVD88. The orthometric height in meters is listed as "ORTHO HEIGHT" and followed by the same in feet. When the height is derived from GPS observation, a geoid model must be used to determine the orthometric height. The model used is given.

Adjustments to NAD27 and NGVD29 datums are a thing of the past. However, these old values may be shown under Superseded Survey Control. Horizontal values may be either scaled, if the station is a benchmark, or adjusted, if the station is indeed a horizontal control point.

When a date is shown in parentheses after NAD83 in the data sheet, it means that the position has been readjusted. There are 13 sources of vertical control values shown on NGS data sheets. Here are a few of the categories. There is adjusted, which are given to three decimal places and are derived from least squares adjustment of precise leveling. Another category is posted, which indicates that the station was adjusted after the general NAVD adjustment in 1991. When a station's elevation has been found by precise leveling but nonrigorous adjustment, it is called computed. Stations' vertical values are given to one decimal place if they are from GPS observation (Obs) or vertical angle measurements (Vert Ang), and they have no decimal places if they were scaled from topographic map, scaled, or found by conversion from NGVD29 values using the program known as VERTCON.

When they are available, Earth-Centered-Earth-Fixed (ECEF) coordinates are shown in X, Y, and Z. These are right-handed system, 3-D Cartesian coordinates and are computed from the position and the ellipsoidal height. They are the same type of X, Y, and Z coordinates presented in Chapter 5. These values are followed by the

The NGS Data Sheet

See file dsdata.txt **for more information about the data sheet.**

```
PROGRAM = datasheet95, VERSION = 8.5
1        National Geodetic Survey ,  Retrieval  Date = July 21, 2014
 KK1696 ************************************************************************
 KK1696  CBN           -   This is a Cooperative Base Network Control Station.
 KK1696  DESIGNATION   -   JOG
 KK1696  PID               KK1696
 KK1696  STATE / COUNTY -  CO / ARAPAHOE
 KK1696  USGS  QUAD    -   PARKER  (1994)
 KK1696
 KK1696
 KK1696                       * CURRENT SURVEY CONTROL
 KK1696  ----------------------------------------------------------------------
 KK1696* NAD 83 (2011)  POSITION  - 39 34 05 . 17515 (N)  104 52 18 . 24505 (W)   ADJUSTED
 KK1696* NAD 83 (2011)  ELLIP HT  -  1779.178   (meters)          (06/27/12)      ADJUSTED
 KK1696* NAD 83 (2011)  EPOCH     -  2010.00
 KK1696* NAVD 88 ORTHO HEIGHT  - 1796.2    (meters)       5893.   (feet)     GPS OBS
 KK1696  ----------------------------------------------------------------------
 KK1696  NAVD 88  orthometric height was determined with geod model          GEOID09
 KK1696  GEOID HEIGHT        -                -17.15 (meters)                 GEOID09
 KK1696  GEOID HEIGHT        -                -17.14 (meters)                 GEOID12A
 KK1696  NAD 83 (2011) x    -        -1,263,970.447 (meters)                  COMP
 KK1696  NAD 83 (2011) y    -        -4,759,798.581 (meters)                  COMP
 KK1696  NAD 83 (2011) z    -         4,042,268.494 (meters)                  COMP
 KK1696  LAPLACE CORR       -                 -5.35 (seconds)                 DEFLEC12A
 KK1696
 KK1696  FGDC Geospatial  Positioning  Accuracy  Standards    (95% confidence, cm)
 KK1696  Type                                     Horiz  Ellip  Dist (km)
 KK1696  -------------------------------------------------------------------
 KK1696  NETWORK                                   0.48   0.78
 KK1696  -------------------------------------------------------------------
 KK1696  MEDIAN LOCAL ACCURACY AND DIST (043 points) 0.50  0.82  16.35
 KK1696  -------------------------------------------------------------------
 KK1696  NOTE: Click here for information on individual local accuracy
 KK1696  values and other accuracy information.
 KK1696
 KK1696
 KK1696.  The horizontal coordinates were established by GPS observations
 KK1696.  and adjusted by the National Geodetic Survey in June 2012.
 KK1696
 KK1696.  NAD 83 (2011) refers to NAD 83 coordinates were the reference
 KK1696.  frame has been affixed to the stable North American tectonic plate.  see
 KK1696.  NA2011 for mor information.
 KK1696
 KK1696.  The horizontal coordinates are valid at the epoch date displayed above
 KK1696.  which is a decimal equivalence of Year/Month/day.
 KK1696
 KK1696.  The orthometric height was determined by GPS observations and a
 KK1696  high-resolution geoid model.
 KK1696
 KK1696.  Photographs are available for this station.
 KK1696
 KK1696
 KK1696.  The X, Y, and Z were computed from the position and the ellipsoidal ht.
 KK1696
 KK1696.  The Laplace correction was computed from DEFLEC12A derived deflections.
 KK1696
 KK1696.  The ellipsoidal height was determined by GPS observation
 KK1696.  and is referenced to NAD 83.
 KK1696
 KK1696.  The following values were computed from the NAD 83 (2011) position.
 KK1696
 KK1696;             North            East      Units     Scale        Converg.
 KK1696; SPC CO C  -  497,563.467    968,386.201   MT   0 . 99996908   +0 23 46.5
 KK1696; SPC CO C  -  1,632,422.81   3,177,113.730  sMT  0 . 99996908   +0 23 46.5
 KK1696; UTM 13    - 4 , 379 , 830 . 668  511 , 017 . 357   MT  0 . 99960149  +0 04 54.1
 KK1696
 KK1696!              Elev Factor x   Scale Factor =  Combined Factor
 KK1696! SPC CO C  -  0.99972095 x    0.99996908  =  0.99969004
 KK1696! UTM 13    -  0.99972095 x    0.99960149  =  0.99932255
 KK1696
 KK1696:             Primary Azimuth Mark            Grid Az
 KK1696: SPC CO C  -  JOG AZ MK                    175  36  49 . 2
 KK1696: UTM 13    -  JOG AZ MK                    175  55  41 . 6
 KK1696
```

FIGURE 6.5   NGS control data sheet.

(Continued)

```
KK1696 |------------------------------------------------------------------------------------|
KK1696 | PID        Reference Object                              Distance        Geod. Az  |
KK1696 |                                                                         dddmmss.s  |
KK1696 | KK1695    DENVER INVERNESS TANK                        66.921 METERS    11309      |
KK1696 | CP8337    JOG RM 1                                     13.428 METERS    13410      |
KK1696 | KK1699    JOG AZ MK                                  APPROX. 0.7 KM   1760035.7    |
KK1696 | KK1701    LITTLETON HONEYWELL CORP TANK              APPROX. 5.2 KM   2840836.7    |
KK1696 | CP8338    JOG RM 2                                     11.912 METERS    32655      |
KK1696 |------------------------------------------------------------------------------------|
KK1696
KK1696                         SUPERSEDED SURVEY CONTROL
KK1696
KK1696   NAD 83 (2007) -   39 34 05 . 17592  (N)        104 52 18 . 24529 (W)   AD (2002 . 00)   0
KK1696   ELLIP H (02/10/07) 1779 . 167       (m)                                GP (2002 . 00)
KK1696   ELLIP H (10/21/02) 1779 . 200       (m)                                GP (            )   5 1
KK1696   NAD 83 (1992) -   39 34 05 . 17172  (N)        104 52 18 . 24505 (W)   AD (            )   B
KK1696   ELLIP H (05/26/92)    1779 . 261    (m)                                GP (            )   4 1
KK1696   NAD 83 (1986) -   39 34 05 . 17172  (N)        104 52 18 . 24385 (W)   AD (            )   1
KK1696   NAD 27        -   39 34 05 . 21252  (N)        104 52 16 . 31971 (W)   AD (            )   1
KK1696   NAVD 88 (06/08/05)   1796 . 4       (m)     GEOI03 model used      GPS OBS
KK1696   NAVD 88 (12/14/94)   1796 . 3       (m)     UNKNOWN model used     GPS OBS
KK1696   NAVD 88 (11/02/93)   1796 . 4       (m)     GEOID93 model used     GPS OBS
KK1696   NAVD 88 (05/26/92)   1796 . 3       (m)     UNKNOWN model used     GPS OBS
KK1696
KK1696.  Superseded values are not recommended for survey control.
KK1696
KK1696.  NGS no longer adjusts projects to the NAD 27 or NGVD 29 datums.
KK1696.  See file dsdata.txt to determine how the superseded data were derived.
KK1696
KK1696.  U.S. NATIONAL GRID SPATIAL ADDRESS: 13SED1101779830 (NAD 83)
KK1696
KK1696_  MARKER: DS = TRIANGULATION STATION DISK
KK1696_  SETTING:   7 = SET IN TOP OF CONCRETE MONUMENT
KK1696_
KK1696_  SP_SET: CONCRETE POST
KK1696_  STAMPING: JOG 1977
KK1696_  MARK LOGO: NGS
KK1696_  PROJECTION: RECESSED 51 CENTIMETERS
KK1696_  MAGNETIC: 0 = OTHER: SEE DESCRIPTION
KK1696_  STABILITY: C = MAY HOLD, BUT OF TYPE COMMONLY SUBJECT TO
KK1696+  STABILITY: SURFACE MOTION
KK1696_  SATELLITE: THE SITE LOCATION WAS REPORTED AS SUITABLE FOR
KK1696+  SATELLITE: SATELLITE OBSERVATIONS - February 26, 2014
KK1696
KK1696   HISTORY    - DATE          Condition          Report By
KK1696   HISTORY    - 1977          MONUMENTED         NGS
KK1696   HISTORY    - 1977          GOOD               NGS
KK1696   HISTORY    - 19910515      GOOD               NGS
KK1696   HISTORY    - 19940307      GOOD               CODOT
KK1696   HISTORY    - 19940307      GOOD               MSAM
KK1696   HISTORY    - 19950206      GOOD               CODOT
KK1696   HISTORY    - 19960415      GOOD               MSAM
KK1696   HISTORY    - 19990415      GOOD               MSAM
KK1696   HISTORY    - 20040228      GOOD               INDIV
KK1696   HISTORY    - 20050308      GOOD               INDIV
KK1696   HISTORY    - 20060605      GOOD               WSSUR
KK1696   HISTORY    - 20080203      GOOD               MSCD
KK1696   HISTORY    - 20100805      GOOD               WOOLPT
KK1696   HISTORY    - 20140226      GOOD               KII
KK1696                         STATION DESCRIPTION
KK1696'  DESCRIBED BY NATIONAL GEODETIC SURVEY 1977 (LHW)
KK1696'  THE STATION IS LOCATED ON THE EAST SIDE OF THE INTERSTATE HIGHWAY 25 AND NORTHEAST
KK1696'  OF THE JUNCTION OF STATE HIGHWAY 470 AND INTERSTATE HIGHWAY 25, ON ROAD
KK1696'  RIGHT-OF-WAY, ON PROPERTY OWNED BY THE STATE OF COLORADO. IT IS JUST SOUTHEAST OF
KK1696'  DENVER CITY LIMITS AND IN THE SOUTHEAST QUARTER OF SECTION 34, T. 5S., R. 67W.
KK1696'
KK1696'  TO REACH THE STATION FROM THE JUNCTION OF INTERSTATE HIGHWAY 25 AND STATE HIGHWAY
KK1696'  88 (ARAPAHOE ROAD) GO SOUTH-SOUTHEAST ON INTERSTATE HIGHWAY 25 FOR 2.5 MILES TO
KK1696'  COUNTY LINE ROAD EXIT. TAKE EXIT RAMP TO A STOP SIGN, TURN LEFT ON COUNTY LINE
KK1696'  ROAD FOR 0.3 MILES TO A FRONTAGE ROAD ON THE RIGHT AND ENTRANCE RAMP FOR
KK1696'  NORTHBOUND INTERSTATE HIGHWAY 25. (TO REACH AZIMUTH MARK FROM HERE TURN RIGHT
KK1696'  ON FRONTAGE ROAD AND GO WESTERLY AND SOUTH FOR 0.4 MILES TO AZIMUTH MARK ON THE
KK1696'  RIGHT). TURN LEFT ONTO ENTRANCE RAMP AND GO NORTHERLY FOR 0.1 MILES TO THE STATION
KK1696'  ON THE RIGHT NEXT TO A CHAIN LINK FENCE.
KK1696'
KK1696'  STATION MARKS, STAMPED---JOG 1977---, ARE STANDARD DISKS. THE SURFACE DISK IS SET IN
KK1696'  THE TOP OF A 12-INCH CYLINDRICAL CONCRETE MONUMENT THAT IS FLUSH WITH THE GROUND.
KK1696'  IT IS 5.5 FEET WEST OF A CHAIN LINK FENCE AND 85 FEET EAST OF THE CENTER OF ENTRANCE
KK1696'  RAMP FOR NORTHBOUND INTERSTATE HIGHWAY 25. THE UNDERGROUND DISK IS SET IN AN
KK1696'  IRREGULAR MASS OF CONCRETE ABOUT 42 INCHES BELOW THE GROUND.
```

FIGURE 6.5 (CONTINUED)   NGS control data sheet.

quantity that when added to an astronomic azimuth yields a geodetic azimuth and is known as the Laplace correction. It is important to note that NGS uses a clockwise rotation regarding the Laplace correction. The ellipsoid height per the NAD83 ellipsoid is shown followed by the geoid height where the position is covered by NGS's GEOID program. The FGDC network accuracy is shown at a 95 percent reliability level. Photographs of the station may also be available in some cases. When the data sheet is retrieved online, one can use the link provided to bring them up. Also, the geoidal model used is noted.

Coordinates

NGS data sheets also provide state plane and Universal Transverse Mercator (UTM) coordinates, the latter only for horizontal control stations. State plane coordinates are given in both U.S. Survey Feet or International Feet and UTM coordinates are given in meters. Azimuths to the primary azimuth mark are clockwise from north and scale factors for conversion from ellipsoidal distances to grid distances. This information may be followed by distances to reference objects. Coordinates are not given for azimuth marks or reference objects on the data sheet.

Station Mark

Along with mark setting information, the type of monument and the history of mark recovery, the NGS data sheets provide a valuable to-reach description. It begins with the general location of the station. Then starting at a well-known location, the route is described with right and left turns, directions, road names, and the distance traveled along each leg in kilometers. When the mark is reached, the monument is described and horizontal and vertical ties are shown. Finally, there may be notes about obstructions to GPS visibility, and so forth.

Significance of the Information

The value of the description of the monument's location and the route used to reach it is directly proportional to the date it was prepared and the remoteness of its location. The conditions around older stations often change dramatically when the area has become accessible to the public. If the age and location of a station increases the probability that it has been disturbed or destroyed, then reference monuments can be noted as alternatives worthy of on-site investigation. However, special care ought to be taken to ensure that the reference monuments are not confused with the station marks themselves.

CONTROL FROM CONTINUOUSLY OPERATING NETWORKS

The requirement to occupy physical geodetic monuments in the field can be obviated by downloading the tracking data available online from appropriate Continuously Operating Reference Stations (CORS) where their density is sufficient. These stations, also known as active stations, comprise fiducial networks that support a variety of GPS applications. While they are frequently administered by governmental organizations, some are managed by public-private organizations and some are commercial ventures. The most straightforward benefit of CORS is the user's ability to do relative positioning without operating his own base station by depending on that role being fulfilled by the network's reference stations.

While CORS can be configured to support differential GPS (DGPS) and real-time kinematic applications, as in Real-Time Networks, most networks constantly collect GPS tracking data from known positions and archive the observations for subsequent download by users from the Internet.

In many instances, the original impetus of a network of CORS was geodynamic monitoring as illustrated by the GEONET established by the Geographical Survey Institute in Japan after the Kobe earthquake. Networks that support the monitoring of the International Terrestrial Reference System (ITRS) have been created around the world by the International GNSS Service (IGS), which is a service of the International Association of Geodesy and the Federation of Astronomical and Geophysical Data Analysis Services originally established in 1993. Also, the Southern California Integrated GPS Network is a network run by a government–university partnership.

Despite the original motivation for the establishment of a CORS network, the result has been a boon for high-accuracy GPS positioning. The data collected by these networks is quite valuable to GPS surveyors around the world. Surveyors in the United States can take advantage of the CORS network administered by the National Geodetic Survey (NGS). The continental NGS system has two components, the Cooperative CORS and the National CORS. Together they comprise a network of hundreds of stations which constantly log multifrequency GPS data and make the data available in the Receiver Independent Exchange (RINEX) format.

NGS CONTINUOUSLY OPERATING REFERENCE STATIONS

NGS manages the National CORS system to support post-processing GPS data. Information is available in both code and carrier phase GPS data from receivers at these stations throughout the United States and its territories. That data can then be conveniently downloaded in its original form from the Internet free of charge for up to 30 days after its collection. It is also available later, but after it has been decimated to a 30 s format.

The Cooperative CORS system supplements the National CORS system. The NGS does not directly provide the data from the cooperative system of stations. Its stations are managed by participating university, public, and private organizations that operate the sites.

Nearly all coordinates provided by NGS for the CORS sites are available in NAD83 (2011) epoch 2010.0. The coordinates of CORS stations are also published in IGS08. However, these positions differ from NAD83 (2011). GPS observations are the foundation of IGS08, which is consistent with ITRF08. The coordinates in both NAD83 (2011) and IGS08 are accompanied by velocities because they are moving. An IGS08 position may differ slightly for a station, but the velocities for both reference frames are identical. These velocities can be used to calculate the stations position at a different date using NGS's Horizontal Time Dependent Positioning (HTDP) utility.

NGS CORS REFERENCE POINTS

At a CORS site, NGS provides the coordinates of the L1 phase center and the Antenna Reference Point (ARP). Generally speaking, it is best to adopt the position that can be physically measured, that means the coordinates given for the ARP. It is the coordinate of the part of the antenna from which the phase center offsets are calculated that is usually the bottom mount.

The phase centers of antennas are not immovable points. They actually change slightly, mostly as the elevation of the satellite's signals change. In any case, the phase centers for L1, L2, and L5 differ from the position of the ARP both vertically and horizontally. NGS provides the position of the phase center on average at a particular CORS site. As most post-processing software will, given the ARP, provide the correction for the phase center of an antenna, based on antenna type, the ARP is the most convenient coordinate value to use.

INTERNATIONAL GLOBAL NAVIGATION SATELLITE SYSTEM (GNSS) SERVICE (IGS)

Like NGS, IGS also provides CORS data. However, it has a global scope. The information on the individual stations can be accessed including the ITRF00 Cartesian coordinates and velocities for the IGS sites, but not all the sites are available from IGS servers. The Scripps Orbit and Permanent Array Center is a convenient access point for IGS data. A virtual map of the available GPS networks can be found there.

STATIC SURVEY PROJECT DESIGN

The selection of satellites to track, start and stop times, mask elevation angle, assignment of data file names, reference position, bandwidth, and sampling rate are some options useful in the static mode, as well as other GPS surveying methods. These features may appear to be prosaic, but their practicality is not always obvious. For example, satellite selection can seem unnecessary when using a receiver with sufficient independent channels to track all satellites above the receiver's horizon without difficulty. However, a good survey project design pays dividends by limiting lost time and maximizing productivity.

HORIZONTAL CONTROL

When geodetic surveying was more dependent on optics than electronic signals from space, horizontal control stations were set with station intervisibility in mind, not ease of access. Therefore, it is not surprising that stations established in that way are frequently difficult to reach. Not only are they found on the tops of buildings and mountains, but they are also in woods, beside transmission towers, near fences, and generally obstructed from GPS signals. The geodetic surveyors that established them could hardly have foreseen a time when a clear view of the sky above their heads would be crucial to high-quality control.

In fact, it is only recently that most private surveyors have had any routine use for NGS stations. Many station marks have not been occupied for quite a long time.

Because the primary monuments are often found deteriorated, overgrown, unstable, or destroyed, it is important that surveyors be well acquainted with the underground marks, reference marks, and other methods used to perpetuate control stations.

Obviously, it is a good idea to propose reconnaissance of several more than the absolute minimum of three horizontal control stations. Fewer than three makes any check of their positions virtually impossible. Many more are usually required in a GPS route survey. In general, in GPS networks, the more well-chosen horizontal control stations available, the better. Some stations will almost certainly prove unsuitable unless they have been used previously in GPS work.

STATION LOCATION

The location of the stations, relative to the GPS project itself, is also an important consideration in choosing horizontal control. For work other than route surveys, a handy rule of thumb is to divide the project into four quadrants and to choose at least one horizontal control station in each. The actual survey should have at least one horizontal control station in three of the four quadrants. Each of them ought to be as near as possible to the project boundary. Supplementary control in the interior of the network can then be used to add more stability to the network (see Figure 6.6).

At a minimum, route surveys require horizontal control at the beginning, the end, and the middle. Long routes should be bridged with control on both sides of the line at appropriate intervals. The standard symbol for indicating horizontal control on the project map is a triangle.

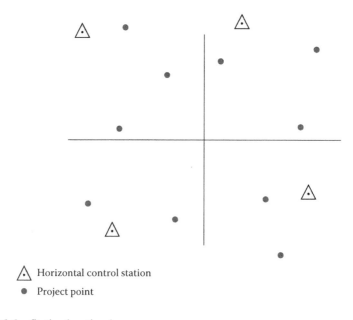

△ Horizontal control station

● Project point

FIGURE 6.6   Station location 1.

VERTICAL CONTROL

Those stations with a published accuracy high enough for consideration as vertical control are symbolized by an open square or circle on the map. Those stations that are sufficient for both horizontal and vertical control are particularly helpful and are designated by a combination of the triangle and square.

A minimum of four vertical control stations are needed to anchor a GPS static network. A large project should have more. In general, the more benchmarks available, the better. Vertical control is best located at the four corners of a project (see Figure 6.7).

Orthometric elevations are best transferred by means of classic spirit leveling; such work should be built into the project plan when it is necessary. When spirit levels are planned to provide vertical control positions, special care may be necessary to ensure that the precision of such conventional work is as consistent as possible with the rest of the survey. Route surveys require vertical control at the beginning and the end. They should be bridged with benchmarks on both sides of the line at intervals from 5 to 10 km.

When the distances involved are too long for spirit leveling to be used effectively, two independent GPS measurements may suffice to connect a benchmark to the project. However, it is important to recall the difference between the ellipsoidal heights available from a GPS observation and the orthometric elevations yielded by a level circuit.

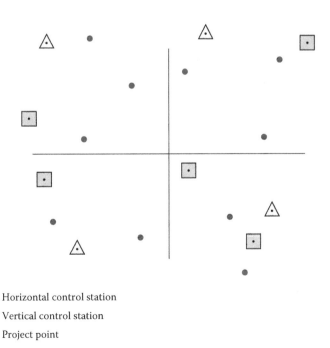

FIGURE 6.7   Station location 2.

PREPARATION

PLOTTING PROJECT POINTS

A solid dot is the standard symbol used to indicate the position of project points. Some variation is used when a distinction must be drawn between those points that are in place and those that must be set (see Figure 6.8).

When its location is appropriate, it is always a good idea to have a vertical or horizontal control station serve double duty as a project point. While the precision of their plotting may vary, it is important that project points be located as precisely as possible even at this preliminary stage.

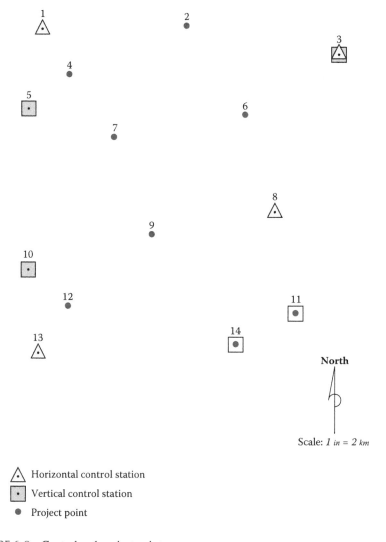

FIGURE 6.8   Control and project points.

The subsequent observation schedule will depend to some degree on the arrangement of the baselines. Also, the preliminary evaluation of access, obstructions, and other information depends on the position of the project point relative to these features.

Evaluating Access

When all potential control and project positions have been plotted and given a unique identifier, some aspects of the survey can be addressed a bit more specifically. If good roads are favorably located, if open areas are indicated around the stations, and if no station falls in an area where special permission will be required for its occupation, then the preliminary plan of the survey ought to be remarkably trouble-free. However, it is likely that one or more of these conditions will not be so fortunately arranged.

The speed and efficiency of transportation from station to station can be assessed to some degree from the project map. It is also wise to remember that while inclement weather does not disturb GPS observations whatsoever, without sufficient preparation, it can play havoc with surveyors' ability to reach points over difficult roads or by aircraft.

Planning Offsets

If control stations or project points are located in areas where the map indicates that topography or vegetation will obstruct the satellite's signals, alternatives may be considered. A shift of the position of a project point into a clear area may be possible where the change does not have a significant effect on the overall network. A control station may also be the basis for a less obstructed position, transferred with a short level circuit or traverse. Of course, such a transfer requires availability of conventional surveying equipment on the project. In situations where such movement is not possible, careful consideration of the actual paths of the satellites at the station itself during on-site reconnaissance may reveal enough windows in the gaps between obstructions to collect sufficient data by strictly defining the observation sessions.

Planning Azimuth Marks

Azimuth marks are a common requirement in GPS projects. They are an accompaniment to static GPS stations when a client intends to use them to control subsequent conventional surveying work. Of course, the line between the station and the azimuth mark should be as long as convenience and the preservation of line of sight allows.

It is wise to take care that short baselines do not degrade the overall integrity of the project. Occupations of the station and its azimuth mark should be simultaneous for a direct measurement of the baseline between them. Both should also be tied to the larger network as independent stations. There should be two or more occupations of each station when the distance between them is less than 2 km.

While an alternative approach may be to derive the azimuth between a GPS station and its azimuth mark with an astronomic observation, it is important to remember that a small error, attributable to the deflection of the vertical, will be present in such an observation. The small angle between the plumb line and a normal to the ellipsoid at the station can either be ignored, because they are likely to be quite similar at both ends, or removed with a Laplace correction.

Obtaining Permissions

Another aspect of access can be considered when the project map finally shows all the pertinent points. Nothing can bring a well-planned survey to a halt faster than a locked gate, an irate landowner, or a government official who is convinced he should have been consulted previously. To the extent that it is possible from the available mapping, affected private landowners and government jurisdictions should be identified and contacted. Taking this precaution at the earliest stage of the survey planning can increase the chance that the sometimes long process of obtaining permissions, gate keys, badges, or other credentials has a better chance of completion before the survey begins.

On the other hand, any aspect of a GPS survey plan derived from examining mapping, virtual or hard copy, must be considered preliminary. Most features change with time, and even those that are relatively constant cannot be portrayed on a map with complete exactitude. Nevertheless, steps toward a coherent workable design can be taken using the information they provide.

SOME GPS SURVEY DESIGN FACTS

Though much of the preliminary work in producing the plan of a GPS survey is a matter of estimation, some hard facts must be considered, too. For example, the number of GPS receivers available for the work and the number of satellites above the observer's horizon at a given time in a given place are two ingredients that can be determined with some certainty.

SOFTWARE ASSISTANCE

Most GPS software packages provide users with routines that help them determine the satellite windows, i.e., the periods of time when the largest numbers of satellites are simultaneously available. Today, observers are virtually assured of 24 hour coverage; however, the mere presence of adequate satellites above an observer's horizon does not guarantee collection of sufficient data. Therefore, despite the virtual certainty that at least four satellites will be available, evaluation of their configuration as expressed in the position dilution of precision (PDOP) is still crucial in planning a GPS survey.

Position Dilution of Precision

The assessment of the productivity of a GPS survey almost always hinges, in part at least, on the length of the observation sessions required to satisfy the survey specifications. The determination of the session's duration depends on several particulars, such as the length of the baseline and the relative position, i.e., the geometry, of the satellites, among others.

Generally speaking, the larger the constellation of satellites, the better the available geometry, the lower the PDOP and the shorter the length of the session needed to achieve the required accuracy. For example, given six satellites and good geometry, baselines of 10 km or less might require a session of 45 min to 1 hour, whereas, under exactly the same conditions, a baseline over 20 km might require a session of 2 hours or more. Alternatively, 45 min of six-satellite data may be worth an hour of four-satellite data, depending on the arrangement of the satellites in the sky.

Stated another way, the GPS receiver's position is derived from the simultaneous solution of vectors between it and the constellation of satellites above the observer's horizon. The quality of that solution depends, in large part, on the distribution of those vectors. For example, any position determined when the satellites are crowded together in one part of the sky will be unreliable, because all the vectors will have virtually the same direction. Fortunately, a computer can predict such an unfavorable configuration if it is given the ephemeris of each satellite, the approximate position of the receiver, and the time of the planned observation. Provided with a forecast of a large PDOP, the GPS survey planner should consider an alternate observation plan.

When one satellite is directly above the receiver and three others are near the horizon and 120° in azimuth from one another, the arrangement is nearly ideal for a four-satellite constellation. The planner of the survey would be likely to consider such a window. However, more satellites would improve the resulting position even more, as long as they are well distributed in the sky above the receiver. In general, the more satellites, the better. For example, if the planner finds eight satellites will be above the horizon in the region where the work is to be done and the PDOP is below two, that window would be a likely candidate for observation.

There are other important considerations. The satellites are constantly moving in relation to the receiver and to each other. Because satellites rise and set, the PDOP is constantly changing. Within all this movement, the GPS survey designer must have some way of correlating the longest and most important baselines with the longest windows, the most satellites, and the lowest PDOP. Most GPS software packages, given a particular location and period of time, can provide illustrations of the satellite configuration.

Polar Plot

One such diagram is a plot of the satellite's tracks drawn on a graphical representation of the upper half of the celestial sphere with the observer's zenith at the center and perimeter circle as the horizon. The azimuths and elevations of the satellites above the specified mask angle are connected into arcs that represent the paths of all available satellites. The utility of this sort of drawing has lessened with the completion of the GPS constellation. In fact, there are so many satellites available that the picture can become quite crowded and difficult to decipher.

Another printout is a tabular list of the elevation and azimuth of each satellite at time intervals selected by the user.

An Example

The position of point Morant in Table 6.2 needed expression to the nearest minute only, a sufficient approximation for the purpose. The ephemeris data were 5 days old when the chart was generated by the computer, but the data were still an adequate representation of the satellite's movements to use in planning. The mask angle

TABLE 6.2
Satellite Azimuth and Elevation Table 1

Point: Morant
Date: Wed., Sept. 29, 2014
24 Satellites: 1 2 3 7 9 12 13 14 15 16 17 18 19 20 21 22 23 24 25 26 27 28 29 31
Sampling Rate: 10 minutes

Lat 36:45:0 N Lon 121:45:0 W
Mask Angle: 15 (deg)

Ephemeris 9/24/2014
Zone: Time Pacific Day (−7)

Time	El	Az	El	Az	El	Az	El	Az	El	Az	El	Az	El	Az	El	Az	PDOP
SV	2		16		18		19		27		28		29		31		
Constellation of 8 SV's																	
0:00	16	219	15	317	77	121	66	330	41	287	23	65	36	129	30	109	1.7
0:10	20	221	18	314	73	131	67	641	44	292	22	60	32	132	33	104	1.8
0:20	24	23	20	310	68	137	68	353	47	297	21	56	28	135	35	99	1.8
0:30	28	226	22	306	64	142	68	5	50	302	20	51	24	138	36	93	1.9
0:40	32	229	23	302	59	146	67	17	52	308	18	48	20	140	37	88	1.8
0:50	36	232	24	297	54	148	66	28	55	314	16	44	16	142	38	82	1.8
SV	2		16		18		19		27		31						
Constellation of 6 SV's																	
1:00	40	235	24	293	49	151	65	39	58	320	38	76					3.0
1:10	43	239	24	288	44	153	63	49	61	328	37	70					3.0
1:20	47	244	24	283	40	155	61	57	634	336	36	64					2.8
1:30	51	249	23	278	35	156	59	65	66	345	34	60					2.6

(Continued)

TABLE 6.2 (CONTINUED)
Satellite Azimuth and Elevation Table 1

Point: Morant   Lat 36:45:0 N Lon 121:45:0 W   Ephemeris 9/24/2014

Date: Wed., Sept. 29, 2014   Mask Angle: 15 (deg)   Zone: Time Pacific Day (−7)

24 Satellites: 1 2 3 7 9 12 13 14 15 16 17 18 19 20 21 22 23 24 25 26 27 28 29 31

Sampling Rate: 10 minutes

Time	El	Az	El	Az	El	Az	El	Az	El	Az	El	Az	El	Az	El	Az	PDOP
								Constellation of 7 SV's									
SV	2		7		16		18		19		27		31				
1:40	54	254	16	186	22	273	30	157	56	73	68	356	32	55			2.3
1:50	57	260	21	186	20	269	26	158	53	79	70	9	29	51			2.2
2:00	60	268	25	186	19	264	22	159	50	85	71	23	26	48			2.0
								Constellation of 8 SV's									
SV	2		7		16		18		19		26		27		31		
2:10	66	276	30	185	16	260	17	160	47	91	15	319	71	38	23	45	1.7

was specified at 15°, so the program would consider a satellite set when it moved below that elevation angle. The zone time was Pacific Daylight Time, 7 hours behind Coordinated Universal Time. The full constellation provided 24 healthy satellites, and the sampling rate indicated that the azimuth and elevation of those above the mask angle would be shown every 10 min.

At 0:00 hour, satellite pseudorandom noise (PRN) 2 could be found on an azimuth of 219° and an elevation of 16° above the horizon by an observer at 36°45′Nφ and 121°45′Wλ. Table 6.2 indicates that PRN 2 was rising and got continually higher in the sky for the 2 hours and 10 min covered by the chart. The satellite PRN 16 was also rising at 0:00 but reached its maximum altitude at about 1:10 and began to set. Unlike PRN 2, PRN 16 was not tabulated in the same row throughout the chart. It was supplanted when PRN 7 rose above the mask angle and PRN 16 shifted one column to the right. The same may be said of PRN 18 and PRN 19. Both of these satellites began high in the sky, unlike PRN 28 and PRN 29. They were just above 15° and setting when the table began and set after approximately 1 hour of availability. They would not have been seen again at this location for about 12 hours.

This chart indicated changes in the available constellation from eight space vehicles, between 0:00 and 0:50, six between 1:00 and 1:30, seven from 1:40 to 2:00, and back to eight at 2:10. The constellation never dipped below the minimum of four satellites, and the PDOP was good throughout. The PDOP varied between a low of 1.7 and a high of 3.0. Over the interval covered by the table, the PDOP never reached the unsatisfactory level of 5 or 6, which is when a planner should avoid observation.

Choosing the Window

Using this chart, the GPS survey designer might well have concluded that the best available window was the first. There was nearly an hour of eight-satellite data with a PDOP below 2. However, the data indicated that good observations could be made at any time covered here, except for one thing: it was the middle of the night.

Ionospheric Delay

It is worth noting that the ionospheric error is usually smaller after sundown. In fact, the FGDC specified dual-frequency receivers for daylight observations for the achievement of the highest accuracies, due, in part, to the increased ionospheric delay during those hours.

Table 6.3 shows data from later in the day. It covers a period of two hours when a constellation of five and six satellites was always available. However, through the first hour, from 6:30 to 7:30, the PDOP hovered around 5 and 6. For the first half of that hour, four of the satellites (PRN 9, PRN 12, PRN 13, and PRN 24) were all near the same elevation. During the same period, PRN 9 and PRN 12 were only approximately 50° apart in azimuth, as well. Even though a sufficient constellation of satellites was constantly available, the survey designer may well have considered only the last 30 to 50 min of the time covered by this chart as suitable for observation.

There is one caution, however. Azimuth-elevation tables are a convenient tool in the division of the observing day into sessions, but it should not be taken for granted

TABLE 6.3

Satellite Azimuth and Elevation Table 2

Point: Morant
Date: Wed., Sept. 29, 2014
24 Satellites: 1 2 3 7 9 12 13 14 15 16 17 18 19 20 21 22 23 24 25 26 27 28 29 31
Sampling Rate: 10 minutes

Lat 36:45:0 N Lon 121:45:0 W
Mask Angle: 15 (deg)

Ephemeris 9/24/2014
Zone: Time Pacific Day (−7)

Time	El	Az	El	Az	El	Az	El	Az	El	Az	El	Az	El	Az	PDOP
SV	7		9		12		13		24						
								Constellation of 5 SV's							
6:30	28	54	60	271	61	319	62	15	48	177						6.3
6:40	24	57	60	261	66	314	57	19	53	176						6.0
6:50	21	60	59	252	70	305	53	22	58	175						5.3
7:00	18	62	57	243	73	292	49	25	63	172						4.6
SV	9		12		13		20		24							
								Constellation of 5 SV's							
7:10	54	235	74	274	44	28	16	308	68	169						4.8
7:20	51	229	74	255	40	32	20	310	72	163						5.7
7:30	47	224	72	238	37	35	23	311	77	153						5.1
7:40	43	219	68	226	33	38	27	313	80	134						4.0

(Continued)

TABLE 6.3 (CONTINUED)
Satellite Azimuth and Elevation Table 2

Point: Morant   Lat 36:45:0 N Lon 121:45:0 W   Ephemeris 9/24/2014

Date: Wed., Sept. 29, 2014   Mask Angle: 15 (deg)   Zone: Time Pacific Day (−7)

24 Satellites: 1 2 3 7 9 12 13 14 15 16 17 18 19 20 21 22 23 24 25 26 27 28 29 31

Sampling Rate: 10 minutes

Time	El	Az	El	Az	El	Az	El	Az	El	Az	El	Az	El	Az	PDOP
SV	9		12		13		16		20		24				
											Constellation of 6 SV's				
7:50	39	215	64	218	29	41	16	149	31	314	81	102			2.1
8:00	35	212	59	213	26	45	19	146	36	314	80	73			2.3
8:10	31	209	54	209	23	48	23	143	40	315	76	57			2.4
8:20	27	207	49	206	19	52	27	140	44	314	72	49			2.5
8:30	23	204	44	204	16	55	30	137	48	314	67	45			2.5

that every satellite listed is healthy and in service. For the actual availability of satellites and an update on atmospheric conditions, it is always wise to check the U.S. Coast Guard's website (http://www.navcen.uscg.gov/?Do=constellationstatus) before and after a project. In the planning stage, the diligence can prevent creation of a design dependent on satellites that prove unavailable. Similarly, after the field work is completed, it can prevent inclusion of unhealthy data in the post-processing.

Supposing that the period from 7:40 to 8:30 was found to be a good window, the planner may have regarded it as a single 50 min session, or divided it into shorter sessions. One aspect of that decision was probably the length of the baseline in question. In static GPS, a long line of 30 km may require 50 min of six-satellite data, but a short line of 3 km may not. Another consideration was probably the approximation of the time necessary to move from one station to another.

Naming the Variables

The next step in the static GPS survey design is drawing the preliminary plan of the baselines on the project map. Once some idea of the configuration of the baselines has been established, an observation schedule can be organized. Toward that end, the FGCC developed a set of formulas, which will be used here.

For illustration, suppose that the project map (Figure 6.9) includes horizontal control, vertical control, and project points for a planned GPS network. They will be symbolized by m. There are four multifrequency GPS receivers available for this project. They will be symbolized by r. There will be five observation sessions each day during the project. They will be symbolized by d. To summarize,

$$m = \text{total number of stations (existing and new)} = 14$$

$$d = \text{number of possible observing sessions per observing day} = 5$$

$$r = \text{number of receivers} = 4 \text{ dual frequency}$$

The design developed from this map must be preliminary. The session for each day of observation will depend on the success of the work the day before. Please recall that the plan must be provisional until the baseline lengths, the obstructions at the observation sites, the transportation difficulties, the ionospheric disturbances, and the satellite geometry are actually known. Those questions can only be answered during the reconnaissance and the observations that follow. Even though these equivocations apply, the next step is to draw the baseline's measurement plan.

Compatible Receivers

Relative static positioning, just as all the subsequent surveying methods discussed here, involves several receivers occupying many sites. Problems can be avoided as long as the receivers on a project are compatible. For example, it is helpful if they have the same number of channels and signal processing techniques, and the Receiver Independent Exchange (RINEX) format, developed by the Astronomical Institute, allows different receivers and post-processing software to work together. Almost all GPS processing software will output RINEX files.

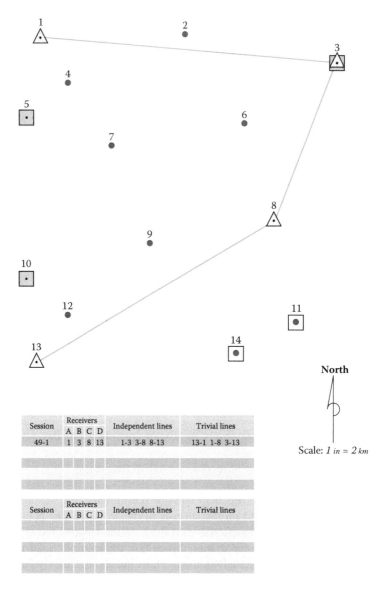

FIGURE 6.9   Drawing the baselines.

Receiver Capabilities and Baseline Length

The number and type of channels available to a receiver are considerations because, generally speaking, the more satellites the receiver can track continuously, the better. Another factor that ought to be weighed is whether a receiver has single or multi-frequency capability. Single-frequency receivers are best applied to relatively short baselines, say, under 25 km. The biases at one end of such a vector are likely to be similar to those at the other. Multifrequency receivers, however, have the capability to nearly eliminate the effects of ionospheric refraction and can handle longer baselines.

DRAWING THE BASELINES

HORIZONTAL CONTROL

A good rule of thumb is to verify the integrity of the horizontal control by observing baselines between these stations first. Vectors can be used to both corroborate the accuracy of the published coordinates and later to resolve the scale, shift, and rotation parameters between the control positions and the new network that will be determined by GPS.

These baselines are frequently the longest in the project, and there is an added benefit to measuring them first. By processing a portion of the data collected on the longest baselines early in the project, the most appropriate length of the subsequent sessions can be found.

This test may allow improvement in the productivity on the job without erosion of the final positions.

Julian Day in Naming Sessions

The table at the bottom of Figure 6.9 indicates that the name of the first session connecting the horizontal control is 49-1. The date of the planned session is given in the Julian system. Taken most literally, Julian dates are counted from January 1, 4713 B.C. However, most practitioners of GPS use the term to mean the day of the current year measured consecutively from January 1.

Under this construction, because there are 31 days in January, Julian day 49 is February 18 of the current year. The designation 49-1 means that this is to be the first session on that day. Some prefer to use letters to distinguish the session. In that case, the label would be 49-A.

Independent Lines

This project will be done with four receivers. The table shows that receiver A will occupy point 1; receiver B, point 3; receiver C, point 8; and receiver D, point 13 in the first session. However, the illustration shows only three of the possible six base lines that will be produced by this arrangement. Only the independent, also known as nontrivial, lines are shown on the map. The three lines that are not drawn are called trivial, and are also known as dependent lines. This idea is based on restricting the use of the lines created in each observing session to the absolute minimum needed to produce a unique solution.

Whenever four receivers are used, six lines are created. However, any three of those lines will fully define the position of each occupied station in relation to the others in the session.

Therefore, the user can consider any three of the six lines independent, but once the decision is made, only those three baselines are included in the network. The remaining baselines are then considered trivial and discarded. In practice, the three shortest lines in a four-receiver session are almost always deemed the independent vectors, and the three longest lines are eliminated as trivial or dependent. That is the case with the session illustrated.

Where r is the number of receivers, every session yields $r - 1$ independent baselines. For example, four receivers used in 10 sessions would produce 30 independent baselines. It cannot be said that the shortest lines are always chosen to be the independent lines. Sometimes there are reasons to reject one of the shorter vectors owing to incomplete data, cycle slips, multipath, or some other weakness in the measurements. Before such decisions can be made, each session will require analysis after the data have actually been collected. In the planning stage, it is best to consider the shortest vectors as the independent lines.

Another aspect of the distinction between independent and trivial lines involves the concept of error of closure or loop closure. Loop closure is a procedure by which the internal consistency of a GPS network is discovered. A series of baseline vector components from more than one GPS session, forming a loop or closed figure, is added together. The closure error is the ratio of the length of the line representing the combined errors of all the vector's components to the length of the perimeter of the figure. Any loop closures that only use baselines derived from a single common GPS session will yield an apparent error of zero, because they are derived from the same simultaneous observations. For example, all the baselines between the four receivers in session 49-1 of the illustrated project will be based on ranges to the same GPS satellites over the same period of time. Therefore, the trivial lines of 13-1, 1-8, and 3-13 will be derived from the same information used to determine the independent lines of 1-3, 3-8, and 8-13. It follows that, if the fourth line from station 13 to station 1 were included to close the figure of the illustrated session, the error of closure would be zero. The same may be said of the inclusion of any of the trivial lines. Their addition cannot add any redundancy or any geometric strength to the lines of the session because they are all derived from the same data. If redundancy cannot be added to a GPS session by including any more than the minimum number of independent lines, how can the baselines be checked? Where does redundancy in GPS work come from?

Redundancy

If only two receivers were used to complete the illustrated project, there would be no trivial lines and it might seem there would be no redundancy at all. However, to connect every station with its closest neighbor, each station would have to be occupied at least twice, and each time during a different session. For example, with receiver A on station 1 and receiver B on station 2, the first session could establish the baseline between them. The second session could then be used to measure the baseline between station 1 and station 4. It would certainly be possible to simply move receiver B to station 4 and leave receiver A undisturbed on station 1. However, some redundancy could be added to the work if receiver A were reset. If it were recentered, replumbed, and its *heights of measurement* (H.I.) remeasured, some check on both of its occupations on station 1 would be possible when the network was completed. Under this scheme, a loop closure at the end of the project would have some meaning.

If one were to use such a scheme on the illustrated project and connect into one loop all of the 14 baselines determined by the 14 two-receiver sessions, the resulting error of closure would be useful. It could be used to detect blunders in the work, such as mis-measured heights of instruments. Such a loop would include many different sessions. The ranges between the satellites and the receivers defining the baselines in such a circuit would be from different constellations at different times. However, if it were possible to occupy all 14 stations in the illustrated project with 14 different receivers simultaneously and do the entire survey in one session, a loop closure would be absolutely meaningless.

In the real world, such a project is not done with 14 receivers nor with 2 receivers, but with 3, 4, or 5. The achievement of redundancy takes a middle road. The number of independent occupations is still an important source of redundancy. In the two-receiver arrangement every line can be independent, but that is not the case when a project is done with any larger number of receivers. As soon as three or more receivers are considered, the discussion of redundant measurement must be restricted to independent baselines, excluding trivial lines.

Redundancy is then partly defined by the number of independent baselines that are measured more than once, as well as by the percentage of stations that are occupied more than once. While it is not possible to repeat a baseline without reoccupying its endpoints, it is possible to reoccupy a large percentage of the stations in a project without repeating a single baseline. These two aspects of redundancy in GPS (i.e., the repetition of independent baselines and the reoccupation of stations) are somewhat separate.

Figure 6.10 shows one of the many possible approaches to setting up the baselines for this particular GPS project. The survey design calls for the horizontal control to be occupied in session 49-1. It is to be followed by measurements between two control stations and the nearest adjacent project points in session 49-2. As shown in the table at the bottom of Figure 6.10, there will be redundant occupations on stations 1 and 3. Even though the same receivers will occupy those points, their operators will be instructed to reset them at different H.I.'s for the new session. A better, but probably less efficient, plan would be to occupy these stations with different receivers than were used in the first session.

Forming Loops

As the baselines are drawn on the project map for a static GPS survey, or any GPS work where accuracy is the primary consideration, the designer should remember that part of their effectiveness depends on the formation of complete geometric figures. When the project is completed, these independent vectors should be capable of formation into closed loops that incorporate baselines from two to four different sessions. In the illustrated baseline plan, no loop contains more than ten vectors, no loop is more than 100 km long, and every observed baseline will have a place in a closed loop.

Finding the Number of Sessions

The illustrated survey design calls for 10 sessions, but the calculation does not include human error, equipment breakdown, and other unforeseeable difficulties. It would be impractical to presume a completely trouble-free project. The FGCC proposed the following formula for arriving at a more realistic estimate:

$$s = \frac{(m \cdot n)}{r} + \frac{(m \cdot n)(p-1)}{r} + k \cdot m$$

where
    s = number of observing sessions
    r = number of receivers
    m = total number of stations involved

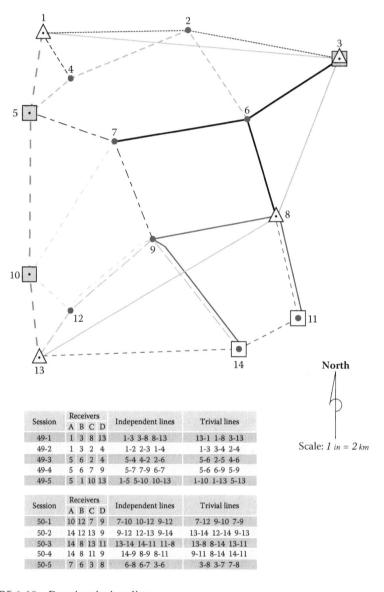

Session	Receivers A B C D	Independent lines	Trivial lines
49-1	1 3 8 13	1-3 3-8 8-13	13-1 1-8 3-13
49-2	1 3 2 4	1-2 2-3 1-4	1-3 3-4 2-4
49-3	5 6 2 4	5-4 4-2 2-6	5-6 2-5 4-6
49-4	5 6 7 9	5-7 7-9 6-7	5-6 6-9 5-9
49-5	5 1 10 13	1-5 5-10 10-13	1-10 1-13 5-13

Session	Receivers A B C D	Independent lines	Trivial lines
50-1	10 12 7 9	7-10 10-12 9-12	7-12 9-10 7-9
50-2	14 12 13 9	9-12 12-13 9-14	13-14 12-14 9-13
50-3	14 8 13 11	13-14 14-11 11-8	13-8 8-14 13-11
50-4	14 8 11 9	14-9 8-9 8-11	9-11 8-14 14-11
50-5	7 6 3 8	6-8 6-7 3-6	3-8 3-7 7-8

North

Scale: 1 *in* $= 2$ *km*

FIGURE 6.10 Drawing the baselines.

The values n, p, and k require a bit more explanation. The variable n is a representation of the level of redundancy that has been built into the network, based on the number of occupations on each station. The illustrated survey design includes more than two occupations on all but 4 of the 14 stations in the network. In fact, 10 of the 14 positions will be visited three or four times in the course of the survey. There are a total of 40 occupations by the four receivers in the 10 planned sessions. By dividing 40 occupations by 14 stations, it can be found that each station will be visited an average of 2.857 times. Therefore, the planned redundancy represented by factor n is equal to 2.857 in this project.

The experience of a firm is symbolized by the variable p in the formula. The division of the final number of actual sessions required to complete past projects by the initial estimation yields a ratio that can be used to improve future predictions. That ratio is the production factor p. A typical production factor is 1.1.

A safety factor of 0.1, known as k, is recommended for GPS projects within 100 km of a company's home base. Beyond that radius, an increase to 0.2 is advised.

The substitution of the appropriate quantities for the illustrated project increases the prediction of the number of observation sessions required for its completion:

$$s = \frac{(m \cdot n)}{r} + \frac{(m \cdot n)(p-1)}{r} + k \cdot m$$

$$s = \frac{(14)(2.857)}{4} + \frac{(14)(2.857)(1.1-1)}{4} + (0.2)(14)$$

$$s = \frac{(40)}{4} + \frac{(4)}{4} + 2.8$$

$$s = 10 + 1 + 2.8$$

$$s = 14 \text{ sessions (rounded to the nearest integer)}$$

In other words, the 2-day, 10-session schedule is a minimum period for the baseline plan drawn on the project map. A more realistic estimate of the observation schedule includes 14 sessions. It is also important to keep in mind that the observation schedule does not include time for on-site reconnaissance.

Ties to the Vertical Control

The ties from the vertical control stations to the overall network are usually not handled by the same methods used with the horizontal control. The first session of the illustrated project was devoted to occupation of all the horizontal control stations. There is no similar method with the vertical control stations. First, the geoidal undulation would be indistinguishable from baseline measurement error. Second, the primary objective in vertical control is for each station to be adequately tied to its closest neighbor in the network.

If a benchmark can serve as a project point, it is nearly always advisable to use it, as was done with stations 11 and 14 in the illustrated project. A conventional level circuit can often be used to transfer a reliable orthometric elevation from vertical control station to a nearby project point.

STATIC GPS CONTROL OBSERVATIONS

The prospects for the success of a GPS project are directly proportional to the quality and training of the people doing it. The handling of the equipment, the on-site reconnaissance, the creation of field logs, and the inevitable last-minute adjustments to the survey design all depend on the training of the personnel involved for their success. There are those who say the operation of GPS receivers no longer requires highly qualified survey personnel. That might be true if effective GPS surveying needed only the pushing of the appropriate buttons at the appropriate time. In fact, when all goes as planned, it may appear to the uninitiated that GPS has made experienced field surveyors obsolete. However, when the unavoidable breakdowns in planning or equipment occur, the capable people, who seemed so superfluous moments before, suddenly become indispensable.

EQUIPMENT

Conventional Equipment

Most GPS projects require conventional surveying equipment for spirit-leveling circuits, offsetting horizontal control stations, and monumenting project points, among other things. It is perhaps a bit ironic that this most advanced surveying method also frequently has need of the most basic equipment. The use of brush hooks, machetes, axes, and so forth, can sometimes salvage an otherwise unusable position by removing overhead obstacles. Another strategy for overcoming such hindrances has been developed using various types of survey masts to elevate a separate GPS antenna above the obstructing canopy.

Flagging, paint, and the various techniques of marking that surveyors have developed over the years are still a necessity in GPS work. The pressure of working in unfamiliar terrain is often combined with urgency. Even though there is usually not a moment to spare in moving from station to station, a GPS surveyor frequently does not have the benefit of having visited the particular points before. In such situations, the clear marking of both the route and the station during reconnaissance is vital.

Despite the best route marking, a surveyor may not be able to reach the planned station, or, has arrived, finds some new obstacle or unanticipated problem that can only be solved by marking and occupying an impromptu offset position for a session. A hammer, nails, shiners, paint, and so forth, are essential in such situations.

In short, the full range of conventional surveying equipment and expertise have a place in GPS. For some, their role may be more abbreviated than it was formerly, but one element that can never be outdated is good judgment.

Safety Equipment

The high-visibility vests, cones, lights, flagmen, and signs needed for traffic control cannot be neglected in GPS work. Unlike conventional surveying operations, GPS observations are not deterred by harsh weather. Occupying a control station in a highway is dangerous enough under the best of conditions, but in the midst of a rainstorm, fog, or blizzard, it can be absolute folly without the proper precautions.

And any time and trouble taken to avoid infraction of the local regulations regarding traffic management will be compensated by an uninterrupted observation schedule.

Weather conditions also affect travel between the stations of the survey, both in vehicles and on foot. Equipment and plans to deal with emergencies should be part of any GPS project. First aid kits, fire extinguishers, and the usual safety equipment are necessary. Training in safety procedures can be an extraordinary benefit, but perhaps the most important capability in an emergency is communication.

Communications

Whether the equipment is handheld or vehicle mounted, two-way radios and cell phones are used in most GPS operations. However, the line of sight that is no longer necessary for the surveying measurements in GPS is sorely missed in the effort to maintain clear radio contact between the receiver operators. A radio link between surveyors can increase the efficiency and safety of a GPS project, but it is particularly valuable when last minute changes in the observation schedule are necessary. When an observer is unable to reach a station or a receiver suddenly becomes inoperable, unless adjustments to the schedule can be made quickly, each end of all of the lines into the missed station will require re-observation.

The success of static GPS hinges on all receivers collecting their data simultaneously. However, it is more and more difficult to ensure reliable communication between receiver operators in geodetic surveys, especially as their lines grow longer.

High-wattage, private-line FM radios are quite useful when line of sight is available between them or when a repeater is available. The use of cell phones may eliminate the communication problem in some areas but probably not in remote locations.

Despite the limitations of the systems available at the moment, achievement of the best possible communication between surveyors on a GPS project pays dividends in the long run.

GPS Equipment

Most GPS receivers capable of geodetic accuracy are designed to be mounted on a tripod, usually with a tribrach and adaptor. However, there is a trend toward bipod- or range-pole-mounted antennas. An advantage of these devices is that they ensure a constant height of the antenna above the station. The mis-measured height of the antenna above the mark is probably the most pervasive and frequent blunder in GPS control surveying.

The tape or rod used to measure the height of the antenna is sometimes built into the receiver and, sometimes a separate device. It is important that the H.I. be measured accurately and consistently in both feet and meters, without merely converting from one to the other mathematically. It is also important that the value be recorded in the field notes and, where possible, also entered into the receiver itself.

Where tribrachs are used to mount the antenna, the tribrach's optical centering should be checked and calibrated. It is critical that the effort to perform GPS surveys to an accuracy of centimeters not be frustrated by inaccurate centering or H.I. measurement. Because many systems measure the height of the antenna to the edge of the ground plane or to the exterior of the receiver itself, the calibration of the tribrach affects both the centering and the H.I. measurement. The resetting of a receiver that occupies the same station in consecutive sessions is an important source of redundancy for many kinds of GPS networks. However, integrity can only be added if the tribrach has been accurately calibrated.

The checking of the carrier phase receivers themselves is also critical to the control of errors in a GPS survey, especially when different receivers or different models of antennas are to be used on the same work. The zero baseline test is a method that may be used to fulfill equipment calibration specifications where a three-dimensional test network of sufficient accuracy is not available. As a matter of fact, the simplicity of this test is an advantage. It is not dependent on special software or a test network. This test can also be used to separate receiver difficulties from antenna errors.

Two or more receivers are connected to one antenna with a *signal* or *antenna splitter*. The antenna splitter can be purchased from specialty electronics shops and is also available online. An observation is done with the divided signal from the single antenna reaching both receivers simultaneously. Because the receivers are sharing the same antenna, satellite clock biases, ephemeris errors, atmospheric biases, and multipath are all canceled. In the absence of multipath, the only remaining errors are attributable to random noise and receiver biases. The success of this test depends on the signal from one antenna reaching both receivers, but the current from only one receiver can be allowed to power the antenna. This test checks not only the precision of the receiver measurements but also the processing software. The results of the test should show a baseline of only a few millimeters. Information is also available on National Geodetic Survey (NGS) calibration baselines throughout the United States.

Auxiliary Equipment

Tools to repair the ends of connecting cables, a simple pencil eraser to clean the contacts of circuit boards, or any of a number of small implements have saved more than one GPS observation session from failure. Experience has shown that GPS surveying requires at least as much resourcefulness, if not more, than conventional surveying.

The health of the batteries is a constant concern in GPS. There is simply nothing to be done when a receiver's battery is drained but to resume power as soon as possible. A backup power source is essential. Cables to connect a vehicle battery, an extra fully-charged battery unit, or both should be immediately available to every receiver operator.

Information

The information every GPS observer carries throughout a project ought to include emergency phone numbers; the names, addresses, and phone numbers of relevant property owners; and the combinations to necessary locks. Each member of the team should also have a copy of the project map, any other maps that are needed to clarify position or access, and, perhaps most important of all, the updated observation schedule.

The observation schedule for static GPS work will be revised daily based on actual production (see Table 6.4). It should specify the start-stop times and station for all the personnel during each session of the upcoming day. In this way, the schedule will not only serve to inform every receiver operator of his or her own expected occupations, but those of every other member of the project as well. This knowledge is most useful when a sudden revision requires observers to meet or replace one another.

TABLE 6.4
Observation Schedule

	Session 1 Start 7:10 Stop 8:10	8:10 to 8:40	Session 2 Start 8:40 Stop 9:50	9:50 to 10:15	Session 3 Start 10:15 Stop 11:15	11:15 to 11:30	Session 4 Start 11:30 Stop 12:30	12:30 to 14:00	Session 5 Start 14:00 Stop 15:00
Svs PRNs	9,12,13, 16,20,24		3,12,13, 16,20,24		3,12,13,16, 17,20,24		3,16,17,20, 22,23,26		1,3,17,21, 23,26,28
Receiver A Dan H.	Station 1 NGS horiz. control	Re-set	Station 1 NGS horiz. control	Move	Station 5 NGS benchmark	Re-set	Station 5 NGS benchmark	Re-set	Station 5 NGS benchmark
Receiver B Scott G.	Station 3 NGS V&H control	Re-set	Station 3 NGS V&H control	Move	Station 6 Project point	Re-set	Station 6 Project point	Move	Station 1 NGS horiz. control
Receiver C Dewey A.	Station 8 NGS horiz. control	Move	Station 2 Project point	Re-Set	Station 2 Project point	Move	Station 7 Project point	Move	Station 10 NGS benchmark
Receiver D Cindy E.	Station 13 NGS horiz. control	Move	Station 4 Project point	Re-Set	Station 4 Project point	Move	Station 9 Project point	Move	Station 13 NGS horiz. control

STATION DATA SHEET

The principles of good field notes have a long tradition in land surveying, and they will continue to have validity for some time to come. In GPS, the ensuing paper trail will not only fill subsequent archives; it has immediate utility. For example, the station data sheet is often an important bridge between on-site reconnaissance and the actual occupation of a monument.

Though every organization develops its own unique system of handling its field records, most have some form of the station data sheet. The document illustrated in Figure 6.11 is merely one possible arrangement of the information needed to recover the station.

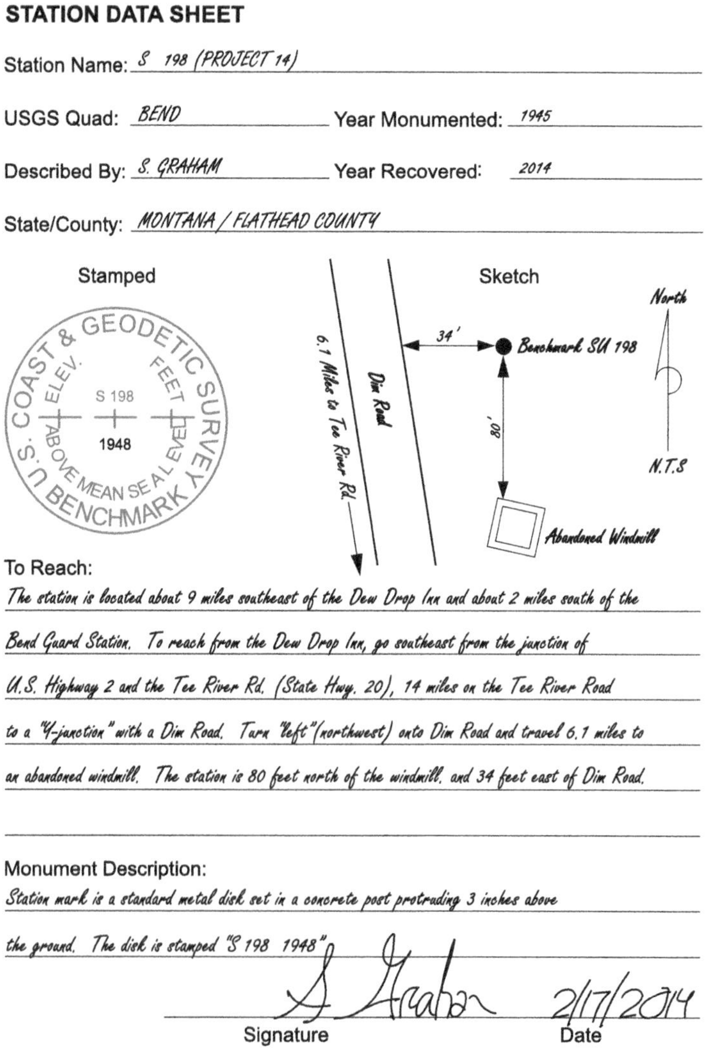

STATION DATA SHEET

Station Name: *S 198 (PROJECT 14)*

USGS Quad: *BEND*   Year Monumented: *1945*

Described By: *S. GRAHAM*   Year Recovered: *2014*

State/County: *MONTANA / FLATHEAD COUNTY*

To Reach:
The station is located about 9 miles southeast of the Dew Drop Inn and about 2 miles south of the

Bend Guard Station. To reach from the Dew Drop Inn, go southeast from the junction of

U.S. Highway 2 and the Tee River Rd. (State Hwy. 20), 14 miles on the Tee River Road

to a "Y-junction" with a Dim Road. Turn "left" (northwest) onto Dim Road and travel 6.1 miles to

an abandoned windmill. The station is 80 feet north of the windmill, and 34 feet east of Dim Road.

Monument Description:
Station mark is a standard metal disk set in a concrete post protruding 3 inches above

the ground. The disk is stamped "S 198 1948".

Signature   Date

FIGURE 6.11   Station data sheet.

The station data sheet can be prepared at any period of the project, but perhaps the most usual times are during the reconnaissance of existing control or immediately after the monumentation of a new project point.

Neatness and clarity, always paramount virtues of good field notes, are of particular interest when the station data sheet is to be later included in the final report to the client. The overriding principle in drafting a station data sheet is to guide succeeding visitors to the station without ambiguity. A GPS surveyor on the way to observe the position for the first time may be the initial user of a station data sheet. A poorly written document could void an entire session if the observer is unable to locate the monument. A client, later struggling to find a particular monument with an inadequate data sheet, may ultimately question the value of more than the field notes.

Station Name

The station name fills the first blank on the illustrated data sheet. Two names for a single monument are far from unusual. In this case the vertical control station, officially named S 198, is also serving as a project point, number 14, but two names purporting to represent the same position can present a difficulty. For example, when a horizontal control station is remonumented, a number 2 is sometimes added to the original name of the station and it can be confusing. For example, it can be easy to mistake station "Thornton 2" with an original station named "Thornton" that no longer exists. Both stations may still have a place in the published record, but with slightly different coordinates. Another unfortunate misunderstanding can occur when inexperienced field personnel mistake a reference mark for the actual station itself. Taking rubbings and/or close-up photographs are widely recommended to avoid such blunders regarding stations names or authority.

Rubbings

The illustrated station data sheet provides an area to accommodate a rubbing. With the paper held on top of the monument's disk, a pencil is run over it in a zigzag pattern producing a positive image of the stamping. This method is a bit more awkward than simply copying the information from the disk onto the data sheet, but it does have the advantage of ensuring the station was actually visited and that the stamping was faithfully recorded. Such rubbings or close-up photographs are often required.

Photographs

The use of photographs is growing as a help for the perpetuation of monuments. It can be convenient to photograph the area around the mark as well as the monument itself. These exposures can be correlated with a sketch of the area. Such a sketch can show the spot where the photographer stood and the directions toward which the pictures were taken. The photographs can then provide valuable information in locating monuments, even if they are later obscured. Still, the traditional ties to prominent features in the area around the mark are the primary agent of their recovery.

Quad Sheet Name

Providing the name of the appropriate state, county, and USGS quad sheet helps to correlate the station data sheet with the project map. The year the mark was monumented, the monument description, the station name, and the "to-reach" description all help to associate the information with the correct official control data sheet and, most importantly, the correct station coordinates.

To-Reach Descriptions

When driving or walking to a position can be aided by computerized turn-by-turn navigation, it is a great tool that may make writing to-reach descriptions unnecessary. However, GPS control work is often done in areas where the roadway mapping in such navigations aids is inadequate. In those situations, the description of the route to the station is one of the most critical documents written during the reconnaissance. Even though it is difficult to prepare the information in unfamiliar territory and although every situation is somewhat different, there are some guidelines to be followed. It is best to begin with the general location of the station with respect to easily found local features.

The description in Figure 6.11 relies on a road junction, guard station, and local business. After defining the general location of the monument, the description should recount directions for reaching the station. Starting from a prominent location, the directions should adequately describe the roads and junctions. Where the route is difficult or confusing, the reconnaissance team should not only describe the junctions and turns needed to reach a station; it is wise to also mark them with lath and flagging, when possible. It is also a good idea to note gates. Even if they are open during reconnaissance, they may be locked later. When turns are called for, it is best to describe not only the direction of the turn, but the new course, too. For example, in the description in Figure 6.11 the turn onto the dim road from the Tee River Road is described to the "left (northwest)." Roads and highways should carry both local names and designations found on standard highway maps. For example in Figure 6.11, Tee River Road is also described as State Highway 20.

The "to-reach" description should certainly state the mileages as well as the travel times where they are appropriate, particularly where packing-in is required. Land ownership, especially if the owner's consent is required for access, should be mentioned. The reconnaissance party should obtain the permission to enter private property and should inform the GPS observer of any conditions of that entry. Alternate routes should be described where they may become necessary. It is also best to make special mention of any route that is likely to be difficult in inclement weather.

Where helicopter access is anticipated, information about the duration of flights from point to point, the distance of landing sites from the station, and flight time to fuel supplies should be included on the station data sheet.

Flagging and Describing the Monument

Flagging the station during reconnaissance may help the observer find the mark more quickly. On the station data sheet, the detailed description of the location of the station with respect to roads, fence lines, buildings, trees, and any other conspicuous features should include measured distances and directions. A clear description of the monument itself is important. It is wise to also show and describe any nearby marks, such as reference marks, that may be mistaken for the station or aid in its recovery. The name of the preparer, a signature, and the date round out the initial documentation of a GPS station.

VISIBILITY DIAGRAMS

Obstructions above the mask angle of a GPS receiver must be taken into account in finalizing the observation schedule. A station that is blocked to some degree is not necessarily unusable, but its inclusion in any particular session is probably contingent on the position of the specific satellites involved.

An Example

The diagram in Figure 6.12 is widely used to record such obstructions during reconnaissance. It is known as a *station visibility diagram*, a *polar plot*, or a *skyplot*. The concentric circles are meant to indicate 10° increments along the upper half of the

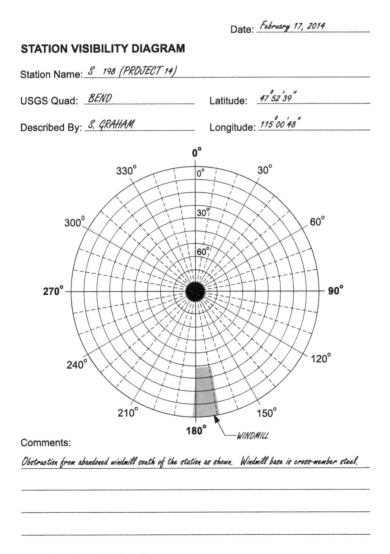

Date: *February 17, 2014*

STATION VISIBILITY DIAGRAM

Station Name: *S 198 (PROJECT 14)*

USGS Quad: *BEND*   Latitude: *47° 52' 39"*

Described By: *S. GRAHAM*   Longitude: *115° 00' 48"*

Comments:

Obstruction from abandoned windmill south of the station as shown. Windmill base is cross-member steel.

FIGURE 6.12   Station visibility diagram.

celestial sphere, from the observer's horizon at 0° on the perimeter, to the observer's zenith at 90° in the center. The hemisphere is cut by the observer's meridian, shown as a line from 0° in the north to 180° in the south.

The prime vertical is signified as the line from 90° in the east to 270° in the west. The other numbers and solid lines radiating from the center, every 30° around the perimeter, are azimuths from north and are augmented by dashed lines every 10°.

Drawing Obstructions

Using a compass and a clinometer, a member of the reconnaissance team can fully describe possible obstructions of the satellite's signals on a visibility diagram. By standing at the station mark and measuring the azimuth and vertical angle of points outlining the obstruction, the observer can plot the object on the visibility diagram. For example, a windmill base is shown in Figure 6.12 as a cross-hatched figure. It has been drawn from the observer's horizon up to 37° in vertical angle from 168°, to about 182° in azimuth at its widest point. This description by approximate angular values is entirely adequate for determining when particular satellites may be blocked at this station.

For example, suppose a 1 hour session from 9:10 to 10:10, illustrated in Table 6.5, was under consideration for the observation on station S 198. The station visibility chart might motivate a careful look at space vehicle (SV) PRN 16. Twenty minutes into the anticipated session, at 9:30 SV 16 has just risen above the 15° mask angle. Under normal circumstances, it would be available at station S 198, but it appears from the polar plot that the windmill will block its signals from reaching the receiver. In fact, the signals from SV 16 will apparently not reach station S 198 until sometime after the end of the session at 10:10.

Working around Obstructions

Under the circumstances, some consideration might be given to observing station S 198 during a session when none of the satellites would be blocked. However, the 9:10 to 10:10 session may be adequate after all. Even if SV 16 is completely blocked, the remaining five satellites will be unobstructed and the constellation still will have a relatively low position dilution of precision (PDOP). Still, the analysis must be carried to other stations that will be occupied during the same session. The success of the measurement of any baseline depends on common observations at both ends of the line. Therefore, if the signals from SV 16 are garbled or blocked from station S 198, any information collected during the same session from that satellite at the other end of a line that includes S 198 will be useless in processing the vector between those two stations.

The material of the base of the abandoned windmill has been described on the visibility diagram as cross-membered steel, so it is possible that the signal from SV 16 will not be entirely obstructed during the whole session. There may actually be more concern of multipath interference from the structure than that of signal availability. One strategy for handling the situation might be to program the receiver at S 198 to ignore the signal from SV 16 completely if the particular receiver allows it.

TABLE 6.5
Satellite Azimuth and Elevation Table 3

Time	SV	El	Az	El	Az	El	Az	El	Az	El	Az	El	Az	El	Az	PDOP
		Constellation of 5 SV's														
SV		3		12		13		20		24						
8:50		54	235	74	274	44	28	16	308	68	169					4.8
9:00		51	229	74	255	40	32	20	310	72	163					5.7
9:10		47	224	72	238	37	35	23	311	77	153					4.9
9:20		43	219	68	226	33	38	27	313	80	134					4.0
		Constellation of 6 SV's														
SV		3		12		13		16		20		24				
9:30		39	215	64	218	29	41	16	179	31	314	81	102			2.1
9:40		35	212	59	213	26	45	19	176	36	314	80	73			2.3
9:50		31	209	54	209	23	48	23	173	40	315	76	57			2.4
10:00		27	207	49	206	19	52	27	170	44	314	72	49			2.5
10:10		23	204	44	204	16	55	30	167	48	314	67	45			2.5

The visibility diagram (Figure 6.12) and the azimuth-elevation table (Table 6.5) complement each other. They provide the field supervisor with the data needed to make informed judgments about the observation schedule. Even if the decision is taken to include station S 198 in the 9:10 to 10:10 session as originally planned, the supervisor will be forewarned that the blockage of SV 16 may introduce a bit of weakness at that particular station.

Approximate Station Coordinates

The latitude and longitude given on the station visibility diagram should be understood to be approximate. It is sometimes a scaled coordinate or it may be taken from another source. In either case, its primary role is as input for the receiver at the beginning of its observation. The coordinate need only be close enough to the actual position of the receiver to minimize the time the receiver must take to lock onto the constellation of satellites it expects to find.

Multipath

The multipath condition is by no means unique to GPS. When a transmitted television signal reaches the receiving antenna by two or more paths, the resulting variations in amplitude and phase cause the picture to have ghosts. This kind of scattering of the signals can be caused by reflection from land, water, or man-made structures. In GPS, the problem can be particularly troublesome when signals are received from satellites at low elevation angles; hence, the general use of a 15° to 20° mask angle. The use of choke ring antennas to mediate multipath may also be considered.

It is also wise, where it is possible, to avoid using stations that are near structures likely to be reflective or to scatter the signal. For example, chain-link fences that are found hard against a mark can cause multipath by forcing the satellite's signal to pass through the mesh to reach the antenna. The elevation of the antenna over the top of the fence with a survey mast is often the best way to work around this kind of obstruction. Metal structures with large flat surfaces are notorious for causing multipath problems. A long train moving near a project point could be a potential problem, but vehicles passing by on a highway or street usually are not, especially if they go by at high speed. It is important, of course, to avoid parked vehicles. It is best to remind new GPS observers that the survey vehicle should be parked far enough from the point to avert any multipath. A good way to handle these unfavorable conditions is to set an offset point.

Point Offsets

An offset must, of course, stand far enough away from the source of multipath or an attenuated signal to be unaffected. However, the longer the distance from the originally desired position the more important the accuracy of the bearing and distance between that position and the offset becomes. Recording the tie between the two correctly is crucial to avoid misunderstanding after the work is completed. Some receivers allow input of the information directly into the observations recorded in the receiver or data logger. However, during a control survey, it is best to also record the information in a field book.

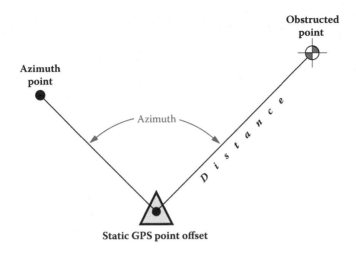

FIGURE 6.13   Static point offset.

Offsets in GPS control surveying are an instance where conventional surveying equipment and expertise are necessary. Clearly, the establishment of the tie requires a position for the occupation of the instrument (i.e., a total station) and a position for the establishment of its orientation (i.e., an azimuth) (see Figure 6.13). It is best to establish three intervisible rather than two points, one to occupy and two azimuth marks. This approach makes it possible to add a redundant check to the tie. The positions on these two, or three, points may be established by setting monuments and performing static observations on them all. Alternatively, azimuthal control may be established by astronomic observations.

Look for Multipath

Both the GPS field supervisor and the reconnaissance team should be alert to any indications on the station visibility diagram that multipath may be a concern. Before the observations are done, there is nearly always a simple solution. Discovering multipath in the signals after the observations are done is not only frustrating but also often expensive.

MONUMENTATION

The monumentation set for GPS projects varies widely and can range from brass tablets to aerial premarks, capped rebar, or even pin flags. The objective of most station markers is to adequately serve the client's subsequent use. However, the time, trouble, and cost in most high-accuracy GPS work warrant the most permanent, stable monumentation.

Many experts predict that GPS will eventually make monumentation unnecessary. The idea foresees GPS receivers in constant operation at well-known master stations that will allow surveyors with receivers to determine highly accurate relative positions with such speed and ease that monumentation will be unnecessary. The idea may prove prophetic, but for now, monumentation is an important part of most GPS projects. The suitability of a particular type of monument is an area still most often left to the professional judgment of the surveyors involved.

LOGISTICS

Scheduling

Once all the station data sheets, visibility diagrams, and other field notes have been collected, the schedule can be finalized for the first observations. There will almost certainly be changes from the original plan. Some of the anticipated control stations may be unavailable or obstructed. Some project points may be blocked, too difficult to reach, or simply not serve the purpose as well as a control station at an alternate location. When the final control has been chosen, the project points have been monumented, and the reconnaissance has been completed, the information can be brought together with some degree of certainty that it represents the actual conditions in the field.

Now that the access and travel time, the length of vectors, and the actual obstructions are more certainly known, the length and order of the sessions can be solidified. Despite all the care and planning that goes into preparing for a project, unexpected changes in the satellites' orbits or health can upset the best schedule at the last minute. It is always helpful to have a backup plan.

The receiver operators usually have been involved in the reconnaissance and are familiar with the area and many of the stations. Even though an observer may not have visited the particular stations scheduled for him, the copies of the project map, appropriate station data sheets, and visibility diagrams will usually prove adequate to their location.

OBSERVATION

When everything goes as planned, a GPS observation is uneventful. However, even before the arrival of the receiver operator at the control or project point the session can get off-track. The simultaneity of the data collected at each end of a baseline is critical to the success of any measurement in static GPS control surveys. When a receiver occupies a master station throughout a project, there need be little concern on this subject, but most static applications depend on the sessions of many mobile receivers beginning and ending together.

Arrival

The number of possible delays that may befall an observer on the way to a station are too numerous to mention. With proper planning and reconnaissance, the observer will likely find that there is enough time for the trip from station to station and that sufficient information is on hand to guide him to the position, but this, too, cannot be guaranteed. When the observer is late to the station, the best course is usually to set up the receiver quickly and collect as much data as possible. The baselines into the late station may or may not be saved, but they will certainly be lost if the receiver operator collects no information at all. It is at times like these that good communication between the members of the GPS team is most useful. For example, some of the other observers in the session may be able to stay on their station a bit longer with the late arrival and make up some of the lost data. Along the same line, it is usually a good policy for those operators who are to remain on a station for two consecutive sessions to collect data as long as possible, while still leaving themselves enough time to reset between the two observation periods.

Setup

Centering an instrument over the station mark is always important. However, the centimeter-level accuracy of static GPS gives the centering of the antenna special significance. It is ironic that such a sophisticated system of surveying can be defeated from such a commonplace procedure. A tribrach with an optical plummet or any other device used for centering should be checked and, if necessary, adjusted before the project begins. With good centering and leveling procedures, an antenna should be within a few millimeters of the station mark.

Unfortunately, the centering of the antenna over the station does not ensure that its phase center is properly oriented. The contours of equal phase around the antenna's electronic center are not themselves perfectly spherical. Part of their eccentricity can be attributed to unavoidable inaccuracies in the manufacturing process. To compensate for some of this offset, it is a good practice to rotate all antennas in a session to the same direction. Many manufacturers provide reference marks on their antennas so that each one may be oriented to the same azimuth. That way they are expected to maintain the same relative position between their physical and electronic centers when observations are made.

The antenna's configuration also affects another measurement critical to successful GPS surveying: the height of the instrument. The frequency of mistakes in this important measurement is remarkable. Several methods have been devised to focus special attention on the height of the antenna. Not only should it be measured in both feet and meters, it should also be measured immediately after the instrument is set up and just before tearing it down to detect any settling of the tripod during the observation.

Height of Instrument

The measurement of the height of the antenna in a GPS survey is often not made on a plumb line. A tape is frequently stretched from the top of the station monument to some reference mark on the antenna or the receiver itself. Some GPS teams measure and record the height of the antenna to more than one reference mark on the ground plane. These measurements are usually mathematically corrected to plumb.

The care ascribed to the measurement of antenna heights is due to the same concern applied to centering. GPS has an extraordinary capability to achieve accurate heights, but those heights can be easily contaminated by incorrect H.I.s.

Observation Logs

Most GPS operations require its receiver operators to keep a careful log of each observation. Usually written on a standard form, these field notes provide a written record of the measurements, times, equipment, and other data that explains what actually occurred during the observation itself. It is difficult to overestimate the importance of this information. It is usually incorporated into the final report of the survey, the archives. However, the most immediate use of the observation log is in evaluation of the day's work by the on-site field supervisor.

An observation log may be organized in a number of ways. The log illustrated in Figure 6.14 is one method that includes some of the information that might be used to document one session at one station.

OBSERVATION LOG

JOB NUMBER   *UL42396*

OBSERVER	STATION	JULIAN DATE	DATE
S. GRAHAM	*S 198 (POINT 14)*	*50*	*2/17/2014*

LATITUDE	LONGITUDE	HEIGHT
47°52'39"	*115°00'48"*	*3,241.09 Feet*

PLANNED OBSERVATION SESSION	SESSION NAME	ACTUAL OBSERVATION SESSION
START TIME: *9:10* STOP TIME: *10:10*	*0014 050 2*	START TIME: *9:10* STOP TIME: *10:10*

ANTENNA TYPE	ANTENNA HEIGHT ABOVE STATION MONUMENT	
ON-BOARD	BEFORE OBSERVATION	AFTER OBSERVATION
	METERS: *1.585*	METERS: *1.585*
MASK ANGLE: *15°*	FEET: *5.20*	FEET: *5.20*

METEOROLOGICAL DATA				
TIME	RELATIVE HUMIDITY	BAROMETER	THERMOMETER (D)	THERMOMETER (W)
9:30	*30%*	*29.94*	*37°F*	*35°F*

VISIBILITY DIAGRAM COMPLETED? (Y) N	TOP OF MONUMENT ABOVE THE SURFACE: *3 IN.*
STATION DATA SHEET COMPLETED? (Y) N	TOP OF MONUMENT BELOW THE SURFACE:

SV PRN TRACKED		COMMENTS:
3	*16*	*DATA FROM SV 16 APPEARS HEALTHY*
12	*20*	*DESPITE WINDMILL OBSERVATION*
13	*24*	

FIGURE 6.14   Observation log.

Of course, the name of the observer and the station must be included, and while the date need not be expressed in both the Julian and Gregorian calendars that information may help in quick cataloging of the data. The approximate latitude, longitude, and height of the station are usually required by the receiver as a reference position for its search for satellites. The date of the planned session will not necessarily coincide with the actual session observed. The observer's arrival at the point may have been late or the receiver may have been allowed to collect data beyond the scheduled end of the session.

There are various methods used to name observation sessions in terminology that is sensible to computers. A widely used system is noted here. The first four digits are the project point's number. In this case it is point 14 and is designated 0014. The next three digits are the Julian day of the session; in this case it is day 50, or 050. Finally, the session illustrated is the second of the day, or 2. Therefore, the full session name is 0014 050 2.

Whether onboard or separate, the type of antenna used and the height of the antenna are critical pieces of information. The relation of the height of the station to the height of the antenna is vital to the station's later utility. The distance that the top of the station's monument is found above or below the surface of the surrounding soil is sometimes neglected. This information cannot only be useful in later recovery of the monument, but it can also be important in the proper evaluation of photocontrol panel points.

Weather

Meteorological data are useful in modeling the atmospheric delay. This information is best recorded at the beginning, middle, and end of each session of projects. Under those circumstances, measurement of the atmospheric pressure in millibars, the relative humidity, and the temperature in degrees Celsius are expected to be included in the observation log. However, the general use is less stringent. The conditions of the day are observed, and unusual changes in the weather are noted. A record of the satellites available during the observation and any comments about unique circumstances of the session round out the observation log.

DAILY PROGRESS EVALUATION

Planned observation schedules of a large GPS project usually change daily. Arrangement of upcoming sessions is often altered based on the success or failure of the previous day's plan. Such a regrouping follows evaluation of the day's data.

This evaluation involves examination of the observation logs as well as the data each receiver has collected. Unhealthy data, caused by cycle slips or any other source, are not always apparent to the receiver operator at the time of the observation. Therefore, a daily quality control check is a necessary preliminary step before finalizing the next day's observation schedule.

Some field supervisors prefer to actually compute the independent baseline vectors of each day's work to ensure that the measurements are adequate. Neglecting the daily check could leave unsuccessful sessions undiscovered until the survey was thought to be completed. The consequences of such a situation could be expensive.

EXERCISES

1. Which of the following information is not available from the NGS data sheet for a particular station?
   a. Type of monument
   b. State plane coordinates
   c. Latitude and longitude of the primary azimuth mark
   d. Permanent Identifier

2. Which of the following is a quantity that when added to an astronomic azimuth yields a geodetic azimuth?
   a. Primary azimuth
   b. Mapping angle, also known as the convergence
   c. Laplace correction
   d. Second term

3. How many nontrivial, or independent, and how many trivial, or dependent, baselines are created in one GPS session using four receivers?
   a. 1 independent baseline and 5 dependent baselines
   b. 3 independent baselines and 1 dependent baseline
   c. 2 independent baselines and 3 dependent baselines
   d. 3 independent baselines and 3 dependent baselines

4. Which of the following statements concerning loop closure in a GPS network is not correct?
   a. A loop closure that uses baselines from one GPS session will appear to have no error at all.
   b. Loop closure is a procedure by which the internal consistency of a GPS network is discovered.
   c. No baselines should be excluded from a loop closure analysis.
   d. At the completion of a GPS control survey, the independent vectors should be capable of formation into closed loops that incorporate baselines from two to four different sessions.

5. How many observation sessions will be required for a GPS control survey involving 20 stations that is to be done with four GPS receivers, a planned redundancy of two, a production factor of 1.1, and a safety factor is 0.2?
   a. 21 observation sessions
   b. 17 observation sessions
   c. 15 observation sessions
   d. 12 observation sessions

6. What of the following options describes a best configuration for the horizontal control of a GPS static control survey?
   a. Divide the project into four quadrants and choose at least one horizontal control station near the project boundary in each quadrant.

b. Divide the project into three parts and choose at least one horizontal control station near the center of the project in each.

c. The primary base station should be located at the center of the project.

d. Draw a north-south line through the center of the project, and choose at least three horizontal control stations near that line.

7. Which of the following data sheet searches is not possible on the NGS Internet site?

a. Rectangular search based on the range of latitudes and longitudes

b. Search using a particular station's geoid height

c. Search using a particular station's Permanent Identifier (PID)

d. Radial search that is defining the region of the survey with one center position and a radius

8. Which of the following is the standard symbol used in this book for project points in a GPS control survey?

a. Solid dot

b. Triangle

c. Square

d. Circle

9. Of the four listed below, what is the most frequent mistake made in GPS control surveying?

a. Failure to use a 15° mask angle

b. Failure to have a fully charged battery

c. Failure to properly measure the H.I. of the antenna

d. Failure to properly center the antenna over the point

10. Which of the following statements about the zero baseline test is not true?

a. Zero baseline test eliminates random noise and receiver biases from the results.

b. Zero baseline test requires that receivers share the same antenna using an antenna splitter.

c. Zero baseline test eliminates satellite clock biases, ephemeris errors, atmospheric biases, and multipath from the results.

d. Zero baseline test is not dependent on special software or a test network.

11. The place of a particular station in a static GPS control survey observation schedule may depend on a number of factors. Which of the following would not be one of them?

a. Obstructions around the station

b. Difficulty involved in reaching the station in bad weather

c. Previous day's successful and unsuccessful occupations

d. Line of sight with other stations in the survey

12. The *to-reach* description in a static GPS control survey would most likely appear on which of the following forms?
    a. Observation log
    b. Station data sheet
    c. Station visibility diagram
    d. Observation schedule

13. The H.I. of the antenna in a static GPS control survey would most likely appear on which of the following forms?
    a. Observation log
    b. Station data sheet
    c. Station visibility diagram
    d. Observation schedule

14. Which of the following forms used in a static GPS control survey would not include information on either the start and stop time of the observation or the approximate latitude and longitude of the station?
    a. Observation log
    b. Station data sheet
    c. Station visibility diagram
    d. Observation schedule

15. What is the overriding principle that ought to guide the preparation of a station data sheet?
    a. Forewarn of obstructed satellites at the site
    b. Finalize the observation schedule
    c. Guide succeeding visitors to the station without ambiguity
    d. Explain what actually occurred during the observation itself

16. Which statement below correctly identifies an advantage of using rubbings in recording the information on an existing monument?
    a. It ensures that the station was actually visited and that the stamping was faithfully recorded.
    b. It makes ties to prominent features in the area unnecessary.
    c. It improves the neatness of the station data sheet.
    d. It increases the efficiency of the actual occupation of the station.

17. What tools are necessary to prepare a visibility diagram at a station that is somewhat obstructed?
    a. Theodolite and EDM
    b. Compass and clinometer
    c. GPS receiver
    d. Level and level rod

18. In planning a kinematic GPS survey, which statement below is the most likely use for control stations placed on either side of a bridge?

a. Control stations may be used to reinitialize a receiver that lost lock while passing under the bridge.

b. Control stations might be occupied during the survey as a check of the accuracy of the survey.

c. Control stations might be used for an antenna swap.

d. Control stations might be set to make the kinematic survey unnecessary in the area.

19. Effect of multipath cannot be reduced or alleviated by

a. Longer observation

b. Choke ring antenna

c. Point offset

d. Mask angle

20. Offsets in GPS control surveying require

a. Second GPS receiver to be used nearby

b. Compass and clinometers

c. Conventional surveying equipment and expertise

d. Connection to base station

21. In order to ameliorate the misalignment of the antenna and its phase center,

a. Antenna height should be measured in meters and feet.

b. Rotate all antennas in a session to the same direction.

c. It is required to calibrate the phase center of the antenna.

d. Type the antenna height in the data collector.

ANSWERS AND EXPLANATIONS

1. Answer is (c)

Explanation: NGS data sheets provide State Plane and UTM coordinates, the latter only for horizontal control stations. State Plane Coordinates are given in either U.S. Survey Feet or International Feet and UTM coordinates are given in meters. They also provide mark setting information, the type of monument, and the history of mark recovery. The data sheet certainly shows the station's designation, which is its name and its Permanent Identifier (PID). Either of these may be used to search for the station in the NGS database. The PID is also found all along the left side of each data sheet record and is always two uppercase letters followed by four numbers. However, it does not show the latitude and longitude of azimuth marks. That information may sometimes be found on a data sheet devoted to the particular azimuth mark.

2. Answer is (c)

Explanation: The Laplace correction is a quantity that, when added to an astronomically observed azimuth, yields a geodetic azimuth. It is important to note that NGS uses a clockwise rotation regarding the Laplace correction. It can contribute several seconds of arc.

3. Answer is (d)

Explanation: Where *r* is the number of receivers, every session yields *r* − 1 independent baselines. Four receivers used in one session would produce six baselines. Of these, three would be independent or nontrivial baselines, and three would be dependent or trivial baselines. It cannot be said that the shortest lines are always chosen to be the independent lines. Sometimes there are reasons to reject one of the shorter vectors owing to incomplete data, cycle slips, multipath, or some other weakness in the measurements. Before such decisions can be made, each session will require analysis after the data have actually been collected. In the planning stage, it is best to consider the shortest vectors as the independent lines, but once the decision is made, only those three baselines are included in the network.

4. Answer is (c)

Explanation: Any loop closures that only use baselines derived from a single common GPS session will yield an apparent error of zero because they are derived from the same simultaneous observations. When the project is completed, the independent vectors in the network, excluding the dependent, or trivial baselines, should be capable of formation into closed loops that incorporate baselines from two to four different sessions.

In GPS the error of closure is valid for orders A and B when three or more independent baselines from three or more GPS sessions are included in the loop. Order AA requires four independent observations be included. Loop closures for first order and lower should include a minimum of two observing sessions.

5. Answer is (c)

Explanation: The appropriate formula is

$$s = \frac{(m \cdot n)}{r} + \frac{(m \cdot n)(p-1)}{r} + k \cdot m$$

$$s = \frac{(20)(2)}{4} + \frac{(20)(2)(1.1-1)}{4} + (0.2)(20)$$

$$s = \frac{(40)}{4} + \frac{(4)}{4} + 4$$

$$s = 10 + 1 + 4$$

$$s = 15 \text{ sessions}$$

where

s = number of observing sessions
r = number of receivers
m = total number of stations involved
n = planned redundancy
p = experience, or production factor
k = safety factor

The variable n is a representation of the level of redundancy that has been built into the network, based on the number of occupations on each station. The experience of a firm is symbolized by the variable p in the formula. A typical production factor is 1.1. A safety factor of 0.1, known as k, is recommended for GPS projects within 100 km of a company's home base. Beyond that radius, an increase to 0.2 is advised. The substitution of the appropriate quantities for the illustrated project increases the prediction of the number of observation sessions required for its completion.

6. Answer is (a)

   Explanation: For work other than route surveys, a handy rule of thumb is to divide the project into four quadrants and to choose at least one horizontal control station in each quadrant. The actual survey should have at least one horizontal control station in three of the four quadrants. Each of them ought to be as near as possible to the project boundary. Supplementary control in the interior of the network can then be used to add more stability to the network.

7. Answer is (b)

   Explanation: It is quite important to have the most up-to-date control information from NGS. A rectangular search based on the range of latitudes and longitudes can now be performed on the NGS Internet site. It is also possible to do a radial search, defining the region of the survey with one center position and a radius. You may also retrieve individual data sheets by the Permanent Identifier (PID) control point name, which is known as the designation, survey project identifier, or USGS quad. However, you cannot retrieve a data sheet for a station based only on its geoid height.

8. Answer is (a)

   Explanation: A solid dot is the standard symbol used to indicate the position of project points. Some variation is used when a distinction must be drawn between those points that are in place and those that must be set. When its location is appropriate, it is always a good idea to have a vertical or horizontal control station serve double duty as a project point. While the precision of their plotting may vary, it is important that project points be located as precisely as possible even at this preliminary stage. A horizontal control point is shown as a triangle, and a vertical control point is shown as a square.

9. Answer is (c)

   Explanation: The mis-measured height of the antenna above the mark is probably the most pervasive and frequent blunder in GPS control surveying.

10. Answer is (a)

    Explanation: The simplicity of the zero baseline test is an advantage. It is not dependent on special software or a test network, and it can be used to separate receiver difficulties from antenna errors. Two or more receivers are connected to one antenna with a signal or antenna splitter.

An observation is done with the divided signal from the single antenna reaching both receivers simultaneously. Because the receivers are sharing the same antenna, satellite clock biases, ephemeris errors, atmospheric biases, and multipath are all canceled. The only remaining errors are attributable to random noise and receiver biases.

11. Answer is (d)

Explanation: In creating an observation, schedule consideration might be given to observing a particular station during a session when none of the satellites would be blocked by obstructions. While GPS is not restricted by inclement weather, particular access routes may not be so immune. Despite best efforts, a planned observation may have been unsuccessful at a required station on a previous day, and it may need to be revisited. However, the line of sight between a particular station and another in the survey is not likely to affect the GPS observation schedule, though such a consideration may be critical in a conventional survey.

12. Answer is (b)

Explanation: The to-reach description in a static GPS control survey would most likely be prepared during reconnaissance and would appear on the station data sheet.

13. Answer is (a)

Explanation: The H.I. of the antenna in a static GPS control survey would most likely be recorded before and after the actual observation on the station and would appear on the observation log.

14. Answer is (b)

Explanation: The station data sheet would not be likely to include either information. The other forms listed would probably include one or both categories.

15. Answer is (c)

Explanation: The station data sheet is often an important bridge between on-site reconnaissance and the actual occupation of a monument. Neatness and clarity, always paramount virtues of good field notes, are of particular interest when the station data sheet is to be later included in the final report to the client. The overriding principle in drafting a station data sheet is to guide succeeding visitors to the station without ambiguity. A GPS surveyor on the way to observe the position for the first time may be the initial user of a station data sheet. A poorly written document could void an entire session if the observer is unable to locate the monument. A client, later struggling to find a particular monument with an inadequate data sheet, may ultimately question the value of more than the field notes.

16. Answer is (a)

   Explanation: Rubbings are performed with paper held on top of the monument's disk; a pencil is run over it in a zigzag pattern producing a positive image of the stamping. This method is a bit more awkward than simply copying the information from the disk onto the data sheet, but it does have the advantage of ensuring the station was actually visited and that the stamping was faithfully recorded. Such rubbings or close-up photographs are required by the provisional FGCC *Geometric Geodetic Accuracy Standards and Specifications for Using GPS Relative Positioning Techniques* for all orders of GPS surveys.

17. Answer is (b)

   Explanation: Using a compass and a clinometer, a member of the reconnaissance team can fully describe possible obstructions of the satellite's signals on a visibility diagram. By standing at the station mark and measuring the azimuth and vertical angle of points outlining the obstruction, the observer can plot the object on the visibility diagram.

18. Answer is (a)

   Explanation: Unavoidable obstructions like bridges and tunnels can be overcome by the placement of control stations on both sides of the barrier. If these control stations are coordinated by some type of static observation, they can later be used to reinitialize the kinematic receivers should they lose lock passing under the bridge.

19. Answer is (a)

   Explanation: In GPS the problem can be particularly troublesome when signals are received from satellites at low elevation angles; hence, the general use of a 15° to 20° mask angle. The use of choke ring antennas to mediate multipath may also be considered. It is best to remind new GPS observers that the survey vehicle should be parked far enough from the point to avert any multipath. A good way to handle these unfavorable conditions is to set an offset point.

20. Answer is (c)

   Explanation: Offsets in GPS control surveying are an instance where conventional surveying equipment and expertise are necessary. Clearly, the establishment of the tie requires a position for the occupation of the instrument (i.e., a total station) and a position for the establishment of its orientation (i.e., an azimuth) (see Figure 6.13). It is best to establish three intervisible rather than two points, one to occupy and two azimuth marks. This approach makes it possible to add a redundant check to the tie. The positions on these two, or three, points may be established by setting monuments and performing static observations on them all. Alternatively, azimuthal control may be established by astronomic observations.

21. Answer is (b)

Explanation: Unfortunately, the centering of the antenna over the station does not ensure that its phase center is properly oriented. The contours of equal phase around the antenna's electronic center are not themselves perfectly spherical. Part of their eccentricity can be attributed to unavoidable inaccuracies in the manufacturing process. To compensate for some of this offset, it is a good practice to rotate all antennas in a session to the same direction. Many manufacturers provide reference marks on their antennas so that each one may be oriented to the same azimuth. That way they are expected to maintain the same relative position between their physical and electronic centers when observations are made.

7 Real-Time Global Positioning System Surveying

REAL-TIME KINEMATIC (RTK) AND DIFFERENTIAL GPS (DGPS)

Most, not all, GPS surveying relies on the idea of differential positioning. The mode of a base or reference receiver at a known location logging data at the same time as a receiver at an unknown location together provide the fundamental information for the determination of accurate coordinates. While this basic approach remains today, the majority of GPS surveying is not done in the static post-processed mode. Post-processing is most often applied to control work. Now, the most commonly used methods utilize receivers on reference stations that provide correction signals to the end user via a data link sometimes over the Internet, radio signal, or cell phone and often in real-time.

In this category of GPS surveying work, there is sometimes a distinction made between code-based and carrier-based solutions. In fact, most systems use a combination of code and carrier measurements so the distinction is more a matter of emphasis rather than an absolute difference.

GENERAL IDEA

Errors in satellite clocks, imperfect orbits, the trip through the layers of the atmosphere, and many other sources contribute inaccuracies to GPS signals by the time they reach a receiver. These errors are variable, so the best to way to correct them is to monitor them as they happen. A good way to do this is to set up a GPS receiver on a station whose position is known exactly; this is called a base station. This base station receiver's computer can calculate its position from satellite data, compare that position with its actual known position, and find the difference. The resulting error corrections can be communicated from the base to the rover. It works well, but the errors are constantly changing so a base station has to monitor them all the time, at least all the time the rover receiver or receivers are working. While this is happening, the rovers move from place to place collecting the points whose positions you want to know relative to the base station, which is the real objective after all. Then all you have to do is get those base station corrections and the rover's data together somehow. That combination can be done over a data link in real-time or applied later in post-processing.

Real-time positioning is built on the foundation of the idea that, with the important exceptions of multipath and receiver noise, GPS error sources are correlated. In other words, the closer the rover is to the base, the more the errors at the ends of the baseline match. The shorter the baseline, the more the errors are correlated. The longer the baseline, the less the errors are correlated.

RADIAL GPS

Such real-time surveying is essentially radial. There are advantages to the approach (Figure 7.1). The advantage is a large number of positions can be established in a short amount of time with little or no planning. The disadvantage is that there is little or no redundancy in positions derived, each of the baselines originates from the same control station. Redundancy can be incorporated, but it requires repetition of the observations so each baseline is determined with more than one GPS constellation. One way to do it is to occupy the project points, the unknown positions, successively with more than one rover. It is best if these successive occupations are separated by at least 4 hours and not more than 8 hours so the satellite constellation can reach a significantly different configuration.

Another more convenient but less desirable approach is to do a second occupation almost immediately after the first. The roving receiver's antenna is blocked or tilted until the lock on the satellites is interrupted. It is then reoriented on the unknown position a second time for the repeat solution. This does offer a second solution, but from virtually the same constellation. It is better to do the second occupation 4 to 8 hours later. Each one of the two occupations may be brief. The time between the two occupations allows the satellites of the second session to reach a configuration different enough to supply a redundant position for the project point.

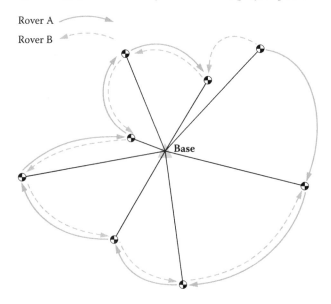

FIGURE 7.1   Radial GPS.

A third way to achieve redundancy is to occupy each point with the same rover but utilize a different base station. This approach allows a solution to be available from two separate control stations. Obviously, this can be done with reoccupation of the project points after one base station has been moved to a new control point or two base stations can be up and running from the very outset and throughout the work, as would be the case using two Continuously Operating Reference Stations (CORS). It is best if there are two occupations on each point and each of the two utilize different base stations.

More efficiency can be achieved by adding additional roving receivers. However, as the number of receivers rises, the logistics become more complicated, and a survey plan becomes necessary. Also, project points that are simultaneously near one another but far from the control station should be directly connected with a baseline between the two of them to maintain the integrity of the survey. Finally, if the base receiver loses lock and that goes unnoticed, it will completely defeat the radial survey for the time it is down.

CORRECTION SIGNAL

The agreed upon protocol for communication between base stations and rovers was first designed for used in marine navigation by an organization known as the Radio Technical Commission for Maritime Services (RTCM). RTCM is an independent not-for-profit organization that is supported by an international membership that includes both governmental and nongovernmental institutions. Its goals are educational, scientific, and professional. Toward those ends, it provides information on maritime radio navigation and radio communication policies and associated regulations to its members. It is also involved in technical standards development.

In 1985, the, RTCM Special Committee (SC-104) created a standard that is still more used than any proprietary formats that have come along since. RTCM is open. In other words, it is a general purpose format and is not restricted to a particular receiver type. The message augments the information from the satellites. It was originally designed to accommodate a slow GPS data rate with a configuration somewhat similar to the Navigation message. The data format has evolved since its inception. For example, RTCM 2.0 supported GPS code only. However, when it became clear in 1994 that including carrier phase information in the message could improve the accuracy of the system, RTCM Special Committee 104 added four new message types to Version 2.1 to fulfill the needs of RTK. RTCM 2.1 supported both code and phase correction but still GPS only. Version 2.2 became available in 1998. RTCM 2.2 added support for GLONASS and version 2.3 included antenna corrections, and the changes continued. In 2007 the Radio Technical Commission for Maritime Services Special Committee 104 published its Version 3. RTCM 3.0 utilizes a more efficient message structure than its predecessors which proves beneficial in the RTK data heavy real-time communications between a base and a rover. Version 3.0 still provides both GPS and GLONASS code and carrier messages, antenna and system parameters. RTCM 3.1 adds a network correction message and version 3.2 announced in 2013 introduces a feature known as multiple signal messages (MSM). MSM includes the capability to handle the European Galileo and the Chinese Beidou Global Navigation Satellite System (GNSS) systems in the RTCM protocol.

The GPS constellation along with the Russian GLONASS system, the European Galileo system, and other systems comprise GNSS currently. It is likely that more systems will become included under the GNSS label in the future. It is also likely that more accuracy of autonomous positions will be available from GNSS than GPS alone. However, in GNSS, as with GPS, even better accuracies can be achieved by broadcasting corrections from reference stations at precisely known locations, and by utilizing RTCM 3.2, it is not only possible to use receivers from different manufacturers together but also to incorporate signals from satellites other than GPS.

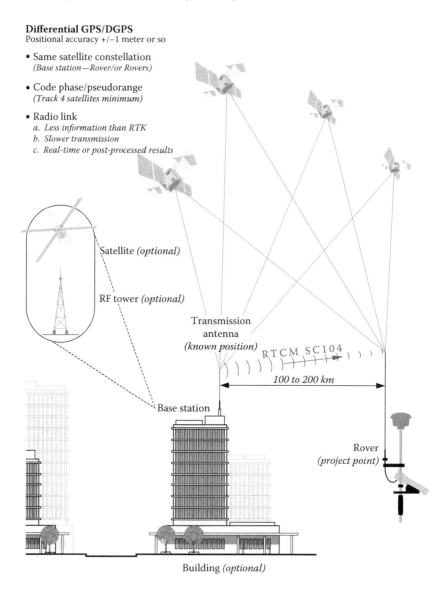

Differential GPS/DGPS
Positional accuracy +/−1 meter or so

- Same satellite constellation
  (Base station—Rover/or Rovers)

- Code phase/pseudorange
  (Track 4 satellites minimum)

- Radio link
  a. *Less information than RTK*
  b. *Slower transmission*
  c. *Real-time or post-processed results*

Satellite *(optional)*

RF tower *(optional)*

Transmission antenna *(known position)*

RTCM SC104

100 to 200 km

Base station

Rover *(project point)*

Building *(optional)*

FIGURE 7.2   DGPS.

DGPS

The term DGPS is sometimes used to refer to differential GPS that is based on pseudoranges, aka code phase (Figure 7.2). Even though the accuracy of code phase applications was given a boost with the elimination of Selective Availability (SA) in May 2000, consistent accuracy better than the 2–5 m range still requires reduction of the effect of correlated ephemeris and atmospheric errors by differential corrections. Though the corrections could be applied in post-processing services that supply these corrections, most often operate in real-time. In such an operation, pseudorange-based versions can offer meter or even submeter results.

Usually, pseudorange corrections are broadcast from the base to the rover or rovers for each satellite in the visible constellation. Rovers with an appropriate input/output port can receive the correction signal and calculate coordinates. The real-time signal comes to the receiver over a data link. It can originate at a project specific base station or it can come to the user through a service of which there are various categories. Some are open to all users, and some are by subscription only. Coverage depends on the spacing of the beacons, aka transmitting base stations, their power, interference, and so forth. Some systems require two-way, some one-way, communication with the base stations. Radio systems, geostationary satellites, low-Earth-orbiting satellites, and cellular phones are some of the options available for two-way data communication.

In any case, most of the wide variety of DGPS services were not originally set up to augment surveying and mapping applications of GPS; they were established to aid GPS navigation.

LOCAL AND WIDE AREA DGPS

Correlation between most of the GPS biases becomes weaker as the rover gets farther from the base. The term Local Area Differential GPS is used when the baselines from a single base station to the roving receivers using the service are less than a couple of hundred kilometers. The term Wide Area Differential GPS is used when the service uses a network of base stations and distributes correction over a larger area, an area that may even be continental in scope. Many bases operating together provide a means by which the information from several of them can be combined to send a normalized or averaged correction tailored to the rover's geographical position within the system. Some use satellites to provide the data link between the service provider and the subscribers. Such a system depends on the network of base stations receiving signals from the GPS satellites and then streaming that data to a central computer at a control center. There the corrections are calculated and uploaded to a geostationary communication satellite. Then the communication satellite broadcasts the corrections to the service's subscribers.

In all cases, the base stations are at known locations and their corrections are broadcast to all rovers that are equipped to receive their particular radio message carrying real-time corrections in the RTCM format. An example of such a DGPS service originated as an augmentation for marine navigation.

Both the U.S. Coast Guard and the Canadian Coast Guard instituted DGPS services to facilitate harbor entrances, ocean mapping, and marine traffic control as well as navigation in inland waterways. Their system base stations beacons broadcast

GPS corrections along major rivers, major lakes, the east coast, and the west coast. The sites use marine beacon frequencies of 255 to 325 kHz, which has the advantage of long range propagation that can be several hundreds of kilometers. Access to the broadcasts is free, and over recent years the service has become very popular outside of its maritime applications, particularly among farmers engaged in GPS-aided precision agriculture. Therefore, the system has been extended beyond waterways across the continental United States and is now known as the Nationwide DGPS (NDGPS). There are currently 85 base stations.

WIDE AREA AUGMENTATION SYSTEMS (WAAS)

Another U.S. DGPS service initiated in 1994 cooperatively by the Department of Transportation and the Federal Aviation Administration is known as WAAS (Figure 7.3).

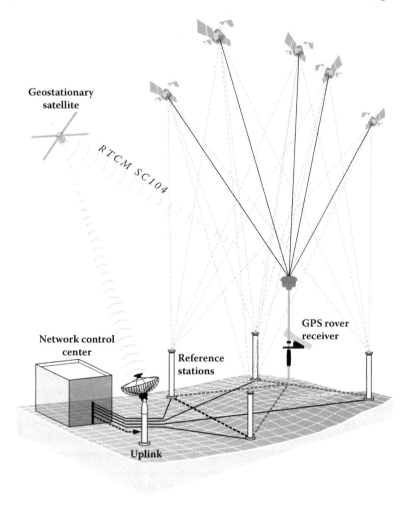

FIGURE 7.3   Wide area differential GPS.

It is available to users with GPS receivers equipped to receive it. The signal is free. The official horizontal accuracy is 7.6 m, but it often delivers better. It utilizes both satellite-based augmentation systems, also known as SBAS, and ground-based augmentations and was initially designed to assist aerial navigation from takeoff through landing. Reference stations at known locations are the bases on the ground. They send their data via processing sites to three Ground Earth Master Stations that upload differential corrections and time to three commercial geostationary satellites devoted to transmission of GPS differential corrections to users on the ground. DGPS requires that all receivers collect pseudoranges from the same constellation of satellites. It is vital that the errors corrected by the base station are common to the rovers. The rover must share its selection of satellites with the base station.

The European Geostationary Navigation Overlay Service has a similar configuration and augments GPS and GLONASS using three geostationary and a network of ground stations. The Japanese system is known by the acronym MSAS, and India's is the GPS And Geo-Augmented Navigation (GAGAN) system.

GEOGRAPHIC INFORMATION SYSTEMS (GIS) APPLICATION

Aerial navigation, marine navigation, agriculture, vehicle tracking, and construction utilize DGPS. DGPS is also useful in land and hydrographic surveying, but perhaps the fastest growing application for DGPS is in data collection, data updating, and even in-field mapping for Geographic Information Systems (GIS). GIS data have long been captured from paper records such as digitizing and scanning paper maps. Photogrammetry, remote sensing, and conventional surveying have also been data sources for GIS. More recently, data collected in the field with DGPS have become significant in GIS.

GIS data collection with DGPS requires the integration of the position of features of interest and relevant attribute information about those features. In GIS it is frequently important to return to a particular site or feature to perform inspections or maintenance. DGPS with real-time correction makes it convenient to load the position or positions of features into a data logger and navigate back to the vicinity. However, to make such applications feasible, a GIS must be kept current. It must be maintained. A receiver configuration including real-time DGPS, sufficient data storage, and graphic display allows verification and updating of existing information.

DGPS allows the immediate attribution and validation in the field with accurate and efficient recording of position. In the past, many GIS mapping efforts have often relied on ties to street centerlines, curb lines, railroads, and so forth. Such dependencies can be destroyed by demolition or new construction, but, meter-level positional accuracy even in obstructed environments such as urban areas, amid high-rise buildings, is possible with DGPS. Therefore, it can provide reliable positioning even if the landscape has changed, and its data can be integrated with other technologies such as laser range finders and so forth in environments where DGPS is not ideally suited to the situation. Finally, loading GPS data into a GIS platform does not require manual intervention. GPS data processing can be automated; the results are digital and can pass into a GIS format without redundant effort, reducing the chance for errors.

REAL-TIME KINEMATIC (RTK)

Kinematic surveying, also known as stop-and-go kinematic surveying, is not new. The original kinematic GPS innovator, Dr. Benjamin Remondi, developed the idea in the mid-1980s. RTK is a method that provides positional accuracy nearly as good as static carrier phase positioning, but faster. RTK accomplishes positioning in real-time (Figure 7.4). It involves the use of at least one stationary reference receiver, the base station, and at least one moving receiver, the rover. All the receivers involved observe the same satellites simultaneously. The base receivers are stationary on

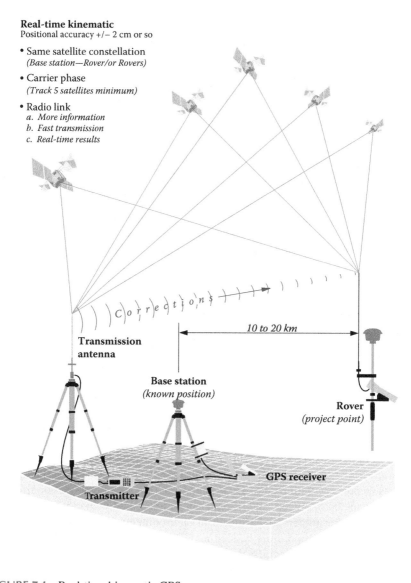

Real-time kinematic
Positional accuracy +/– 2 cm or so

• Same satellite constellation
  (Base station—Rover/or Rovers)

• Carrier phase
  (Track 5 satellites minimum)

• Radio link
  a. More information
  b. Fast transmission
  c. Real-time results

Corrections

10 to 20 km

Transmission antenna

Base station
(known position)

Rover
(project point)

GPS receiver

Transmitter

FIGURE 7.4   Real-time kinematic GPS.

control points. The rovers move from project point to project point, stopping momentarily at each new point, usually briefly. The collected data provide vectors between themselves and the base receivers as shown in Figure 7.1 in real-time.

RTK has become routine in development and engineering surveys where the distance between the base and roving receivers can most often be measured in thousands of feet. When compared with the other relative positioning methods, there is little question that the very short sessions of the real-time kinematic method can produce the largest number of positions in the least amount of time. The remarkable thing is that this technique can do so with only slight degradation in the accuracy of the work.

INTEGER CYCLE AMBIGUITY FIXING

RTK receivers can be single- or multifrequency receivers with GPS antennas, but multifrequency receivers are usual because RTK relies on carrier phase observations corrected in real-time. In other words, it depends on the fixing of the integer cycle ambiguity and that is most efficiently accomplished with a multifrequency GPS receiver capable of making both carrier phase and precise pseudorange measurements.

Here is one way it can be done. A search area is defined in the volume of the possible solutions, but that group is narrowed down quite a bit by using pseudoranges. If the number of integer combinations to be tested is greatly reduced with precise pseudoranges, the search can be quickly limited. The possible solutions in that volume are tested statistically, according to a minimal variance criterion, and the best one is found. This candidate is verified, that is, compared with the second best candidate. The process can take less than 10 s under the best circumstances where the receivers are tracking a large constellation of satellites, the position dilution of precision (PDOP) is small, the receivers are multifrequency, there is no multipath, and the receiver noise is low. This technique relies on multifrequency information. Observations are combined into a wide lane, which has an ambiguity of about 86 cm, and the integer ambiguity is solved in a first pass. This information is used to determine the kinematic solution on L1. Therefore, it is a good idea to restrict RTK to situations where there is good correlation of atmospheric biases at both ends of the baseline. In other words, RTK is best used when the distance between the base and rover is between 10 and 20 km (6 and 12 miles) or less. It is fortunate that GPS receivers with virtually instantaneous carrier phase-based positioning are available. These techniques of integer cycle ambiguity resolution, validation, and quality control are being further improved to apply to GNSS data processing.

WIRELESS LINK

RTK also requires a real-time wireless connection be maintained between the base station and the rover. The radio receiving antennas for the rovers will either be built into the GPS antenna or be present as separate units. It is usual that the radio antenna for the data transmitter and the rover are omnidirectional whip antennas; however, at the base it is usually on a separate mast and has a higher gain than those at the rovers. The position of the transmitting antenna affects the performance of the system significantly. It is usually best to place the transmitter antenna as high as is practical for maximum coverage, and the longer the antenna, the better its transmission characteristics. It is also best if the base station occupies a control station that has no overhead obstructions, is unlikely to be affected by multipath, and is somewhat away from the action if the work is on a construction site. It is also best if the base station is within line of sight of the rovers. If line of sight is not practical, as little obstruction as possible along the radio link is best.

The data radio transmitter consists of an antenna, a radio modulator, and an amplifier. The modulator converts the correction data into a radio signal. The amplifier increases the signal's power, which determines how far the information can travel. Well, not entirely; the terrain and the height of the antenna have something to do with it, too. RTK work requires that a great deal of information be successfully communicated from the base station to the receivers. The base station transmitter ought to be VHF, UHF, or spread spectrum-frequency hopping or direct to have sufficient capacity to handle the load. UHF spread spectrum radio modems are the most popular for DGPS and RTK applications. The typical gain on the antenna at the base is 6 dB. However, while DGPS operations may need no more than 200 bits per second (bps), updated every 10 s or so, RTK requires at least 2400 bps updated about every 1/2 s or less. Like the power of the transmission, the speed of the link between the base and rover, the data rate, can be a limiting factor in RTK performance.

It is also important to note that it takes some time for the base station to calculate corrections, and it takes some time for it to put the data into packets in the correct format and transmit them. Then the data make their way from the base station to the rover over the data link. They are decoded and must go through the rover's software. The time this takes is called the *latency* of the communication between the base station and the rover. It can be as little as a quarter of a second or as long as a couple of seconds, and because the base stations corrections are only accurate for the moment they were created, the base station must send a range rate correction along with them. Using this rate correction, the rover can back date the correction to match the moment it made that same observation.

As mentioned in the previous section, RTK is at its best when the distance between the base station and the rovers is 6–12 miles or less. However, the baselines length may be further limited by the effective range of the radio data link. In areas with high radio traffic it can be difficult to find an open channel. It is remarkable how often the interference emanates from other surveyors in the area doing RTK as well. Most radios connected to RTK GPS surveying equipment operate between UHF 400–475 MHz or VHF 170–220 MHz, and emergency voice communications also tend to operate in this same range, which can present problems from time to time. That is why most radio data transmitters used in RTK allow the user several frequency options within the legal range.

The usual data link configuration operates at 4800 baud or faster. The units communicate with each other along a direct line of sight. The transmitter at the base station is usually the larger and more powerful of the two radios. However, the highest wattage radios, 35 W or so, cannot be legally operated in some countries. Lower power radios, from 1/2 to 2 W, are sometimes used in such circumstances. The radio at the rover has usually lower power and is smaller. The Federal Communications Commission (FCC) is concerned with some RTK GPS operations interfering with other radio signals, particularly voice communications. It is important for GPS surveyors to know that voice communications have priority over data communications.

The FCC requires cooperation among licensees who share frequencies. Interference should be minimized. For example, it is wise to avoid the most typical community voice repeater frequencies. They usually occur between 455–460 MHz and 465–470 MHz. Part 90 of the Code of Federal Regulations, 47 CFR 90, contains the complete text of the FCC Rules including the requirements for licensure of radio spectrum for private land mobile use. The FCC does require application be made for licensing a radio transmitter. Fortunately, when the transmitter and rover receivers required for RTK operations are bought simultaneously, radio licensing and frequency selection are often arranged by the GPS selling agent. Nevertheless, it is important that surveyors do not operate a transmitter without a proper license. Please remember that the FCC can levy fines for several thousand dollars for each day of illegal operation. More can be learned by consulting the FCC Wireless Fee Filing Guide. There are also other international and national bodies that govern frequencies and authorize the use of signals elsewhere in the world. In some areas, certain bands are designated for public use, and no special permission is required. For example, in Europe, it is possible to use the 2.4 GHz band for spread spectrum communication without special authorization with certain power limitations. Here in the United States the band for spread spectrum communication is 900 MHz.

It is vital, of course, that the rover and the base station are tuned to the same frequency for successful communication. The receiver also has an antenna and a demodulator. The demodulator converts the signal back to an intelligible form for the rover's receiver. The data signal from the base station can be weakened or lost at the rover from reflection, refraction, atmospheric anomalies, or even being too close. A rover that is too close to the transmitter may be overloaded and not receive the signal properly, and, of course, even under the best circumstances, the signal will fade as the distance between the transmitter and the rover grows too large.

VERTICAL COMPONENT IN RTK

The output of RTK can appear to be somewhat similar to that of optical surveying with an electronic distance measuring (EDM) and a level. Nevertheless, it is not a good idea to consider the methods equivalent. RTK offers some advantages and some disadvantages when compared with more conventional methods. For example, RTK can be much more productive because it is available 24 hours a day and is not really affected by weather conditions. However, when it comes to the vertical component of surveying, RTK and the level are certainly not equal.

GPS can be used to measure the differences in ellipsoidal height between points with good accuracy. However, unlike a level, unaided GPS cannot be used to measure differences in orthometric height. Orthometric elevations are not directly available from the geocentric position vectors derived from GPS measurements. The accuracy of orthometric heights in GPS is dependent on the veracity of the geoidal model used and the care with which it is applied.

Fortunately, ever improving geoid models have been, and still are, available from NGS. Because geoidal heights can be derived from these models, and ellipsoidal heights are available from GPS, it is certainly feasible to calculate orthometric heights especially when a geoid model is onboard the RTK systems. However, it is important to remember that without a geoid model, RTK will only provide differences in ellipsoid heights between the base station and the rovers.

It is not a good idea to presume that the surface of the ellipsoid is sufficiently parallel to the surface of the geoid and ignore the deviation between the two. They may depart from one another as much as a meter, approximately 3 feet, in 4 or 5 km (2.5 to 3 miles).

SOME PRACTICAL RTK SUGGESTIONS

In RTK, generally speaking, the more satellites that are available, the faster the integer ambiguities will be resolved. A multifrequency receiver is a real benefit in doing RTK. Using a multifrequency receiver instead of a single-frequency receiver is almost as if there were one and a half more satellites available to the observer.

It is best to set up the base station over a known position first, before configuring the rover. After the tripod and tribrach are level and over the point, attach the GPS antenna to the tribrach and, if possible, check the centering again. Set up the base station transmitter in a sheltered location at least 10 feet (about 3 meters) from the GPS antenna and close to the radio transmitter's antenna. It is best if the airflow of the base station transmitter's cooling fan is not restricted.

The radio transmitting antenna is often mounted on a range pole attached to a tripod. Set the radio transmitting antenna as far as possible from obstructions and as high as stability will allow. Be certain there are no power lines in the vicinity before setting up the radio transmitting antenna to eliminate the danger of electrocution.

The base station transmitter's power is usually provided by a deep-cycle battery. Even though the attendant power cable is usually equipped with a fuse, it is best to be careful to not reverse the polarity when connecting it to the battery. It is also best to have the base station transmitter properly grounded and avoid bending or kinking any cables.

After connecting the base station receiver to the GPS antenna, to the battery and the data collector, if necessary, carefully measure the GPS antenna height. This measurement is often the source of avoidable error, both at the base station and the rovers. Many surveyors measure the height of the GPS antenna to more than one place on the antenna, and it is often measured in both meters and feet for additional assurance.

Select a channel on the base station transmitter that is not in use, and be sure to note the channel used so that it may be set correctly on the rovers as well.

When the RTK work is done, it is best to review the collected data from the data logger. Whether or not fixed height rods have been used, it is a good idea to check the antenna heights. Incorrect antenna heights are a very common mistake. Another bulwark against blunders is the comparison of different observations of the same stations. If large discrepancies arise, there is an obvious difficulty. Along the same line, it is worthwhile to check for discrepancies in the base station coordinates. Clearly, if the base coordinate is wrong, the work created from that base is also wrong. Finally, look at the residuals of the final coordinates to be sure they are within reasonable limits. Remember that multipath and signal attenuation can pass by the observer without notice during the observations but will likely affect the residuals of the positions where they occur.

A base station on a coordinated control position must be available. Its observations must be simultaneous with those at the roving receivers, and it must observe the same constellation of satellites. It is certainly possible to perform a differential survey in which the position of the base station is either unknown or based on an assumed coordinate at the time of the survey. However, unless only relative coordinates are desired, the position of the base station must be known or determined in the end. In other words, the base station must occupy a control position, even if that control is established later. Utilization of the DGPS techniques requires a minimum of four satellites for three-dimensional positioning. RTK ought to have at least five satellites for initialization. Tracking five satellites provides insurance against losing one abruptly; also, it adds considerable strength to the results. While cycle slips are always a problem, it is imperative in RTK that every epoch contains a minimum of four satellite data without cycle slips. This is another reason to always track at least five satellites when doing RTK. Both methods most often rely on real-time communication between the base station and roving receivers.

There is an alternative to the radio link method of RTK; the corrections can be carried to the rover using a cell phone. The cell phone connection does tend to ameliorate the signal interruptions that can occur over the radio link, and it offers a somewhat wider effective range in some circumstances. The use of cell phones in this regard is also a characteristic of Real-Time Network (RTN) solutions.

REAL-TIME NETWORK SERVICES

There is no question that RTK dominates the GPS surveying applications. It is applicable to much of engineering, surveying, air-navigation, mineral exploration, machine control, hydrography, and a myriad of other areas that require centimeter-level accuracy in real-time. However, the requirements of setting up a GPS reference station on a known position, the establishment of a radio frequency transmitter and all attendant components before a single measurement can be made are both awkward and expensive. This, along with the baseline limitation of short baselines, 10 to 20 km for centimeter-level work, has made RTK both more cumbersome and less flexible than most surveyors prefer.

In an effort to alleviate these difficulties, services have arisen around the world to provide RTCM real-time corrections to surveyors by a different means. RTN (see Figure 7.5) have been implemented by both governments and commercial interests. The services are sometimes free and sometimes require subscription or the payment of a fee before the surveyor can access the broadcast corrections over a data link via a modem such as a cell phone or some other device. Nevertheless, there are definite advantages including the elimination of individual base station preparation and the measurement of longer baselines without rapid degradation of the results.

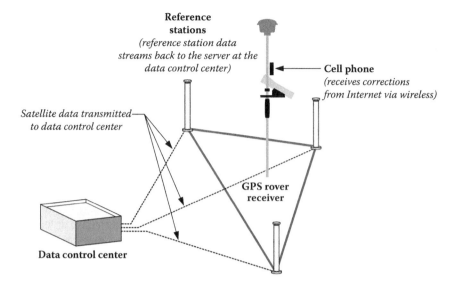

FIGURE 7.5   Real-time network GPS.

These benefits are accomplished by the services gleaning corrections from a whole network of Continuously Operating Reference Stations (CORS) rather than just a single base. In this way, quality control is facilitated by the ability to check corrections from one CORS with those generated from another and should a CORS go off-line or give incorrect values other CORS in the network can take up the slack with little accuracy loss.

The central idea underlying RTN differential corrections is the combination of observations from several CORS at known positions used to derive a model of an entire region. So rather than being considered as isolated beacons with each covering its own segregated area, the CORS are united into a network. The data from the network can then be used to produce a virtual model of the area of interest. From this model, distance-dependent biases such as ionospheric, tropospheric, and orbit errors can be calculated. Once the roving receiver's place within that network is established, it is possible to predict the errors at that position with a high degree of certainty. Not only can the CORS network be used to model errors in a region more correctly, but the multibase solution also can improve redundancy. Solving several baselines that converge on a project point simultaneously rather than relying on just one from typical RTK adds more certainty to resulting coordinates.

Implementing an RTN requires data management and communication. The information from the CORS must be communicated to the central master control station where all the calculations are done. Their raw measurement data, orbits, and so forth, must be managed as they are received in real-time from each of the CORS that make up the network. Along with the modeling of the distance-dependent errors all the integer ambiguities must be fixed for each CORS in real-time. This is probably the most significant data processing difficulty required of an RTN, especially considering that there are usually large distances between the CORS. To accomplish it, post-computed ephemerides, antenna phase center corrections, and all other available information are brought to bear on the solution such as tropospheric modeling and ionospheric modeling. Modeling is subject to variation in both space and time. For example, ionospheric and orbit biases are satellite specific, whereas tropospheric corrections can be estimated station by station. However, the ionospheric, dispersive, biases change more rapidly than the tropospheric and orbit biases, which are nondispersive. Therefore, ionospheric corrections must be updated more frequently than orbit and tropospheric corrections, and while it is best to keep the modeling for ionosphere within the limited area around three or so CORS, when it comes to tropospheric and orbital modeling, the more stations used the better.

Finally, the pseudorange and/or carrier phase residuals must be determined for the L1, L2, and/or L5 by using one of many techniques to interpolate the actual distance-dependent corrections for the surveyor's particular position within the network. Then the subsequent corrections must be communicated to the surveyor in the field, which typically requires the transmission of a large amount of data. There is more than one way the correction can be determined for a particular position within an RTN. So far, there is no clear, best method. One approach is the creation of a position sometimes known as a virtual reference station (VRS) and the attendant corrections. This approach requires a two-way communication link. Users must send their approximate positions to the master control center, usually as a string in the standard format that was defined by the National Marine Electronics Association. The master control center returns corrections for an individual VRS, via RTCM and then the baseline processing software inside the rover calculates its position using the VRS, which seems to the receiver to be a single nearby reference station. Another method involves sending basic RTK-type corrections or the system may broadcast raw data for all the reference stations.

REAL-TIME GPS TECHNIQUES

Station diagrams, observation logs, and to-reach descriptions would rarely be necessary in real-time GPS surveying. However, some components of static GPS control methods are useful. One such technique is offsetting points to avoid multipath and signal attenuation.

OFFSETS

The need to offset points is prevalent in real-time GPS (see Figure 7.6). For example, an offset point must often be established far enough from the original position to avoid an obstructed signal but close enough to prevent unacceptable positioning error. While the calculation of the allowable vertical and horizontal measurement errors can be done trigonometrically, the measurements themselves will be different than those for an offset point in a static survey. For example, rather than the total station point and an azimuth point used in static work, a magnetic fluxgate digital compass and laser may be used to measure the tie from the offset point to the original point in real-time work. It is worth noting that magnetic declination must be accommodated and metal objects avoided when using a magnetically

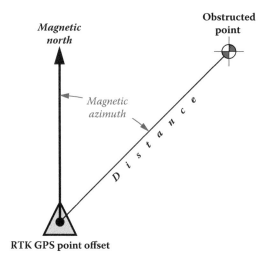

FIGURE 7.6   DGPS or RTK point offset.

determined direction. Such internal compasses should be carefully checked before they are relied upon.

The length of the tie may be measured by an external laser, a laser cabled directly into the GPS receiver, or even a tape and clinometer. Lasers are much more convenient because they can be used to measure longer distances more reliably and taping requires extra field crew members.

Rather than recording the bearing and distance in a field book for post-processing, the tie is usually stored directly in the data collector. In fact, often the receiver's real-time processor can combine the measured distance and the direction of the *sideshot* with the receiver's position and calculate the coordinate of the originally desired position.

DYNAMIC LINES

A technique that is used especially in mobile GPS application is the creation of dynamic lines. The GPS receiver typically moves along a route to be mapped logging positions at predetermined intervals of time or distance. These points can then be joined together to create a continuous line. Obstructions along the route present a clear difficulty for this procedure. Points may be in error or lost completely owing to multipath or signal attenuation. Also, in choosing the epoch interval the capacity of the receiver's memory must be considered, especially when long lines are collected. If the interval chosen is too short, the receiver's storage capacity may be overwhelmed. If the interval is too long, important deflections along the way may be missed.

Where it is impossible or unsafe to travel along the line to be collected in the field, the dynamic line may be collected with a consistent offset (see Figure 7.7). This technique is especially useful in the collection roads and railroads where it is possible to

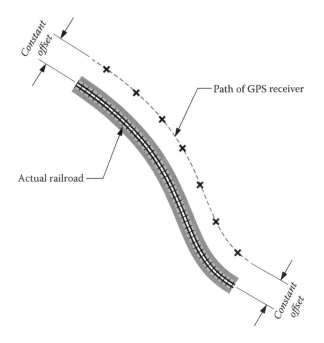

Path of GPS receiver

Actual railroad

Constant offset

Constant offset

FIGURE 7.7   Line offset.

estimate the offset with some certainty due to the constant width of the feature. It is also possible, of course, to collect routes with individual discrete points with short occupations where that approach recommends itself.

PLANNING

Multipath and signal attenuation are particularly troublesome for the real-time GPS. While visibility diagrams are not directly applicable it is nevertheless prudent to plan the work so that at least five GPS satellites are available above the mask angle in the area where data are to be collected. It is also useful to lower the mask angle when conditions warrant it. There are, of course, trade-offs to such strategies that may involve a reduction in positional integrity. The balance between accuracy and productivity is always a consideration.

A FEW RTK PROCEDURES

Redundancy in RTK work can be achieved by occupying each newly established position twice, and it is best if the second occupation is done using a different base station than was used to control the first. If this technique is used, the control points occupied by base stations should not be too close to one another. A minimum of 300 m is a good rule of thumb. Each time the base is set up and before it is taken down, it is best to do a check shot on at least one known control point to verify the work. In order to ensure that the GPS constellation during the second occupation differs substantially from that of the first, it is best if the second occupation takes place not less than 4 hours and not more than 8 hours later or earlier than the first.

To ensure that the centering is correct during the short occupations of RTK, it is best if a bipod is used with a fixed height rod to eliminate the possibility of incorrect height of instrument measurement corrupting the results. Concerning heights, if orthometric heights in real-time are desired, a geoidal model is required and it is best if it is the most recent. However, note that work retraced with a different geoidal model than was used initially will likely show vertical differences at the reoccupied points.

Some rover configurations facilitate *in-fill* surveys. In other words, when the correction signals from the base station fail to reach the rover, the collected data are stored in the memory of the receiver for post-processing after the work is completed.

SITE CALIBRATION

The area of interest that is the project area, covered by an RTK survey, is usually relatively small and defined. Typically a *site calibration*, aka *localization*, is performed to prepare such a GPS project to be done using plane coordinates. A site calibration establishes the relationship between geographical coordinates (i.e., latitude, longitude, and ellipsoidal height) with plane coordinates (i.e., northing, easting) and orthometric heights across the area. In the final analysis the relationship is expressed in three dimensions: translation, rotation, and scale. Because of the inevitable distortion that a site calibration must model, one of the prerequisites for such localization is the enclosure of the area by the control stations that will be utilized during the work.

In the horizontal plane the method of using plane coordinates on an imaginary flat reference surface with northings and eastings, or x and y coordinates, assumes a flat Earth. That is incorrect, of course, but a viable simplification if the area is small enough and the distortion is negligible. Such local tangent planes fixed at discrete points, control points, by GPS site calibration have been long used by land surveyors. Such systems demand little if any manipulation of the field observations, and once the coordinates are derived, they can be manipulated by straightforward plane trigonometry. In short, Cartesian systems are simple and convenient.

However, there are difficulties as the area grows as was mentioned in Chapter 5. For example, typically each of these planes has a unique local coordinate system derived from its own unique site calibration. The axes, the scale, and the rotation of each one of these individual local systems will not be the same as those elements of its neighbor's coordinate system. Therefore, when a site calibration is done and a local flat plane coordinate system is created, it is important to keep all of the work in that system inside limits created by the control points used in its creation. In the simplest case, a single-point calibration, a flat plane is brought tangent to the Earth at one point, but a more typical approach is the utilization of three or four points enclosing the area of interest to be covered by the independent *local coordinate system*. Working outside of the limits created by those points should be avoided as it involves working where the distortion has not been modeled.

It might be said that a site calibration is a *best fit* of a plane onto a curved surface in which the inevitable distortion is distributed in both the horizontal and vertical planes. The vertical aspect is particularly important. It is called on to adjust the measured GPS ellipsoid heights to a desired local vertical datum. Therefore, it must account for undulations in the geoid because the separation between the ellipsoid and geoidal models is seldom if ever consistent over the project area. The separation is not consistent and usually can be modeled approximately as a trend across the area of interest so that the site calibration typically produces an inclined plane in the vertical aspect. Toward that end, the set of control points used to establish the site calibration must have both geographical coordinates (i.e., latitude, longitude, and ellipsoidal height) and plane coordinates (i.e., northing, easting, and orthometric heights in the desired local system). It is best if these control points are from the National Spatial Reference System when possible that enclose the project and are distributed evenly around its boundary.

PRECISE POINT POSITIONING (PPP)

Real-time or static differential GPS and differential GNSS have long been the preferred methods of data processing for surveyors and geodesists. Typically, the differential processing technique depends on one of at least two receivers standing at a control station whose position is known (i.e., the base). It follows that the size of the positional error of the base receiver is knowable. By finding the *difference* between the biases at the base and the biases at the rover, the positional error at the other end of the baseline can be estimated. Through the process of differencing, corrections are generated that reduce the three-dimensional positional error at the unknown point by reducing the level of the biases there. The approach can generally provide up to submeter position from single-frequency pseudorange observations. Differentially processed carrier phase observations can typically reach accuracies of a few centimeters. These facts have led to the construction of networks of Continuously Operating Reference Stations (CORS) on control stations and around the world to support differential processing. There are many regional networks such as the Australian Fiducial Network (AFN) administered by the Australian Surveying and Land Information Group (AUSLIG), EUREF with its EUREF Permanent Network (EPN), the CORS administered by the NGS, AFREF, NAREF, SIRGAS, as well as many commercial networks, and the list is constantly growing.

Some of these networks stream real-time differential corrections to users. These real-time networks (RTNs) support the now well-known real-time kinematic (RTK) methods. The convenience of these RTNs has contributed substantially to the extraordinary expansion of relatively high-accuracy GPS applications that rely on differential processing. However, there is an alternative in both real-time and post-processed work. It is known as Precise Point Positioning (PPP).

Single-point positioning can be a real-time solution using a single receiver measuring to a minimum of four satellites simultaneously. There is no question that this is the most common GPS solution outside of the geodesy and surveying disciplines, and it is in a sense the fulfillment of the original idea of GPS. However, its weakness is that the receiver must rely on the information it collects from the satellite's Navigation message to learn the positions of the satellites, the satellite clock offset, the ionospheric correction, etc. These data contain substantial errors. Under such circumstances, the typical pseudorange or carrier phase single-point position cannot be highly accurate, but what if the positions of the satellites, the satellite clock offset, the ionospheric correction, etc. were not derived from the Navigation message?

There is such a source. It contains much more accurate data about the satellites orbits and the clocks and they enable single-point positioning to achieve higher accuracy and thereby present some advantages over differential methods. For example, the user need not establish control stations or have access to corrections from reference stations operated by others. It ameliorates the limits on the baseline lengths imposed by differential processing, and the solution is global. It can work anywhere in the world. The PPP corrections are expressed in a global reference frame, the *International Terrestrial Reference Frame 2008* (ITRF08), which offers better overall consistency than does a local or regional solution.

The International GNSS Service (IGS) is a collaboration between more than 200 organizations in more than 80 countries. As a public service, it collects and archives GPS and GLONASS data from a worldwide network of more than 300 continuously operating reference stations. It formulates precise satellite ephemerides and clock solutions from these data. Up to eight IGS analysis centers are involved in the processing, and then IGS freely distributes the results. In other words, these data allow users to process their observations using the positions of the satellites and the state of the clocks derived from the period of time the satellites were being tracked. That period of time includes the moment the user actually made the observations. Because these data reflect the precise position of the satellites and clock offsets during the actual measurements, it stands to reason that these data are more precise than the broadcast ephemeris and clock corrections can be. There are several categories of these data available.

POST PROCESSED (PP-PPP)

The *observed Ultra-Rapid* ephemeris and clock data are available online 3 to 9 hours after an observation is completed. It is posted 4 times daily at 03hr, 09hr, 15hr, and 21hr UTC. There is also a *predicted Ultra-Rapid ephemeris* and clock data that are available ahead of time and are posted at the same times as the observed Ultra-Rapid. It takes longer for the *Rapid* to come online, from 17 to 41 hours. It is posted each day at 17hr UTC. The *Final* product takes the longest of all, 12–18 days. It is posted each week on Thursday. As would be expected, the accuracy of the ephemeris and the clock data of each increment increase.

These post-computed data have been available for more than a decade, and there are free PPP post-processing services. Users may upload their data files to these service websites (in the RINEX format) and be served automatically computed GNSS receiver positions at the centimeter level. The services do require that the submitted data be derived from long observation times.

In the United States, the National Oceanic and Atmospheric Administration (NOAA) and, more specifically, the National Geodetic Survey (NGS) have worked with IGS to provide accurate GPS satellite ephemerides or orbits.

REAL-TIME SERVICE (RTS-PPP)

Recently, near real-time information has also become available from IGS. This real-time service (RTS) was created in partnership with other organizations, specifically Natural Resources Canada (NRCan), the German Federal Agency for Cartography and Geodesy (BKG), and the European Space Agency's Space Operations Centre in Darmstadt, Germany (ESA/ESOC). The RTS's precise ephemeris and clock data are available on the Internet every 25 s via an open source protocol known as the Network Transport of RTCM (NTRIP), which has been an RTCM standard for the real-time collection and distribution of GNSS information since 2004.

These precise satellite orbits and clocks along with code and phase observations from a dual-frequency receiver provide the data from which the PPP algorithm derives accurate positions. This is currently a GPS-only service, but there are plans for it to be expanded to the Russian GLONASS system and other GNSS constellations.

PPP DISADVANTAGE

PPP currently has disadvantages. One of the most persistent is the time necessary to resolve the cycle ambiguity. The time necessary to move from a float to a fixed solution is extended because the ambiguity cannot be assumed to be an integer as it is in a differenced solution. As things stand, the convergence can take 20 min or more. The situation may be improved with the increase in satellite observables that will come with the inclusion of GLONASS in the IGS PPP solution.

You may recall that ionospheric delays are significantly reduced in differenced solutions. In PPP, dual-frequency receivers are needed to mitigate the ionospheric delay.

If data about precise satellites orbits and the clocks are available, a single receiver can achieve accuracies commensurate with those available from differenced measurements. This can be done without the user establishing control stations or using corrections directly from continuously operating reference stations. PPP is more consistent over large areas than are the usual results from difference observations, and it works over longer baselines.

EXERCISES

1. What is the meaning of latency as applied to DGPS and RTK?
   a. Baud rate of a radio modem in real-time GPS
   b. Time taken for a system to compute corrections and transmit them to users in real-time GPS
   c. Frequency of the RTCM SC104 correction signal in real-time GPS
   d. Range rate broadcast with the corrections from the base station

2. Which of the following errors are not reduced by using DGPS or RTK methods?
   a. Atmospheric errors
   b. Satellite clock bias
   c. Ephemeris bias
   d. Multipath

3. Who was the original kinematic GPS innovator?
   a. Ann Bailey
   b. Benjamin Remondi
   c. George F. Syme
   d. R. E. Kalman

4. Which of the following statements about rules governing RTK radio communication is correct?
   a. According to the FCC regulations, voice communications have priority over data communications.
   b. Code of Federal Regulations, 48 CFR 91, contains the complete text of the FCC Rules.
   c. Most typical community voice repeater frequencies are about 900 MHz.
   d. In Europe it is possible to use the 4.2 GHz band for spread spectrum communication without special authorization.

5. Which of the following data rates is closest to that required for RTK?
   a. 200 bps updated every 10 s
   b. 1400 bps updated about every 5 s
   c. 2400 bps updated about every 1/2 s
   d. 3500 bps updated about every 0.05 of a second

6. How many satellites is the minimum for the best initialization with RTK?
   a. 4 satellites
   b. 5 satellites
   c. 6 satellites
   d. 8 satellites

7. Which of the following is not a part of a typical data radio transmitter used in RTK?
   a. Radio modulator
   b. Amplifier
   c. Demodulator
   d. Antenna

8. Why is it advisable that successive occupations in a GPS survey be separated by a period of time?
   a. To eliminate multipath when it corrupted the first occupation of the station
   b. To allow the satellite constellation to reach a significantly different configuration than it had during the first occupation of the station

   c.   To overcome an overhead obstruction during the first occupation of the
       station

   d.   To eliminate receiver clock errors

9. Which of the following is the weakness of the PPP?
   a.   Length of time necessary to resolve the cycle ambiguity
   b.   PPP does not use base stations
   c.   RTCM message is not received by the receiver
   d.   The PPP uses only one receiver

ANSWERS AND EXPLANATIONS

1. Answer is (b)

   Explanation: In DGPS and RTK it takes some time for the base station to
calculate corrections and it takes some time for it to transmit them. The base
station's data are put into packets in the correct format. The data make their way
from the base station to the rover over the data link. They must then be decoded
and go through the rover's software. The time it takes for all of this to happen is
called the latency of the communication between the base station and the rover.
It can be as little as a quarter of a second or as long as a couple of seconds.

2. Answer is (d)

   Explanation: In autonomous point positioning, all the biases discussed
in this chapter affect the GPS position. For example, in autonomous point
positioning, the satellite clock bias normally contributes about 3 m to the
final GPS positional error, ephemeris bias contributes about 5 m and solar
radiation, and so forth, contribute about another 1.5 m. However, with
DGPS and RTK, when errors are well correlated, all these biases are virtually zero, as are the atmospheric errors, unlike autonomous point positioning where the ionospheric delay contributes about 5 m and the troposphere
about 1.5 m. While most other errors in GPS discussed in this chapter can
be mediated, or canceled, owing to the relatively short distance between
receivers, the same cannot be said for multipath.

3. Answer is (b)

   Explanation: Kinematic surveying, also known as stop-and-go kinematic
surveying, is not new. The original kinematic GPS innovator, Dr. Benjamin
Remondi, developed the idea in the mid-1980s.

4. Answer is (a)

   Explanation: In the United States, most transmitters connected to RTK
GPS surveying equipment operate between 450–470 MHz, and voice
communications also operate in this same range. It is important for GPS
surveyors to know that voice communications have priority over data communications. The FCC requires cooperation among licensees that share frequencies. Interference should be minimized. For example, it is wise to avoid
the most typical community voice repeater frequencies. They usually occur

between 455–460 MHz and 465–470 MHz. Part 90 of the Code of Federal Regulations, 47 CFR 90, contains the complete text of the FCC Rules.

There are also other international and national bodies that govern frequencies and authorize the use of signals elsewhere in the world. In some areas, certain bands are designated for public use, and no special permission is required. For example, in Europe it is possible to use the 2.4 GHz band for spread spectrum communication without special authorization with certain power limitations. Here in the United States the band for spread spectrum communication is 900 MHz.

5. Answer is (b)

Explanation: DGPS operations may need no more than 200 bits per second (bps), updated every 10 s or so. RTK requires at least 2400 bps updated about every 1/2 s or less.

6. Answer is (b)

Explanation: Utilization of the DGPS techniques requires a minimum four satellites for three-dimensional positioning. RTK ought to have at least five satellites for initialization. Tracking five satellites is a bit of insurance against losing one abruptly; also, it adds considerable strength to the results. While cycle slips are always a problem, it is imperative in RTK that every epoch contains a minimum of four satellite data without cycle slips. This is another reason to always track at least five satellites when doing RTK. However, most receivers allow the number to drop to four after initialization.

7. Answer is (c)

Explanation: A data radio transmitter consists of a radio modulator, an amplifier, and an antenna. The modulator converts the correction data into a radio signal. The amplifier increases the signal's power, and then the information is transmitted via the antenna.

8. Answer is (b)

Explanation: Successive occupations ought to be separated by a period of time so the satellite constellation can reach a significantly different configuration than that which it had during the first occupation. Recall that GPS measurements are not actually made between the occupied stations but directly to the satellites themselves.

9. Answer is (a)

Explanation: PPP currently has disadvantages. One of the most persistent is the time necessary to resolve the cycle ambiguity. The time necessary to move from a float to a fixed solution is extended because the ambiguity cannot be assumed to be an integer as it is in a differenced solution. As things stand, the convergence can take 20 min or more. The situation may be improved with the increase in satellite observables that will come with the inclusion of GLONASS in the IGS PPP solution.

8 Global Positioning System Modernization and Global Navigation Satellite System

GLOBAL POSITIONING SYSTEM (GPS) MODERNIZATION

Configuration of the GPS Space Segment is well known. The satellites are in orbit at a nominal height of about 20,000 km above the Earth. There are three carriers L1 (1575.42 MHz), L2 (1227.60 MHz), and L5 (1176.42 MHz). A minimum of 24 GPS satellites ensure 24 hour worldwide coverage, but there are more than that minimum on orbit. There are a few spares on hand in space. The redundancy is prudent because GPS is critical to positioning, navigation, and timing, of course. It is also critical to the smooth functioning of financial transactions, air traffic, ATMs, cell phones, and modern life, in general, around the world. This very criticality requires continuous modernization.

GPS was put in place with amazing speed, considering the technological hurdles and reached its fully operational capability on July 17, 1995. The oldest satellites in the current constellation were launched in the early 1990s. If you imagine using a personal computer of that vintage today, it is not surprising that there are plans in place to alter the system substantially. In 2000, U.S. Congress authorized the GPS III effort. The project involves new ground stations and satellites, additional civilian and military navigation signals, and improved availability. This chapter is about some of the changes in the modernized GPS, its inclusion in the Global Navigation Satellite System, and more.

SATELLITE BLOCKS

BLOCK I, BLOCK II/IIA, BLOCK IIR, AND BLOCK III SATELLITES

Here is an illustration that summarizes the improvements made in the satellite blocks that have made up the GPS constellation over the years (Table 8.1).

Block I

The first of the 11 successful Block I satellites was launched in 1978 from Vandenberg Air Force Base. The last of them was launched in 1985. One launch failed, Navstar 7. They were all retired by late 1995. None of the Block I satellites are in orbit now.

These satellites needed frequent help from the Control Segment. They could operate independently for only 3½ days. The Control Segment handled the necessary

TABLE 8.1
GPS Modernization

			Launch Dates			
1978–1985	**1989–1990**	**1990–1997**	**1997–2004**	**2005–2009**	**2010–?**	**Planned**
Block I	**Block II**	**Block IIA**	**Block IIR**	**Block IIR-M**	**Block IIF**	**Block III**
Demostrated GPS – L1 (CA) navigation signal – L1 and L2 (P code) navigation signal – 4.5-year design life	Some of the improvements over Block I – 14 days operation without Control Segment – 7.3-year design life	Some of the improvements over Block II – 6 months operation without Control Segment (with degradation) – 7.5-year design life – Radiation hardened	Some of the improvements over Block IIA – 6 months operation without Control Segment (without on-board clock monitoring degradation) – Intersatellite cross-link ranging (AutoNav) – 3 rubidium frequency standards – 7.8-year design life – Distress Alerting Satellite System (DASS) proof of concept	Some of the improvements over Block IIR – Second civil signal L2 (L2C) – M code on L1/L2 – L5 Demo – Flexible power levels for military signals – Anti-jam flex power	Some of the improvements over Block IIR-M – Third civil signal L5 – 12-year design life – Improved rubidium frequency standards – Direct orbit insertion – Operational Distress Alerting Satellite System (DASS) repeaters	Some of the expected improvements over Block IIF IIIA – Increased Earth coverage power – 15-year design life – Fourth civil signal (L1C) – On-board Laser Retro-reflector Arrays (LRA) IIIB – 2-4 crosslink antennas for more efficient uploads IIIC – Spot beam for anti-jam (AJ) – Navigation integrity

momentum dumping for the satellites and maintained their attitudes using hydrazine thrusters. The inclination of these satellites relative to the equator was 63° instead of the inclination of 55° used for subsequent GPS blocks.

They had a design life of 4.5 years, though some operated for double that. They were powered by 7.25 m² of solar panels, and they also had three rechargeable nickel-cadmium batteries. In subsequent blocks of satellites, design lives increased and dependence on the Control Segment decreased. However, some of the features of Block I were carried forward into the subsequent blocks of GPS satellites. They carried onboard nuclear detonation detection sensors a feature that has continued in future GPS satellite blocks.

It was clear from the beginning that atomic frequency standards, clocks, were necessary for the proper functioning of the system. Therefore, the Block I satellites had cesium and rubidium frequency standards onboard, a feature that future GPS satellites share. The first three Block I satellites carried three rubidium clocks. Unfortunately, they stopped working after about a year in space. The three rubidium standards were improved. Equipment was added to keep the frequency standards at a constant temperature during flight and 1 cesium frequency standard was added to subsequent satellites in this block.

Block II

The Block II satellites were about twice as heavy as the Block I satellites. The first of them was launched in 1989. The Block II satellites often exceeded their 7.3-year design life. The last was decommissioned in 2007 after 17 years of operation. They could be autonomous, without contact with the Control Segment, for up to 14 days. Uploads from the Control Segment to the Block II satellites were encrypted unlike uploads during the Block I. The signals from the Block II satellites were periodically and purposely disrupted. Specifically, the onboard clocks were intentionally dithered in a procedure known as Selective Availability (SA).

Block IIA

Block IIA satellites are an improved version of the Block II. The first of 19 Block IIA satellites was launched in 1990. While none of the Block II satellites are functioning today, some of the Block IIA satellites are still healthy. There are six Block IIA satellites in orbit and operational. These survivors are now the oldest of the GPS satellites operating in orbit.

They are radiation hardened against cosmic rays, built to provide SA, anti-spoofing (AS) capability, and onboard momentum dumping. This SA continued until May 2, 2000, when it was discontinued.

Block IIA satellites can store more of the Navigation message than the Block II satellites could and can, therefore, operate without contact with the Control Segment for 6 months. However, if that were actually done, their broadcast, ephemeris, and clock correction would degrade.

Two Block IIA satellites, SVN 35 (PRN 05) and SVN 36 (PRN 06), have been equipped with Laser Retro-reflector Arrays (LRA). The second of these was launched in 1994 and is still in service. The retro-reflectors facilitate satellite laser ranging (SLR). Such ranging can provide a valuable independent validation of GPS orbits.

Like the Block II satellites, the Block IIA satellites are equipped with two rubidium and two cesium frequency standards. They are expected to have a design life of 7.3 years. While the design life has obviously been exceeded in most cases, Block IIA satellites do wear out.

Block IIR

The first launch of the next Block, Block IIR satellites in January 1997, was unsuccessful. The following launch in July 1997 succeeded. There are 12 Block IIR satellites in orbit and operational. There are some differences between the Block IIA and the Block IIR satellites. The Block IIR satellites have a design life of 7.8 years and can determine their own position using intersatellite crosslink ranging called AutoNav. This involves their use of reprogrammable processors onboard to do their own fixes in flight.

They can operate in that mode for up to 6 months and still maintain full accuracy. The Control Segment can also change their software while the satellites are in flight and, with a 60 day notice, move them into a new orbit. Unlike some of their direct predecessors, these satellites are equipped with three rubidium frequency standards. Some of the Block IIR satellites also have an improved antenna panel that provides more signal power. They are more radiation hardened than their predecessors, and they cost about a third less than the Block II satellites did.

Despite their differences, Block IIA and Block IIR satellites are very much the same in some ways. They both broadcast the same fundamental GPS signals that have been in place for a long time. Their frequencies are centered on L1 and L2. The Coarse/Acquisition code or C/A code is carried on L1 and has a chipping rate of 1.023 million chips per second. It has a code length of 1023 chips over the course of a millisecond before it repeats itself. There are actually 32 different code sequences that can be used in the C/A code, more than enough for each satellite in the constellation to have its own. The Precise code or P code on L1 and L2 has a chipping rate that is 10× faster than the C/A code at 10.23 million chips per second. The P code has a code length of about a week, approximately 6 trillion chips, before it repeats. If this code is encrypted, it is known as the P(Y) code, or simply the Y code.

Nine of the Block IIR satellites carry Distress Alerting Satellite System (DASS) repeaters. These DASS repeaters are used to relay distress signals from emergency beacons and were part of a proof of the concept of satellite-supported search and rescue that was completed in 2009. Twelve additional IIR satellites will carry them, too.

Block IIR-M

In the current constellation, there are seven Block IIR-M satellites in orbit and operational. These are IIR satellites that were modified before they were launched. The modifications upgraded these satellites so that they radiate two new codes: a new military code, the M code, and a new civilian code, the L2C code. The modifications also demonstrate a new carrier, L5. The L2C code is broadcast on L2 only, and the M code is on both L1 and L2. The L2C code helps in the correction of the ionospheric delay, and the M code improves the military anti-jamming efforts through flexible power capability. One of the Block IIR-M satellites, SVN 49, transmits on L1, L2, and L5. L5 is a frequency intended for safety-of-life applications. The first of these

Block IIR-M satellites was launched in the summer 2005 and the last in the summer 2009.

Block IIF

The first Block IIF satellite was launched in the summer of 2010. As of 2014, there are seven Block IIF satellites in orbit and 33 are planned. GPS IIF-7 was launched August 1, 2014. Their design life is 12 to 15 years. Block IIF satellites have faster processors and more memory onboard. They broadcast all of the previously mentioned signals, and one more, a new carrier known as L5. This is a signal that was demonstrated on the Block IIR-M. It will be available from all of the Block IIF satellites. The L5 signal is within the Aeronautical Radio Navigation Services (ARNS) frequency and can service aeronautical applications. The improved rubidium frequency standards on Block IIF satellites have a reduced white noise level. The Block IIF satellite's launch vehicles can place the satellites directly into their intended orbits so they do not need the apogee kick motors their predecessors required. All of the Block IIF satellites will carry DASS repeaters. The Block IIF satellites will replace the Block IIA satellites as they age. Their onboard *navigation data units* support the creation of new Navigation messages with improved broadcast ephemeris and clock corrections. Like the Block IIR satellites, the Block IIF can be reprogrammed on orbit.

Block III

Block III satellites will replace the older Block IIR satellites as they are taken out of service. As yet, there are no Block III satellites in orbit. This block will be deployed in three increments. The first of these is known as Block IIIA. It will be resistant to hostile jamming. The next two increments are Block IIIB and Block IIIC. Higher power is planned for the signals broadcast by the IIIB satellites. The IIIB and IIIC satellites will also carry Distress Alerting Satellite System (DASS) repeaters.

When the whole GPS constellation has DASS repeaters on board, there will be global coverage for satellite-supported search and rescue and at least four DASS-equipped satellites will always be visible from anywhere on Earth. This system will enhance the international Cospas-Sarsat satellite-aided search and rescue (SAR) system and will be interoperable with the similarly planned Russian (SAR/GLONASS) and European (SAR/Galileo) systems.

Block III satellites will have cross-link capability to support intersatellite ranging and transfer, telemetry, tracking, and control capability. Block IIIB satellites will have from two- to four-directional crosslink antennas. This means they can be updated from a single ground station instead of requiring each satellite to be in the range of a ground antenna to be updated. This and their high-speed upload and download antennas could help increase the upload frequency from once every 12 hours to once every 15 min.

Each Block III satellite will have three enhanced rubidium frequency standards (clocks) and a fourth slot will be available for a new clock, i.e., a hydrogen maser.

It was established in 2010 that all Block III satellites will have onboard Laser Retro-reflector Arrays (LRA) (aka *retro-reflectors*). The satellite laser tracking available with this payload will provide data from which it will be possible to distinguish between clock error and ephemeris error. Similar LRA are planned for the Russian (GLONASS) and European (Galileo) systems.

There is a plan for these satellites that includes the broadcast of a new civil signal, known as L1C, on the L1 carrier. This signal was designed with international cooperation to maximize interoperability with Galileo's Open Service Signal and Japan's Quazi-Zenith Satellite System (QZSS).

Codes available from earlier blocks (i.e., the M code, L5, the P code, and the C/A code) will be broadcast with increased power from the Block III satellites. The broadcast of the M code will change in an interesting way. It will continue to be radiated with a *wide angle* to cover the full Earth just as in the Block IIR-M satellites, but the Block IIIC M code will also have a rather large deployable high-gain antenna to produce a directional *spot beam*. The spot beam will have approximately 100× more power (−138 dBW) compared with (−158 dBW) the wide angle M code broadcast. It will have the anti-jam capability to be aimed at a region several hundreds of kilometers in diameter. A side-effect of having two antennas is that the GPS satellite will appear to be two GPS satellites occupying the same position to those inside the spot beam. While the full Earth M code signal is available on the Block IIR-M satellites, the spot beam antennas will not be available until the Block III satellites are deployed (Figure 8.1).

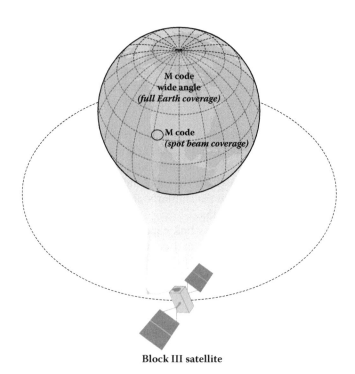

Block III satellite

FIGURE 8.1   Spot beam.

POWER SPECTRAL DENSITY DIAGRAMS

Many of the improvements in GPS are centered on the broadcast of new signals. Therefore, it is pertinent to have a convenient way to visualize all the GPS and GNSS signals that illustrates the differences in the new signals and a good deal of signal theory as well. It is a diagram of the power spectral density (PSD) function. They graphically illustrate the signals power per bandwidth in Watts per Hertz as a function of frequency.

In GPS and GNSS literature the PSD diagram is often represented with the frequency in MHz on the horizontal axis and the density, the power, represented on the perpendicular axes in decibels relative to one Hertz per Watt or dBW/Hz as shown in Figure 8.2.

dBW/Hz

Perhaps a bit of background is in order to explain those units. A bel unit originated at Bell Labs to quantify power loss on telephone lines. A decibel is a tenth of a bel. A decibel, dB, is a dimensionless number. In other words, it's a ratio that can acquire dimension by being associated with measured units. Here are some of the quantities with which it is sometimes associated: seconds of time, symbolized dBs, bandwidth measured in Hertz, symbolized dBHz; and temperature measured in Kelvins, symbolized dBK. Because signal power is of interest here, dB will be described with respect to 1 Watt; the symbol used is dBW.

The dBW is a short concise number that can conveniently express the wide variation in GPS signal power levels. It can represent quite large and quite small amounts

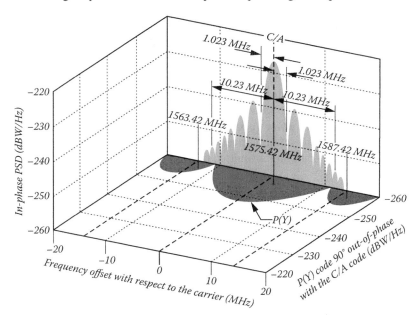

FIGURE 8.2   Legacy L1 signal.

of power more handily than other notations. For example, consider a value of interest in GPS signals. The value is very small, one tenth of a millionth billionth of a watt. Expressing it as 0.0000000000000001 W is a bit exhausting. It would be more convenient expressed in dBW, a value that can be derived using

$$P_{dBW} = 10\log_{10}\frac{P_W}{1\ W}$$

where P_W is the power of the signal.

$$P_{dBW} = 10\log_{10}\frac{10^{-16}\ W}{1\ W}$$

$$P_{dBW} = 10\log_{10}10^{-16}\ W$$

$$-160\ dBW = 10\log_{10}10^{-16}\ W$$

The expression −160 dBW is immediately useful. Here's an example. A change in 3 dB is always an increase or a decrease of 100 percent in power level. Stated another way, a 3 dB increase indicates a doubling of signal strength and a 3 dB decrease indicates a halving of signal strength. Therefore, it is easy to see that a signal of −163 dBW has half the power of a signal of −160 dBW. Considering the broadcasts from the current constellation of satellites, the minimum power received from the P code on L1 by a GPS receiver on the Earth's surface is about −163 dBW and the minimum power received from the C/A code on L1 is about −160 dBW. This difference between the two received signals is not surprising because at the start of their trip to Earth they are transmitted by the satellite at power levels that are also 3 dB apart. The P code on L1 is transmitted at a nominal +23.8 dBW (240 W), whereas the nominal transmitted power of the C/A code on L1 is +26.8 dBW (479 W). It is interesting to note that the minimum received power of the P code on L2 is even less at −166 dBW and its nominal transmitted power is +19.7 dBW (93 W).

$$P_{dBW} = 10\log_{10}\frac{P_W}{1\ W}$$

$$P_{dBW} = 10\log_{10}\frac{93\ W}{1\ W}$$

$$+19.7\ dBW = 10\log_{10}93\ W$$

One might wonder why there are such differences between the power of the transmitted GPS signal, called the effective isotropic radiated power, and the power of the received signal. The difference is large. It is 186 to 187 dB, nearly 10 quintillion decibels. The loss is mostly because of the 20,000 km distance from the satellite to a GPS receiver on the Earth. There is also an atmospheric loss and a polarization mismatch loss, but the biggest loss by far, about 184 dB, is along the path in free space.

Much of this loss is a function of the spreading out of the GPS signal in space as described by the inverse square law. The intensity of the GPS signal varies inversely to the square of the distance from the satellite. In other words, by the time the signal makes that trip and reaches the GPS receiver, it is pretty weak (see Figure 8.3). It follows that GPS signals are easily degraded by vegetation canopy, urban canyons, and other interference.

A GPS signal has power, of course, but it also has bandwidth. PSD is a measure of how much power a modulated carrier contains within a specified bandwidth. That

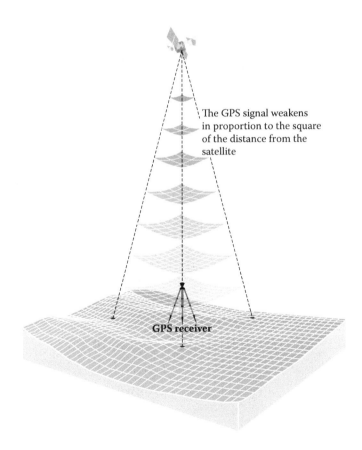

The GPS signal weakens in proportion to the square of the distance from the satellite

GPS receiver

FIGURE 8.3   Inverse square law.

value can be calculated using the following formula and allowing that there is an even distribution of 10^{-16} W over the 2.046 MHz C/A bandwidth of the C/A code:

$$\text{Power density}\left(\frac{\text{dBW}}{\text{Hz}}\right) = 10\log_{10}\left[\frac{\text{power (W)}}{\text{bandwidth (Hz)}}\right]$$

$$\text{Power density}\left(\frac{\text{dBW}}{\text{Hz}}\right) = 10\log_{10}\left[\frac{10^{-16}\text{ W}}{2.046\times10^{6}\text{ Hz}}\right]$$

$$\text{Power density}\left(\frac{\text{dBW}}{\text{Hz}}\right) = 10\log_{10}\left[4.888^{-23}\right]$$

$$\text{Power density}\left(\frac{\text{dBW}}{\text{Hz}}\right) = 10(-22.3)$$

$$\text{Power density} = -223\text{ dBW/Hz}$$

The calculation is also frequently normalized and done presuming an even distribution of 1 W over the 2.046 MHz C/A bandwidth of the C/A code. In other words, the following calculation presumes an even distribution of the power over 1 W instead of the 10^{-16} W used in the previous calculation:

$$\text{Power density}\left(\frac{\text{dBW}}{\text{Hz}}\right) = 10\log_{10}\left[\frac{\text{power (W)}}{\text{bandwidth (Hz)}}\right]$$

$$\text{Power density}\left(\frac{\text{dBW}}{\text{Hz}}\right) = 10\log_{10}\left[\frac{1\text{ W}}{2.046\times10^{6}\text{ Hz}}\right]$$

$$\text{Power density}\left(\frac{\text{dBW}}{\text{Hz}}\right) = 10\log_{10}\left[4.888^{-7}\right]$$

$$\text{Power density}\left(\frac{\text{dBW}}{\text{Hz}}\right) = 10(-6.31)$$

$$\text{Power density} = -63\text{ dBW/Hz}$$

L1 LEGACY SIGNALS

The PSD diagrams show the increase or decrease, in decibels, of power, in Watts with respect to frequency in Hertz. Figure 8.2 illustrates the PSD diagrams of the well-known codes on L1. In the diagram the C/A code on the L1 signal is centered on the frequency 1575.42 MHz and a portion of the bandwidth over which it is spread, approximately 20.46 MHz, is shown evenly split, 10.23 MHz, on each side of the center frequency. The horizontal scale shows the offset in MHz from 1575.42 center frequency. Other scales show the decibels relative to 1 Watt per Hertz (dBW/Hz).

The P(Y) code is in quadrature that is 90° from the C/A code. In both cases the majority of the power is close to the center frequency. The C/A code has many lobes, but the P code with the same bandwidth with 10× the clock rate has just the one main lobe.

NEW SIGNALS

An important aspect of GPS modernization is the advent of some new and different signals that are augmenting the old reliable codes. In GPS a dramatic step was taken in this direction on September 21, 2005, when the first Block IIR-M satellite was launched. One of the significant improvements coming with the Block IIRM satellites is increased L-band power on both L1 and L2 by virtue of the new antenna panel. The Block IIR-M satellites will also broadcast new signals, such as the M code.

M Code

Eight to twelve of these replenishment satellites are going to be modified to broadcast a new military code, the M code. This code will be carried on both L1 and L2 and will probably replace the P(Y) code eventually. It has the advantage of allowing the Department of Defense (DoD) to increase the power of the code to prevent jamming. There was consideration given to raising the power of the P(Y) code to accomplish the same end, but that strategy was discarded when it was shown to interfere with the C/A code.

The M code was designed to share the same bands with existing signals, on both L1 and L2, and still be separate from them. See those two peaks in the M code in Figure 8.4. They represent a split-spectrum signal about the carrier. Among other

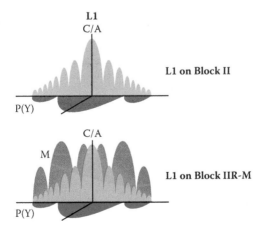

FIGURE 8.4   L1 with M code.

things, this allows minimum overlap with the maximum power densities of the P(Y) code and the C/A code, which occur near the center frequency. That is because the actual modulation of the M code is done differently. It is accomplished with binary offset carrier (BOC) modulation, which differs from the binary phase shift key (BPSK) used with the legacy C/A and P(Y) signals. As a result of this BOC modulation, the M code has its greatest power density at the edges, which is at the nulls, of the L1 away from P(Y) and C/A. This architecture both simplifies implementation at the satellites and receivers and also mitigates interference with the existing codes. Suffice it to say that this aspect and others of the BOC modulation strategy offer even better spectral separation between the M code and the older legacy signals.

The M code is unusual in that a military receiver can determine its position with the M code alone, whereas with the P(Y) it must first acquire the C/A code to do so. It is also spread across 24 MHz of the bandwidth.

Perhaps it would also be useful here to mention the notation used to describe the particular implementations of the binary offset carrier. It is characteristic for it to be written BOC (α, β). Here the α indicates the frequency of the square wave modulation of the carrier, also known as the subcarrier frequency factor. The β describes the frequency of the pseudorandom noise modulation, also known as the spreading code factor. In the case of the M code the notation BOC (10, 5) describes the modulation of the signal. Both here are multiples of 1.023 MHz. In other words, their actual values are

$$\alpha = 10 \times 1.023 \text{ MHz} = 10.23 \text{ MHz and } \beta = 5 \times 1.023 \text{ MHz} = 5.115 \text{ MHz (Betz)}$$

The M code is tracked by direct acquisition. This means that as mentioned in Chapter 1, the receiver correlates the signal coming in from the satellite with a replica of the code that it has generated itself.

L2 SIGNAL

In Figure 8.5 the L2 signal diagram is centered on 1227.60 MHz. As you can see, it is similar to the L1 diagram except for the absence of the C/A code, which is, of course, not

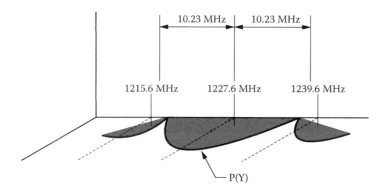

FIGURE 8.5   Legacy L2 signal.

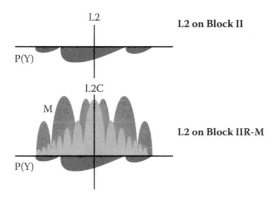

FIGURE 8.6 L2 with M code and L2C.

carried on the L2 frequency. As well-known as these are, this state of affairs is changing. We have been using the L2 carrier since the beginning of GPS, of course, but now there will be two new codes broadcast on the carrier, L2, that previously only carried one military signal exclusively, the P(Y) code. Now L2 will carry a new military signal, the M code, just discussed, and a new civil signal as well (Figure 8.6).

L2C

A new military code on L1 and L2 may not be terribly exciting to civilian users, but these IIR-M satellites have something else going for them. They broadcast a new civilian code. We have been using the L2 carrier since the beginning of GPS, of course, but now there will be two new codes broadcast on the carrier, L2, that previously only carried one military signal, the P(Y) code. Now L2 will carry the new military signal, the M code, and a new civil signal as well. This is a code that was first announced back in March 1998. It is transmitted by all block IIR-M satellites and subsequent blocks and is known as L2C. The *C* is for civil.

Even though its 2.046 MHz from null to null gives it a very similar power spectrum to the C/A code, it is important to note that L2C is a copy of the C/A code even though that was the original idea. The original plan was that it would be a replication of the venerable C/A code but carried on L2 instead of L1. This concept changed when Colonel Douglas L. Loverro, Program Director for the GPS Joint Program Office, was asked if perhaps it was time for some improvement of C/A. The answer was yes. The C/A code is somewhat susceptible to both waveform distortion and narrowband interference and its cross-correlation properties are marginal at best. So the new code on L2, known as L2 civil, or L2C, was announced. It is more sophisticated than C/A.

Civil-Moderate (CM) and Civil-Long (CL)

L2C is actually composed of two pseudorandom noise signals: the civil-moderate length code, CM, and the civil long code, CL. They both utilize the same modulation scheme, binary phase shift key (BPSK), as the legacy signals and both signals are broadcast at 511.5 kilobits per second (Kbps). This means that CM repeats its 10,230 chips every 20 ms and CL repeats its 767,250 chips every 1.5 s.

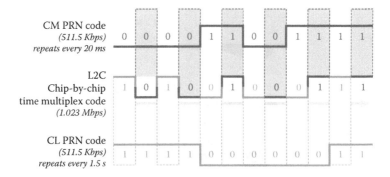

FIGURE 8.7   L2C signal structure.

But wait a minute, how can you do that? How can you have two codes in one? L2C achieves this by time multiplexing. Because the two codes have different lengths, L2C alternates between chips of the CM code and chips of the CL code as shown in Figure 8.7. It is called chip-by-chip time multiplexing. So even though the actual chipping rate is 511.5 KHz, half the chipping rate of the C/A code, with the time multiplexing it still works out that taken together L2C ends up having the same overall chip rate as L1 C/A code, 1.023 MHz. This provides separation from the M code.

L2C has better autocorrelation and cross-correlation protection than the C/A code because both of the CM and CL codes are longer than the C/A code. Longer codes are easier to keep separate from the background noise. In practice, this means these signals can be acquired with more certainty by a receiver that can maintain lock on them more surely in marginal situations where the sky is obstructed. There is also another characteristic of L2C that pays dividends when the signal is weak. While CM carries newly formatted navigation data and is, therefore, known as the *data channel*, CL does not. It is dataless and known as a *pilot channel*. A pilot channel can support longer integration when the signal received from the satellite is weak. This is an idea that harks all the way back to Project 621B at the very beginnings of GPS. The benefits of a pilot channel distinct from the data components carried by a signal was known in the 1960s and 1970s but was not implemented in GPS until recently.

Phase-Locked Loop

Even though the L2C signals' transmission power is 2.3 dB weaker than is C/A on L1 and even though it is subject to more ionospheric delay than the L1 signal, L2C is still much more user friendly. The long dataless CL pilot signal has 250× (24 dB) better correlation protection than C/A. This is due in large part to the fact that the receiver can track the long dataless CL with a phase-locked loop instead of a squaring Costas loop that is necessary to maintain lock on CM, C/A, and P(Y). This allows for improved tracking from what is, in fact, a weaker signal and a subsequent improvement in protection against continuous wave interference. As a way to illustrate how this would work in practice, here is one normal sequence by which a receiver would lock onto L2C. First, there would be acquisition of the CM code with a frequency locked or Costas loop; next, there would be testing of the 75 possible phases of CL, and, finally, acquisition of CL. The CL as mentioned can be then tracked with a basic phase-locked loop. Using this strategy, even though L2C is weaker than C/A, there is actually an improvement in the threshold of nearly 6 dB by tracking the CL with the phase-locked loop. Compared to the C/A code, L2C has 2.7 dB greater data recovery and 0.7 dB greater carrier tracking.

PRACTICAL ADVANTAGES

Great, so what does all that mean in English? Having two civilian frequencies being transmitted from one satellite affords the ability to model and lessen the ionospheric delay error for that satellite while relying on code phase pseudorange measurements alone. In the past, ionospheric modeling was only available to multi-carrier frequency observations, or by reliance on the atmospheric correction in the Navigation message. Before May 2, 2000, with Selective Availability on a code-based receiver could get you within 30 to 100 m of your true position, when SA was turned off that was whittled down to 15 to 20 m or so under very good conditions. However, with just one civilian code C/A on L1, there was no way to remove the second largest source of error in that position, the ionospheric delay. Now with two civilian signals, one on L1 (C/A) and one on L2 (L2C), it becomes possible to effectively model the ionosphere using code phase. In other words, it may become possible for an autonomous code-phase receiver to achieve positions with a 5–10 m positional accuracy with some consistency.

The L2C signal also ameliorates the effect of local interference. This increased stability means improved tracking in obstructed areas like woods, near buildings, and urban canyons. It also means fewer cycle slips.

So, even if it is the carrier phase that ultimately delivers the wonderful positional accuracy we all depend on, the codes get us in the game and keep us out of trouble every time we turn on the receiver. The codes have helped us to lock on to the first satellite in a session and allowed us to get the advantage of cross-correlation techniques almost since the beginning of GPS. In other words, our receivers have been combining pseudorange and carrier phase observables in innovative ways for some time now to measure the ionospheric delay, detect multipath, do wide laning, and so forth. Those techniques, however, can be improved, because while the current methods work, the results can be noisy and not quite as stable as they might be, especially over long baselines. It will be cleaner to get the signal directly once there are two clear civilian codes, one on each carrier. It may also help reduce the complexity of the chipsets inside our receivers and might just reduce their cost as well.

Along that line, it is worthwhile to recall that the L2C has an overall chip rate of 1.023 MHz, just like L1 C/A. Such a slow chip rate can seem to be a drawback until you consider that that rate affects the GPS chipset power consumption. In general, the slower the rate, the longer the battery life and the improvement in receiver battery life could be very helpful, and not only that, but the slower the chip rate, the smaller the chipset. That could mean more miniaturization of receiver components.

L2C is clearly going to be good for the GPS consumer market, but it also holds promise for surveyors. Nevertheless, there are a few obstacles to full utilization of the L2C signal. It will be some time before the constellation of Block IIR-M necessary to provide L2C at an operational level is up and functioning. Additionally, aviation authorities do not support L2C. It is not in an *Aeronautical Radionavigation*

Service (ARNS) protected band. It happens that L2 itself occupies a radiolocation band that includes ground-based radars.

CNAV

The navigation data on the CM portion of the L2C signal is improved over the legacy Navigation message. It is called CNAV. It was tested using some of the message types, MT, shown in Figure 8.8 in June 2013.

You may recall that the legacy Navigation message was mentioned in Chapter 1. It is also known by the acronym NAV. The information in CNAV is fundamentally the same as that in the original NAV message, but there are differences. It includes the almanac, ephermides, time, and satellite health. CNAV contains 300-bit messages like the legacy NAV message, but CNAV packages its 12-s 300-bit messages differently. Instead of the repetition of frames and subframes in a fixed pattern as is the case with the legacy NAV, CNAV utilizes a *pseudo-packetized* message protocol. One of every four of these packets includes clock data; two of every four contains ephemeris data, and so on. This makes CNAV more flexible than NAV. The order and the repetition of the individual messages in CNAV can be varied. CNAV can

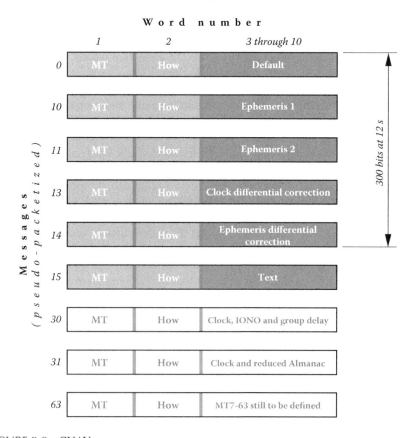

FIGURE 8.8   CNAV.

accommodate the transmission of the information in support of 32 satellites using 75 percent or less of its bandwidth. CNAV is also more compact than NAV and that allows a receiver to get to its first fix on a satellite much faster.

A packet in message 35 of CNAV was assigned to the time offset between GPS and GNSS in the June 2013 test. Such a message would be a boon for interoperability between GPS, Galileo, and GLONASS. Also, each packet contains a flag that can be toggled on within a few seconds of the moment a satellite is known to be unhealthy. This is exactly the sort of quick access to information necessary to support safety-of-life applications. Only a fraction of the available packet types are being used at this point. CNAV is designed to grow and can accommodate 63 satellites as the system requires. For example, there could be packets that would contain differential correction like that available from satellite-based augmentation systems.

There is also a very interesting aspect to the data broadcast on CM known as *forward error correction* (FEC). An illustration of this technique is to imagine that every individual piece of data is sent to the receiver twice. If the receiver knows the details of the protocol to which the data ought to conform, it can compare each of the two instances it has received to that protocol. If they both conform, there is no problem. If one does and one does not, the piece of data that conforms to the protocol is accepted and the other is rejected. If neither conforms, then both are rejected. Using FEC allows the receiver to correct transmission errors itself, on the fly. FEC will ameliorate transmission errors and reduce the time needed to collect the data.

L5

L5 Carrier

Alright, L2C is fine, but what about L5, the new carrier being broadcast on the Block IIF satellites? It is centered on 1176.45 MHz, 115× the fundamental clock rate.

As you see from Figure 8.9, the basic structure of L5 looks similar to that of L1. There are two pseudorandom noise (PRN) codes on this 20 MHz carrier. The two codes are modulated using quad phase shift key (QPSK), and they are broadcast in quadrature to each other. However, borrowing a few pages from other recent developments the in-phase (I) signal carries a data message that is virtually identical to the CNAV on L2. The other, the quadraphase signal (Q), is dataless and L5 utilizes chip-by-chip time multiplexing in broadcasting its two codes as does L2C in broadcasting CM and CL.

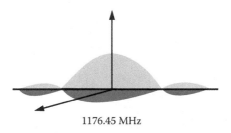

1176.45 MHz

FIGURE 8.9   L5.

Both L5 codes have a 10.23 MHz chipping rate, the same as the fundamental clock rate. This is the same rate that has been available on the P(Y) code from the beginning of the system. However, this is the fastest chipping rate available in any civilian code. L5 has the only civilian codes that are both 10× longer and 10× faster than the C/A code. Because the maximum resolution available in a pseudorange is typically about 1 percent of the chipping rate of the code used, the faster the chipping rate the better the resolution.

L5 has about twice as much power as L1, and because L5 does not carry military signals, it achieves an equal power split between its two signals. In this way, L5 lowers the risk of interference and improves multipath protection. It also makes the dataless signal easier to acquire in unfavorable and obstructed conditions.

Unlike L2C, L5 users will benefit from its place in a band designated by the International Telecommunication Union for the Aeronautical Radio Navigation Services worldwide. Therefore, it is not prone to interference with ground-based navigation aids and is available for aviation applications. While no other GPS signal occupies this band, L5 does share space with one of the Galileo signals, E5. L5 does also incorporate FEC.

GPS MODERNIZATION IS UNDERWAY

GPS modernization is no longer a future development; it is underway. New spacecraft with better electronics, better Navigation messages, and newer and better clocks are just part of the story. Beginning with the launch of the first IIR-M satellite, new civil signals began to appear, starting with L2C, which was followed by L5 on the Block IIF satellites.

These signals tend to have longer codes, faster chipping rates, and more power than the C/A and P(Y) codes have. In practical terms, these developments lead to faster first acquisition, better separation between codes, reduced multipath, and better cross-correlation properties.

Figure 8.10 provides a general outline of all of the signals that are available on the current satellite blocks.

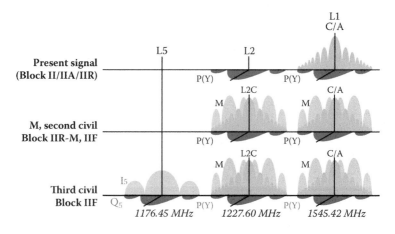

FIGURE 8.10 GPS modernized signal evolution. (Adapted from Lazar, S., *Crosslink* 3(2):42–46, 2002.)

IONOSPHERIC BIAS

As you know ionospheric delay is inversely proportional to frequency of the signal squared. So it is that L2's atmospheric bias is about 65 percent larger than L1, and it follows that the bias for L5 is the worst of the three at 79 percent larger than L1. L1 exhibits the least delay as it has the highest frequency of the three.

CORRELATION PROTECTION

Where a receiver is in an environment where it collects some satellite signals that are quite strong and others that are weak, such as inside buildings or places where the sky is obstructed, correlation protection is vital. The slow chipping rate, short code length, and low power of L1 C/A means it has the lowest correlation protection of the three frequencies L1, L2, and L5. That means that a strong signal from one satellite can cross correlate with the codes a receiver uses to track other satellites. In other words, the strong signal will actually block collection of the weak signals. To avoid this, the receiver is forced to test every single signal so to avoid incorrectly tracking the strong signal it does not want instead of a weak signal that it does. This problem is much reduced with L2. It has a longer code length and higher power than L1. It is also reduced with L5 as compared to L1. L5 has a longer code length, much higher power, and a much faster chipping rate than L1. In short, both of the civilian codes on L2 and L5 have much better cross-correlation protection and better narrowband interference protection than L1, but L5 is best of them all.

L1C ANOTHER CIVIL SIGNAL

Another civil signal will be broadcast by the Block III satellites. It is known as L1C. As a result of an agreement between the United States and the European Union reached in June 2004, this signal will be broadcast by both GPS and Galileo's L1 Open Service signals. It will also be broadcast by the Japanese Quasi-Zenith Satellite System (QZSS), the Indian Regional Navigation Satellite System (IRNSS), and China's BeiDou system. These developments open extraordinary possibilities of improved accuracy and efficiency when one considers there may eventually be a combined constellation of 50 or more satellites all broadcasting this same civilian signal. All this is made possible by the fact that each of these different satellite systems utilizes carrier frequencies centered on the L1, 1575.42 MHz. Perhaps that has something to do with the fact that L1, having the highest frequency, experiences the least ionospheric delay of the carrier frequencies. There are many signals on L1, but looking at GPS alone, there is the C/A code, the P(Y) code, the M code, and eventually the coming L1C code. It is a challenge to introduce yet another code on the crowded L1 frequency and still maintain separability.

As shown in Figure 8.11, the L1C design shares some of the M code characteristics, i.e., binary offset carrier. It will also have some similarities with L2C. For example, it will have a dataless pilot signal (L1C$_P$) and a signal with a data message (L1C$_D$) whose codes will be of the same length as CM on L2C 10,230 chips and be broadcast at 1.023 Mbps. The data signal will use BOC (1, 1) modulation, and the

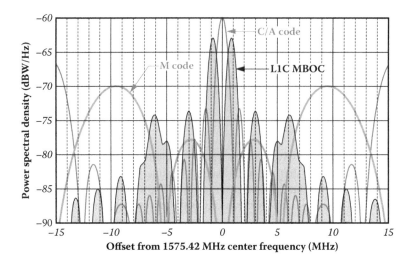

FIGURE 8.11   L1 with M code and L1C.

pilot will use time multiplexed binary offset carrier (TMBOC). The TMBOC will be BOC (1, 1) for 29 of its 33 cycles, but switch to BOC (6, 1) for 4 of them. The pilot component will have 75 percent of the signal power and the data portion of the signal will have 25 percent. This means that L1C will have good separation from the other signals on L1, a good tracking threshold and a receiver will reach its first fix to the satellite broadcasting L1C faster. The utility of L1C will be further enhanced by the fact that it will have double, 1.5 dB, the power of C/A.

The message carried by $L1C_D$ will be known as CNAV-2. Unlike CNAV, CNAV-2 will have frames. Each frame will be divided into three subframes. Both CNAV and CNAV-2 have 18 percent more satellite ephemeris information than NAV. The structure of CNAV-2 will virtually ensure that once a receiver acquires the satellite the time to first fix will not exceed 18 s.

GLOBAL NAVIGATION SATELLITE SYSTEM (GNSS)

The GPS system is one component of the worldwide effort known as the *Global Navigation Satellite System* (GNSS). Another component of GNSS is the GLONASS system of the Russian Federation and a third is the Galileo system administered by the EU. It is likely that more constellations will be included in GNSS, such as the Japanese Quasi-Zenith Satellite System (QZSS), the Indian Regional Navigation Satellite System (IRNSS), and the Chinese BeiDou Satellite Navigation and Positioning System. They will be augmented by both *ground-based augmentation systems* and *space-based augmentation systems* deployed by the United States, Europe, Japan, China, and Australia. One immediate effect of GNSS is the substantial growth of the available constellation of satellites; the more signals that are available for positioning and navigation, the better. The concept is that these networks of satellites and others will begin to work together to provide positioning, navigation, and timing solutions to users around the world.

The objectives of this cooperation are interoperability and compatibility. Interoperability is the idea that properly equipped receivers will be able to obtain useful signals from all available satellites in all the constellations and have their solutions improved rather than impeded by the various configurations of the different satellite broadcasts. Compatibility refers to the ability of U.S. and foreign space-based positioning, navigation, and timing services to be used separately or together without interfering with use of each individual service or signal. The two systems that are fully operational are GPS and GLONASS.

GLONASS

Russia's Global Orbiting Navigation Satellite System (Globalnaya Navigationnaya Sputnikovaya Sistema), known as GLONASS, did not reach full operational status before the collapse of the Soviet Union. The first of the Uragan satellites reached orbit in October 1982, a bit more than 4 years after the GPS constellation was begun. There were 87 Uragan satellites launched and a nearly full constellation of 24 made up of 21 satellites in three orbital planes, with three on-orbit spares achieved in 1996.

However, only about seven healthy satellites remained in orbit about 1000 km lower than the orbit of GPS satellites in 2001, and the remaining seven were only expected to have a design life of 3 years. The situation was not helped by the independence of Kazakhstan, subsequent difficulties over the Baikonur Cosmodrome launch facility, and lack of funds. The system was in poor health when a decision was made in August 2001, outlining a program to rebuild and modernize GLONASS. Improvements followed.

Today, Russia's GLONASS is operational and has worldwide coverage. A complete GLONASS constellation of improved satellites is in place at the altitude of 19,100 km inclined 64.8° toward the Equator (see Figure 8.12).

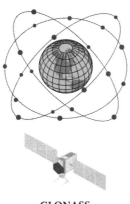

GLONASS
3 Orbital planes
21 Satellites + 3 spares
64.8° Inclination angle
Altitude 19,100 km

FIGURE 8.12   GLONASS constellation.

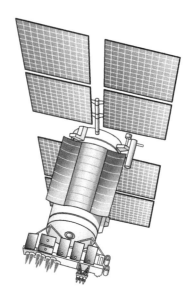

FIGURE 8.13   GLONASS Uragan-M.

URAGAN-M

Compared with the Uragan satellites, the Uragan-M have several new attributes (Figure 8.13). These satellites have an increased lifetime of 7 years, improved solar array orientation, better clock stability, and better maneuverability. They are three-axis stabilized and have onboard cesium clocks. Launched between 2001 and 2014, they were augmented with an L2 frequency for civilian users in 2004. These satellites comprise most of the current GLONASS constellation.

GLONASS-K

The GLONASS K satellite is lighter than the Uragan-M. It has an unpressurized bus, a 12 year service life, and costs less to produce. These satellites also carry the international search and rescue instrument COSPAS-SARSAT. As the Uragan-M satellites age, they will be replaced by the smaller GLONASS-K satellites, which will be followed by further improved versions known as GLONASS-K2 and later GLONASS-KM. As shown in Figure 8.14, the K version of the GLONASS has a transmitter for a third L-band civilian signal.

GLONASS SIGNALS

Regarding the signals broadcast by these satellites, the original objective was similar to the plan embraced by GPS, a system that would provide 100 m accuracy with a deliberately degraded standard C/A signal and 10 to 20 m accuracy with its P signals available exclusively to the military. However, that changed at the end of 2004, when

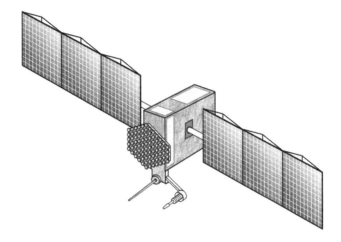

FIGURE 8.14   GLONASS K.

the Federal Space Agency announced a plan to provide access to the high-precision navigation data to all users whose foundation is a right-hand circular polarized code-based solution.

Code Division Multiple Access

A receiver collecting signals from GPS, or from most other GNSS constellations for that matter, collects a unique segment of the P code, the C/A code, or other PRN code from each satellite. For example, a particular segment of the 37 week long P code is assigned to each GPS satellite; i.e., SV14 is so named because it broadcasts the fourteenth week of the P code. Also, each GPS satellite broadcasts its own completely unique segment of the C/A code. Even though the segments of the P code and the C/A code coming into a receiver on L1 are unique to their satellite or origin, they all arrive at the same frequency, 1575.42 MHz. The same is true of the P code that arrives on L2. They all arrive at the same frequency, 1227.60 MHz.

This approach is known as *Code Division Multiple Access* (CDMA). CDMA technology was originally developed by the military during World War II. Researchers were looking for ways of communicating that would be secure in the presence of jamming. CDMA does not use frequency channels or time slots. The technique is called multiple access because it serves many simultaneous users and CDMA does this over the same frequency. As in GPS, CDMA usually involves a narrowband message multiplied by a wider bandwidth PRN (pseudorandom noise) signal. The increased bandwidth is wider than necessary to broadcast the data information and is called a spread spectrum signal.

To make this work, it is important that each of the PRN codes, C/A, P, and all the others, have high autocorrelation and low cross-correlation properties. High autocorrelation promotes efficient de-spreading and recovery of the unique code coming from a particular satellite that includes matching it with the PRN code available for that satellite inside the receiver. Low cross-correlation means that the autocorrelation process for a particular satellite's signal will not be interfered with by any of the other satellite's signals that are coming in from the rest of the constellation at the same time.

Frequency Division Multiple Access (FDMA)

GLONASS uses a different strategy. As shown in Figure 8.15, the satellites transmit L-band signals, and unlike GPS each code a GLONASS receiver collects from any one of the GLONASS satellites is exactly the same. Also unlike GPS each GLONASS satellite broadcasts its codes at its own unique assigned frequency. This is known as *Frequency Division Multiple Access* (FDMA). This ensures signal separation known as an improved spectral separation coefficient. However, the system does require more complex hardware and software development.

The three GLONASS L-bands have a range of frequencies to assign to satellites. GLONASS uses carriers in three areas. The first is L1 (~1602 MHz) in which the separation between individual carriers is 0.5625 MHz; the range is between ~1598.0625 and ~1607.0625 MHz. The second is L2 (~1246 MHz), in which the separation between individual carriers of 0.4375 MHz; the range is between ~1242.9375 and ~1249.9375 MHz. The third is L3. This third civil signal on L3 is available on the K satellites and within a new frequency band (~1201 MHz) that includes 1201.743 to 1208.511 MHz and will overlap Galileo's E5b signal. On L3, there will be a separation between individual carriers is 0.4375 kHz. However, within those ranges, there can be up to 25 channels of L-band signals; currently, there are 16 channels on each to accommodate the available satellites. In other words, each GLONASS satellite broadcasts the same code, but each satellite gets its own frequencies.

While there are some differences in the signals available from GPS, the EU's Galileo system and Russia's GLONASS, they are surmountable. Russia has also discussed development and use of GLONASS in parallel with the American GPS and European Galileo systems.

Changes to FDMA

There may be some changes to the FDMA approach in the future. Recently, Russia agreed to alter the architecture a bit. In order to use only half as many bands, GLONASS will now assign the same frequency to satellites that are in the same orbital plane but are always on opposite sides of the Earth.

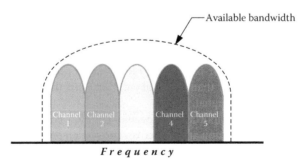

FIGURE 8.15   Frequency Division Multiple Access (FDMA).

This will not only reduce the amount of the radio spectrum used by GLONASS; it may actually improve its broadcast ephemeris information. Utilizing so many frequencies makes it difficult to accommodate the wide variety of propagation rates and keep the ephemeris information sent to the receivers within good limits. There are a number of receiver manufacturers that have GPS/GLONASS receivers available, but the differences between FDMA and CDMA signals increases the technical difficulty and costs of such equipment. In the last few months of 2006, it was mentioned that GLONASS probably will be able to implement CDMA signals on the third frequency and at L1. This could make it easier for GPS and Galileo to be interoperable with GLONASS and will probably improve GLONASS's commercial viability.

GLONASS TIME

In fact, there are many efforts underway to improve the GLONASS accuracy. The stability of the satellites' onboard clocks has improved from 5×10^{-13} to 1×10^{-13} over 24 hours with precision thermal stabilization. The GLONASS Navigation message will include the difference between GPS Time and GLONASS Time, which is significant.

There are no leap seconds introduced to GPS Time. The same may be said of Galileo and BeiDou. However, things are different in GLONASS. Leap seconds are incorporated into the Time standard of the system. Therefore, there is no integer-second difference between GLONASS Time and UTC as there is with GPS. Still that is not the whole story. The version of UTC used by GLONASS is the Coordinated Universal Time of Russia, known as UTC (SU). The epoch and rate of Russian time, relative to UTC (BIH), is monitored and corrected periodically by the Main Metrological Center of Russian Time and Frequency Service (VNIIFTRI) at Mendeleevo. They establish the regional version of UTC, which is known as UTC (SU). There is a constant offset of 3 hours between GLONASS Time and UTC (SU). GLONASS Central Synchronizer, CS, time is the foundation of GLONASS Time, GLONASST. The GLONASS-M satellites are equipped with cesium clocks that are kept within 8 ns of GLONASST.

GLONASS EPHEMERIS

The phase center of the satellite's transmitting antennas is provided in the PZ-90.02 Earth-Centered Earth-Fixed reference frame in the same right-handed three-dimensional Cartesian coordinate system described in Chapter 5.

GALILEO

The European Union's civilian controlled Galileo system is approaching operational status (Figure 8.16). The satellites are in orbit at a nominal height of about 23,222 km above the Earth. The full constellation will include 30 satellites in 3 planes, 10 in each plane higher than the GPS constellation. There are four carriers E5a (1176.45 MHz), E5b (1191.795 MHz), E6 (1278.75 MHz), and E2-L1-E1 (1575.42 MHz).

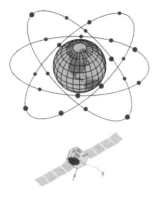

Galileo
3 Orbital planes
27 Satellites + 3 spares
56° Inclination angle
Altitude 23,222 km

FIGURE 8.16   Galileo constellation.

Galileo is intended to have worldwide coverage. It uses CDMA as its access scheme, and will be interoperable with GPS and GLONASS.

Just over a dozen years after the idea was first proposed, the work on Galileo culminated in the launch of GIOVE-A (Galileo In Orbit Validation Experiment-A) on December 28, 2005. The name GIOVE, Italian for Jupiter, was a tribute to Galileo Galilei, discoverer of Jupiter's moons. The first follow-on satellite, GIOVE-B, was launched in 2008. It was more like the satellites that will eventually comprise the Galileo constellation than was GIOVE-A. One of the motivations for launching GIOVE-A and GIOVE-B was to allow European government authorities to register its Galileo frequencies with international regulators at the International Telecommunications Union. Registration is necessary to prevent the frequency registration from expiring. They did their job and bought time for Europe to build additional satellites without facing a confiscation of its frequency reservations. The experimental satellites also facilitated investigation of the transmitted Galileo signals and provided measurements of the radiation environment. GIOVE-B also demonstrated the utilization of a hydrogen maser frequency standard. Both satellites did their job and were retired in 2012.

Two In-Orbit Validation (IOV) satellites were launched from French Guiana in 2011 and two more in 2012. They are Galileo-IOV PFM (GSAT0101, Thijs), Galileo-IOV FM2 (GSAT0102, Natalia), Galileo-IOV FM3 (GSAT0103, David), and Galileo-IOV FM4 (GSAT0104, Sif). The first ground location using these four satellites was accomplished at the European Space Agency's, Navigation Laboratory in Noordwijk, the Netherlands in 2013. The fifth and sixth Galileo satellites were also launched from French Guiana in 2014. The satellites initially entered elliptical orbits instead of the required circular orbits. These IOV Galileo satellites have been launched aboard Russian Soyuz rockets, and it is possible that the orbit difficulty was caused by a malfunction of the third stage of the Soyuz launch vehicle.

GALILEO'S GROUND SEGMENT

Galileo's ground segment has two centers: one at the Fucino Control Centre in Italy and one at the Oberpfaffenhofen Control Centre in Germany. There is also a network of sensor stations that provide data to the centers through the dedicated data dissemination network and the uplink stations communicate the calculated integrity, time, and other information back up to the satellites. There are also two telemetry, tracking, and command stations: one at Kourou, French Guiana, and one at Kiruna, Sweden. The stations are all located in areas controlled by European nations.

GALILEO'S SIGNALS AND SERVICES

There are four Galileo signals: E5a, E5b, E6, and E2-L1-E1. E5a and E2-L1-E1 overlap the existing L1 and L5 GPS signals. The minimum power received from the Galileo signals is −152 dBW, more than double the power of the C/A code from GPS. There is a pilot, dataless, as well as a data component in all the Galileo signals. They are broadcast in quadrature. The pilot signal enhances correlation and allows longer integration as it does on L2C and L5 in GPS. The frequency standards in the Galileo satellites are rubidium and passive hydrogen masers.

Galileo has defined five levels of service that will be provided by the system. They include the *Open Service* (OS), which uses the basic signals and is quite similar to GPS and GLONASS. The OS is free and available for timing and positioning applications. The *Safety-of-Life Service* (SOL) is along the same line but provides increased guarantees including integrity monitoring, meaning that users are warned if there are signal problems. SOL will be a global service and will include both a critical level service and a less accurate noncritical level. Both SOL and OS are on the E5a, E5b, and E2-L1-E1 carriers. Their availability on separate frequencies presents the ability to ameliorate the ionospheric bias.

The *Public Regulated Service* (PRS) is encrypted and is meant to assist public security and civil authorities. The PRS is under government control and provides significant jamming protection. PRS will be provided on E6 and E2-L1-E1. Likely applications will include emergency services, law enforcement, intelligence services, and customs.

The *Search and Rescue Service* (SAR) is intended to enhance space-based services and improve response time to distress beacons and alert messages. Galileo is another constellation in the COSPAS-SARSAT effort already mentioned in the discussions of GPS and GLONASS. Transponders on the satellites transfer the distress signals broadcast by a user to centers that initiate rescue operations. In the Galileo application the user also receives a notification that help has been dispatched.

The availability of the OS, SAR, and PRS will coincide with initial operational capability. Other services, such as the encrypted custom solutions and unique applications of the *Commercial Service* (CS) will follow as full operational capability is achieved.

GOVERNANCE OF GALILEO

The Galileo Joint Undertaking was established in 2002 by the European Commission and the European Space Agency to oversee Galileo's development phase. Its duties were transferred to the European GNSS Agency (GSA) in 2007. The governance of the Galileo system is now divided between the GSA, the European Commission and the European Space Agency (ESA). The GSA is in charge of the exploitation of the system. The European Commission owns the public assets, the physical system, the ground stations, and satellites and supervises the program. ESA handles the deployment.

INTEROPERABILITY BETWEEN GPS, GLONASS, AND GALILEO

Any discussion of interoperability between GPS and Galileo must consider the overlapping signals. It is helpful that the signals center on the same frequency if they are to be used in a combined fashion. For example, the third GLONASS civil reference signal on L3 that is available from the K satellites will be within a new frequency band that includes 1201.743–1208.511 MHz and will overlap Galileo's E5b signal.

In Figure 8.17, the Galileo signals are shown on the top and the GPS signals on the bottom. The Galileo satellites broadcast signals in several frequency ranges including 1176–1207 MHz, near GPS L5. Galileo's E5a signal is centered exactly at 1176.45 MHz as is L5. The other overlapping signals can be seen at 1575.42 MHz where both Galileo's L1 and the GPS L1 frequency are both centered. There the GPS signal is based on the binary phase shift key and the Galileo signal is accomplished with the binary offset carrier (BOC) method. The compatibility of these methods can be seen graphically in Figure 8.17. An important characteristic of BOC modulation is that the code's greatest power density is at the edges that is at the *nulls*, which, as it did with the M code on GPS, mitigates interference with the existing codes. In this case, not only will there not be interference between the codes on Galileo and GPS where they overlap, they can actually be used together. Galileo also has a signal E6 at 1278.75 MHz. As you can see, this band does not overlap any GPS frequency; however, it does happen to coincide with the band that Russia is considering for L3 on GLONASS.

E6 is part of the Radio Navigation Satellite Service (RNSS) allocation for Galileo. The Galileo signal E2-L1-E1 from 1559 to 1592 MHz is also part of the Radio Navigation Satellite Service. This signal is often known as simply L1. That is a convenient name because the GPS L1 is right there, too. Spectral separation of GPS and Galileo L1 signals is accomplished by use of different modulation schemes. This strategy allows jamming of civil signals, if that should prove necessary, without affecting GPS M code or the Galileo service. You can see the modulation method, BOC or BPSK, chipping rates, data rates in Figure 8.17. Also, please note the places where the carrier frequencies and frequency bands are common between GPS, GLONASS, and Galileo.

There are two signals on E6 with encrypted ranging codes, including one dataless channel that is only accessible to users who gain access through a given CS provider. Last, there are two signals, one in E6 band and one in E2–L1–E1, with encrypted ranging codes and data that are accessible to authorized users of the PRS.

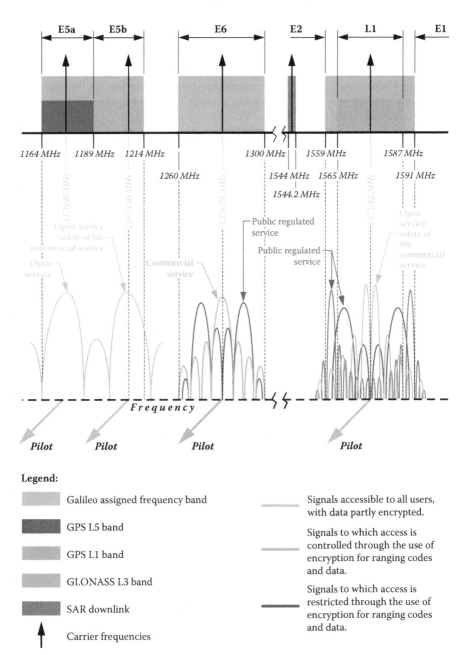

FIGURE 8.17 GALILEO–GPS–GLONASS signals. (Courtesy of Hein, Guenter W., Director of the Institute of Geodesy and Navigation, University FAF Munich, http://www .ifen.unibw-muenchen.de/research/signal.htm.)

Fortuitous coincidences of frequencies between GPS and Galileo did not happen without discussion. As negotiations proceeded between the United States and the European Union, one of the most contentious issues arose just as the European Union was moving to get Galileo off the ground. They announced their intention to overlay Galileo's PRS code on the U.S. Military's M code. The possibility that this would make it difficult for the DoD to jam the Galileo signal in wartime without also jamming the U.S. signal was considered. It became known as the M code overlay issue. In June 2004, the United States and the European Union reached an agreement that ensured the Galileo's signals would not harm the navigation warfare capabilities of the United States and the North Atlantic Treaty Organization (NATO).

BeiDou

The fourth GNSS system, joining those undertaken by the United States (GPS), Russia (GLONASS), and Europe (Galileo) will be the Chinese *BeiDou*. The system is named after the Big Dipper (Figure 8.18).

The first generation of the system is known as *BeiDou-1*. It is a regional satellite navigation system that services a portion of the Earth from 70°E longitude to 140°E longitude and from latitude 5°N to 55°N. It relies on three satellites with one backup. The first satellites were launched into geostationary (GEO) orbits in 2000; BeiDou-1A at 140°E longitude and BeiDou-1B 80°E longitude. A third satellite, BeiDou-1C, joined them 3 years later at 110.5°E longitude. With the launch of the fourth, BeiDou-1D in 2007, the first BeiDou-1 system was operational, regionally.

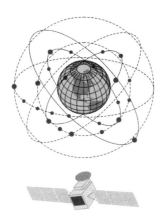

BeiDou
6 Orbital planes
35 Satellites + 5 GEO + 27 MEO + 3 IGSO
55° Inclination angle
Altitude 38,300 km, 21,500 km

FIGURE 8.18   BeiDou constellation.

BeiDou-2 (aka *Compass*) is the second generation of the BeiDou Navigation Satellite System. In 2007, Xinhua, the government news agency, announced that the People's Republic of China National Space Administration would launch two more GEO satellites to open the way to a global Chinese Satellite Navigation system to replace the regional BeiDou-1.

The first BeiDou-2 satellite, a medium Earth orbit (MEO) satellite named Compass-M1, was launched into a circular orbit at 21,500 km at an inclination of 55.5°. Similar satellites followed between 2007 and 2012. There were five MEO satellites with sequential names from Compass-M1 to M6 (without an M2) launched. During the period from 2009 to 2012 the six GEO BeiDou-2 satellites with sequential names from Compass-G1 to G6 were launched. Their positions are at 58.75°E longitude (G5), 80.0°E longitude (G6), 110.5°E longitude (G3), 140.0°E longitude (G1), and 160.0°E longitude (G4). The G2 satellite is inactive. From 2010 to 2011 the five high Earth orbit BeiDou-2 satellites with sequential names from Compass-IGS01 to IGS05 were launched and achieved an altitude of approximately 38,000 km. The acronym IGOS means inclined geosynchronous orbit satellites. The IGS01, IGS02, and IGS03 satellite are at ~120°E. IGS04 and IGS05 are at ~95°E. All the IGSO satellites have an inclination of 55° and are arranged so that one of them is always over the Chinese region. The system began trial operations in late 2011 and followed with service to the region bounded by 55°E longitude to 180°E longitude and latitude 55°S to 55°N.

These launches are harbingers of a full BeiDou-2 constellation, which will eventually include 35 satellites; 5 of them will be geostationary satellites and 30 will be nonstationary. The constellation will occupy six orbital planes. The 30 nonstationary satellites will include 27 in MEO and 3 in inclined geosynchronous orbit (IGSO). All the satellites will have a phased array antenna, onboard retro-reflector, a C-band horn antenna, and an S/L-band dish antenna.

BEIDOU'S SIGNALS AND SERVICES

BeiDou satellites currently transmit signals in three bands: B1 (1561.098 MHz), B1-2 (1589.742 MHz), B2 (1207.14 MHz), and B3 (1268.52). These bands overlap Galileo on E2-L1-E1, E5b, and E6, respectively (Figure 8.19). Each of the BeiDou signals in these bands have an I (in-phase) and Q (quadraphase) component in quadrature with one another. BeiDou's multiple access system is CDMA. The modulation scheme is quadraphase shift key (QPSK) (Table 8.2).

There are two service types. The Open Service (OS) is available to the public and offers an autonomous (not differentially corrected) positional accuracy of 10 m, 0.2 m/s velocity accuracy, and timing accuracy within 20 ns. The authorized service is not available to the public. It is available to the military, specifically the militaries of China and Pakistan.

There are two Navigation messages available from the BeiDou satellites. The MEO satellites and the IGSO satellites transmit the D1 NAV message, which is similar to the GPS NAV message. The D1 message has 30 bit words and 10-word subframes, i.e., 300 bits. It is broadcast at 50 bps and has adequate capacity for almanac data for 30 satellites. It also includes time offset information for UTC and other GNSS system clocks.

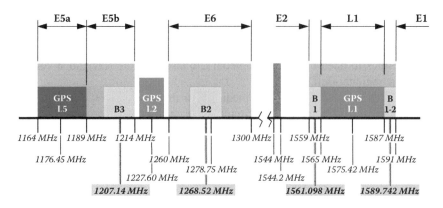

Legend:

    Galileo assigned frequency band

    GLONASS L3 band

    BeiDou

    SAR downlink

FIGURE 8.19   BeiDou bands.

TABLE 8.2
BeiDou Signals

Signals	Carrier Frequency (MHz)	Chip Rate (cps)	Multiple Access Scheme	Bandwidth (MHz)	Modulation Type	Service Type	Minimum Received Power
B1(I)	1561.098	2.046	CDMA	4.092	QPSK	Open	−163 dbW
B1(Q)		2.046				Authorized	
B1-2(I)	1589.742	2.046	CDMA	4.092	QPSK	Open	−163 dbW
B1-2(Q)		2.046				Authorized	
B2(I)	1207.14	2.046	CDMA	24	QPSK	Open	−163 dbW
B2(Q)		10.23				Authorized	
B3	1268.52	10.23	CDMA	24	QPSK	Authorized	−163 dbW

The GEO satellites transmit the D2 message. It also has 30-bit words and 10-word subframes but is broadcast at 500 bps. The D2 message includes pseudorange corrections satellites in subframes 2 and 3 with enough capacity to accommodate corrections for 18 satellites. D2 has ionospheric corrections and clock corrections to other GNSS systems in subframe 5. This provides a unique service planned for the BeiDou Control/Ground Segment that has not been incorporated into other GNSS systems, a wide area differential correction available directly from the constellation rather than from a separate system. The BeiDou Ground-Based Enhancement System network includes 150 reference stations; information from these stations is processed in the BeiDou Control/Ground Segment and the resulting corrections are sent to the BeiDou GEO satellites. Those corrections are then broadcast by the GEO satellites via the D2 NAV message to the BeiDou user's *terminals* (receivers). In other words, the GEO satellites provide satellite based augmentation for users in the region between 70°E longitude to 145°E longitude and 5°N latitude to 55°N latitude. While the BeiDou's open service is said to offer positional accuracy of ~10 m, the differential corrected is expected to produce ~1 m. Within the same region, BeiDou also provides a short message service. Users are enabled to send up to 120 Chinese characters in each message.

BeiDou's Control/Ground Segment

The BeiDou Control/Ground Segment is comprised of a Master Control Station (MCS), two upload stations and a network of 30 widely distributed monitoring stations. Similar to the control of other GNSS constellations the MCS receives data from the Monitoring Stations which track the constellation continuously. The upload stations send the data generated information by the MCS to the satellites.

The MCS is responsible for the operational control of the system, including orbit determination, Navigation messages, and ephemerides, which are based on the China Geodetic Coordinate System 2000 (CGCS2000). CGCS2000 is the coordinate framework of the BeiDou System and is within a few centimeters of ITRF. The MCS also coordinates mission planning, scheduling and time synchronization with BeiDou time (BDT). BDT is synchronized within 100 ns of UTC as maintained by National Time Service Center, China Academy of Science. BDT does not incorporate leap seconds. The leap second offset is broadcast by the BeiDou satellites in the Navigation message. The initial epoch of BDT is 00:00:00 UTC on January 1, 2006. The offset between BDT and GPST/GST is also to be measured and broadcast in the Navigation message.

QUASI-ZENITH SATELLITE SYSTEM (QZSS)

The first demonstration satellite of the Japanese Quasi-Zenith Satellite System (QZSS), QSZ-1 was launched in 2010 by the Japan Aerospace Exploration Agency (JAXA) from the Tanegashima Space Center (Figure 8.20). It is expected to have a design life of 10 years. The system is nicknamed Michibiki, meaning *guide*. There will be three more satellites launched in phase two. It is likely that they will include two additional quasi-zenith satellites, QZ-2 and QZ-3, and one geostationary satellite, QZ-4, which will be on orbit at the equator. Thereafter a four-satellite constellation is planned that will increase to seven satellites in the future.

QZSS is intended to provide satellites in highly elliptical orbits that are inclined and geosynchronous. The orbits are designed so the satellites will always be available at high elevation angles that are almost directly overhead in Japan, Oceania, and East Asia. This is the origin of the term *quasi-zenith*. They will orbit at ~32,000 to ~40,000 km and will all follow the same asymmetrical figure-eight ground track in the region.

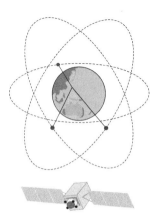

Quasi-Zenith Satellite System (QZSS)
3 Orbital planes
3 Satellites
~40° Inclination angle
Altitude 32,000 km to 40,000 km

FIGURE 8.20   QZSS constellation.

QZSS is primarily a multisatellite regional augmentation system. The objective is to improve satellite positioning, navigation, and timing services in urban canyons and mountainous areas in the region. The system is designed to improve the positioning of GPS and Galileo receivers. The system transmits six signals; L1-C/A, L1C, L2C, L5, L1-SAIF, and LEX. The first four on the list are the familiar GPS signals. The others are unique to QZSS. L1-SAIF (Submeter-class Augmentation with Integrity Function) is broadcast at the L1 frequency, 1575.42 MHz. It is interoperable with GPS and is intended to provide a submeter correction signal to users. Another unique signal to be broadcast by QZSS is LEX (L-band Experiment) at 1278.75 MHz. LEX is being developed to provide high-accuracy positioning that is interoperable with Galileo E6. QZSS will broadcast multiple-frequency signals and also provide a short message service as does the Chinese BeiDou system. There may be user fees developed for these signals and services.

The commercial portion of the QZSS operation will be managed by Quasi-Zenith Satellite System Services Inc. The system integration, research, and development of the QZS bus, research, the ground segment, etc. will be under JAXA's control.

QZSS Control/Ground Segment

The Master Control Station (MCS) develops the ephemerides, time, and Navigation messages that are uploaded to the QZS satellite constellation by the main telemetry, tracking, and command ground station in the Okinawa prefecture. Other monitoring stations on Japanese territory are at Ogasawara, Koganei, and Sarobetsu. However, there are also monitoring stations in areas governed by other nations. They are on Hawaii, Guam, Bangkok, Bangalore, and Canberra. The QZSS ground segment also includes laser ranging and tracking control stations (TCS). The main TCS station is at JAXA's Tsukuba Space Center.

The reference system for QZSS is the Japanese Satellite Navigation Geodetic System, which is quite near the International Terrestrial Reference System. The time reference for QZSS is known as Quasi-Zenith Satellite System Time (QZSST). The system does not use leap seconds. The duration of the second in this system is the same as that in International Atomic Time.

IRNSS

The building of the Indian Regional Navigation Satellite System (IRNSS) was authorized by the Indian government in 2006 (Figure 8.21). When fully developed by the Indian Space Research Organization, the constellation will provide position, navigation, and timing service in a region from 30°S latitude to 50°N latitude and from 30°E longitude to 130°E longitude. The region embraced is approximately 1500 km around India.

The space segment of IRNSS will be comprised of seven satellites. All the satellites will be continuously visible to the extensive Indian Control Segment's 21 stations located across the country including the Master Control Center (MCC) at Hassan, Karnataka. They will all be named with the prefix IRNSS-1. The constellation will include three of the seven satellites and will be in geostationary orbits at 32.5°E longitude, 83°E longitude, and 131.5°E longitude. There will be four of the seven satellites that will be geosynchronous and in orbit of 24,000 km apogee with an inclination of 29°. The small inclination is appropriate to the coverage of India as the nation is located in the low latitudes. The equator crossing of two of the geosynchronous satellites will be at 111.75°E longitude and two will cross the equator at 55°E longitude. The first IRNSS satellites launched have been these geosynchronous satellites.

IRNSS-1A was launched in 2013. IRNSS-1B and IRNSS-1C were launched in 2014. All three were launched from the Satish Dhawan Space Centre at Sriharikota, India. They carry rubidium clocks, corner cube retro-reflectors for laser ranging, and C-band transponders. They will broadcast in the L5 band (1176.45 MHz) with a

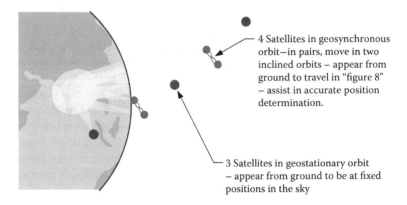

4 Satellites in geosynchronous orbit—in pairs, move in two inclined orbits – appear from ground to travel in "figure 8" – assist in accurate position determination.

3 Satellites in geostationary orbit – appear from ground to be at fixed positions in the sky

FIGURE 8.21   IRNSS constellation.

bandwidth of 24 MHz and S-band (2492.028 MHz) with a bandwidth of 16.5 MHz. They have a 10-year design life.

IRNSS will provide two levels of service, a public Standard Positioning Service (SPS) and an encrypted Restricted Service (RS). Both will be available on L5 and on the S-band; however, the SPS signal will be modulated by BPSK at 1 MHz and the RS will use BOC (5, 2). The navigation signals will be broadcast on the S-band.

THE FUTURE

So what is coming? Someday there may be as many as 130 navigation satellites aloft. They will be from GPS, GLONASS, Galileo, BeiDou, QZSS, IRNSS, and perhaps others. If so, the systems will provide users with quite a variety of signals and codes. The availability of many more satellites will enable new applications in areas where their scarcity has been a hindrance. For civil users, new signals will provide more protection from interference, ability to compensate for ionospheric delays with pseudoranges, and wide-laning or even tri-laning capabilities. For military users, there will be greater anti-jam capability and security. For everybody, there will be improvement in accuracy, availability, integrity, and reliability. So it is no surprise that there is great anticipation from a business perspective, but from a user's point of view the situation is not unlike the advent of GPS years ago. Much is promised but little assured. New capabilities will be available, but exactly what and exactly when is by no means certain.

More satellites will be above an observer's horizon. GPS and GLONASS together now provide the user with ~2× the satellites than does GPS alone. If GPS, GLONASS, and Galileo are considered together, there will be ~3× more. If BeiDou is added to the mix, ~4× more satellites will be available to a user. In a sense, the more satellites the better the performance. This is particularly highlighted where multipath abounds among trees, in urban canyons and those places where signals bounce and scatter. For example, with the six satellites in a window available to a GPS receiver in Figure 8.22, the user may be able to increase the mask angle to decrease the multipath and still have four satellites to observe. Imagine if there were 12, 18, or even 24 satellites in the picture and you can see how more satellites can mean better accessibility in restricted environments.

More satellites also means more measurements in shorter time and that means observation periods can be shortened without degrading accuracy in part because interference can be ameliorated more easily. With more satellites available, the time to first fix for carrier phase receivers, the period when the receiver is solving for the integers, downloading the almanac, and so forth, also known as *initialization*, will be shortened significantly. Additionally, fixed solution accuracy will be achieved more quickly. Today dual-frequency carrier phase solutions are accurate but noisy, but with the new signals available on L2C, L5, and other GNSS signals, dual-frequency solutions will be directly enhanced. While a GNSS capable receiver may offer a user improved availability and reliability, it may not necessarily offer higher accuracy than is available from GPS. However, the achievement of high accuracy more conveniently and in more places seems to be within reach with GNSS. It also means better ionospheric correction. Remember the ionospheric delay is frequency dependent.

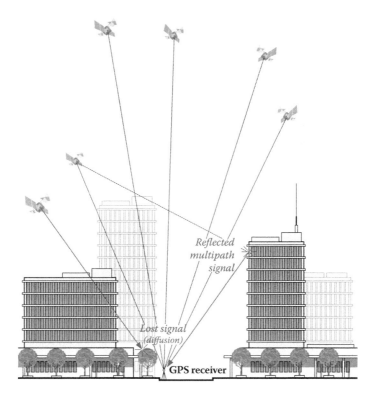

FIGURE 8.22 Urban canyon.

The algorithms currently necessary for the achievement of high accuracy with carrier phase ranging may be simplified because many of the new GNSS signals will be carrying a civilian code. Generally speaking, code correlation is a more straightforward problem than is carrier differencing. This may lead to less complicated receivers. This presents the possibility that they will be less expensive.

Also, the more diverse the maintenance of the components of GNSS, the less chance of overall system failure; the United States, Russia, the European Union, and China all have infrastructure in place to support their contribution to GNSS. Under such circumstances, simultaneous outages across the entire GNSS constellation are extremely unlikely.

INTEROPERABILITY

GPS–GALILEO–GLONASS CONSTELLATIONS

Inconsistency

Despite similarities, there are some issues in the consistency issues with GNSS. For example, the roll out of the new systems is not coordinated. In other words, GPS modernization, GLONASS replenishment, and Galileo deployment have not been synchronized. Despite much cooperation, there is no clear agreement among nations that launches and operational capabilities will happen in the same time frame. Also, there are some differences in multiple access schemes, time standards, and other GNSS system design parameters. For the full potential of the systems to be realized, the components of GNSS need to be interoperable. In other words, as many satellites as possible delivering signals that can be used in conjunction with one another any time, any place without interfering with one another. Fortunately, there is signal compatibility. Interoperability is achieved by partial frequency overlap using different signal structures and/or different code sequences for spectral separation.

In fact, high accuracy and interoperability are not only a matter of convenience, but robust, reliable solutions are becoming a business necessity. Consider safety-of-life uses for things such as routing of emergency vehicles or the GPS-based automated machine control system now used in construction. Mining, agriculture, aircraft control, and so forth, are depending more and more on satellite navigation systems. These industries have high costs and high risks and not only require high accuracy but reliability as well. If GNSS can deliver inexpensive receivers tracking the maximum number of satellites broadcasting the maximum number of signals, it will live up to the fondest hopes of not only many individuals but also many industries as well.

In fact, the goal of a single receiver that can track all the old and new GNSS satellite signals with a significant performance improvement looks possible. After all, the main attraction of interoperability between these systems is the greatly increased number of satellites and signals, better satellite availability, better dilution of precision, immediate ambiguity resolution on long baselines with three-frequency data, better accuracy in urban settings, and fewer multipath worries. Those are some of the things we look forward to. It looks like these things are achievable.

EXERCISES

1. How many GPS satellites are needed for worldwide 24 hour coverage?
   a. 18 satellites
   b. 12 satellites
   c. 20 satellites
   d. 24 satellites

2. Which of the following statements is true of the current GPS constellation?
   a. There are four orbital planes.
   b. There are eight satellites in each orbital plane.
   c. Satellites orbit approximately 20,000 nautical miles above the Earth.
   d. Each satellite completes an orbit in 12 sidereal hours.

3. Which of the following statements about L2C is not correct?
   a. It is broadcast by the IIR-M satellites.
   b. It was originally announced in 1998.
   c. It is carried on L1 and L2.
   d. It carries CNAV.

4. What characteristic of L5 is an improvement over both L1 and L2?
   a. It has a higher chipping rate.
   b. Its frequency is a multiple of the GPS fundamental clock rate.
   c. Its codes are broadcast in quadrature.
   d. It has two pseudorandom noise (PRN) codes.

5. The L5 is a bit stronger than L1. How much stronger?
   a. L5 has about 2× as much power as L1.
   b. L5 has about 3× as much power as L1.
   c. L5 has about 4× as much power as L1.
   d. L5 has about 5× as much power as L1.

6. Which satellites broadcast L5?
   a. Block I
   b. Block II
   c. Block IIR
   d. Block IIF

7. On the Block IIF satellites, what code is carried on L1 and not carried on L2?
   a. C/A code
   b. P(Y) code
   c. L2C code
   d. M code

8. On the Block IIF satellites, what code is carried on L2 and not carried on L1?
   a. C/A code
   b. P(Y) code
   c. L2C code
   d. M code

9. Which of the following statements is not true of the Block IIR satellites?
   a. Block IIR satellites can determine their own position.
   b. Block IIR satellites have programmable processors onboard to do their own fixes in flight.
   c. Block IIR satellites can be moved into a particular orbit with a 60 day advance notice.
   d. Block IIR satellites cost about one-third more than the earlier Block II satellites did.

ANSWERS AND EXPLANATIONS

1. Answer is (d)
   Explanation: The basics of its configuration are well known. Twenty-four GPS satellites are needed for 24 hour worldwide coverage, but there are actually more than 24 birds up at any one time to assure that the minimum will always be available.

2. Answer is (d)

   Explanation: The satellites are in six orbital planes, four or more satellites per plane in orbit about 11,000 nautical miles above the Earth. Each completes an orbit every 12 hours sidereal time.

3. Answer is (c)

   Explanation: A new military code may not be terribly exciting to civilian users like us, but these IIR-M satellites have something else going for them. They broadcast a new civilian code. This is a code that was announced way back in March 1998. It is on L2 and is known as L2C. Also, this new code is not a merely a copy of the C/A-code. L2C is a bit more sophisticated.

4. Answer is (a)

   Explanation: Alright, L2C is fine, but what about the new carrier, L5? It is centered on 1176.45 MHz, 115× the fundamental clock rate, in the Aeronautical Radio Navigation Services (ARNS) band. The L5 codes have a higher chipping rate. L5 was first introduced on the fourth-generation GPS satellites, Block IIF. Even though both codes are broadcast on L1, they are distinguishable from one another by their transmission in *quadrature*. That means that the C/A code modulation on the L1 carrier is phase shifted 90° from the P code modulation on the same carrier. L5 uses the same strategy.

5. Answer is (a)

   Explanation: L5 has about twice as much power as L1, and because L5 does not carry military signals, it achieves an equal power split between its two signals. In this way, L5 lowers the risk of interference and improves multipath protection. It also makes the dataless signal easier to acquire in unfavorable and obstructed conditions.

6. Answer is (d)

   Explanation: L5 is broadcast by the fourth-generation GPS satellites, Block IIF.

7. Answer is (a)

   Explanation: The C/A code.

8. Answer is (c)

   Explanation: L2C.

9. Answer is (d)

   Explanation: There are some significant changes with the Block IIR satellites. Block IIR satellites can determine their own position. They have programmable processors onboard to do their own fixes in flight. The Block IIR satellites can be moved into a particular orbit with a 60 day advance notice. In short, they have more autonomy. They are also more radiation hardened than their predecessors, and they cost about a third less than the Block II satellites did.

Glossary

Absolute Positioning: (*also known as* Point Positioning, Point Solution, or Single Receiver Positioning) A single receiver position, defined by a coordinate system. Most often the coordinate system used in absolute positioning is geocentric. In other words, the origin of the coordinate system is intended to coincide with the center of mass of the Earth.

Accuracy: Agreement of a value, whether measured or computed, with the standard or accepted true value. In the absolute sense, the true value is unknown, and, therefore, accuracy can only be estimated. Nevertheless, in measurement, accuracy is considered to be directly proportional to the attention given to the removal of systematic errors and mistakes. In GPS specifically, the values derived are usually the position, time, or velocity at GPS receivers.

Almanac: Data file containing a summary of the orbital parameters of all GPS satellites. The almanac is found in subframe 5 of the Navigation message. This information can be acquired by a GPS receiver from a single GPS satellite. It helps the receiver find the other GPS satellites.

Analog: Representation of data by a continuous physical variable. For example, the modulated carrier wave used to convey information from a GPS satellite to a GPS receiver is an analog mechanism.

Antenna: Resonant device that collects and often amplifies a satellite's signals. It converts the faint GPS signal's electromagnetic waves into electric currents sensible to a GPS receiver. Microstrip, also known as patch antennas, are the most often used with GPS receivers. Choke-ring antennas are intended to minimize multipath error.

Antenna Splitter: (*see also* Zero Baseline Test) Attachment that divides a single GPS antenna's signal in two. The divided signal goes to two GPS receivers. This is the foundation of Zero Baseline Test. The two receivers are using the same antenna so the baseline length should be measured as zero; because perfection remains elusive, it is usually a bit more.

Anti-Spoofing (AS): Encryption of the P code to render spoofing ineffective. Spoofing is generation of false transmissions masquerading as the precise code (P code). This countermeasure, anti-spoofing, is actually accomplished by the modulation of a W code to generate the more secure Y code that replaces the P code. Commercial GPS receiver manufacturers are not authorized to use the P code directly. Therefore, most have developed proprietary techniques both for carrier wave and pseudorange measurements on L2 indirectly. Dual-frequency GPS receivers must also overcome AS. Anti-spoofing was first activated on all Block II satellites on January 31, 1994.

Anywhere Fix: Receiver's ability to achieve lock-on without being given a somewhat correct beginning position and time.

Atomic Clock: (*see also* Cesium Clock) Clock regulated by the resonance frequency of atoms or molecules. In GPS satellites the substances used to regulate atomic clocks are cesium, hydrogen, or rubidium.

Attosecond: One-quintillionth (10^{18}) of a second.

Attribute: Information about features of interest. Attributes such as date, size, material, and color are frequently recorded during data collection for Geographic Information Systems (GIS).

Availability: Period, expressed as a percentage, when positioning from a particular system is likely to be successful.

Bandwidth: Frequency range of a signal. It is a measurement of the difference between the highest frequency and the lowest frequency expressed in Hertz. For example, the bandwidth of a voice signal is about 3 kHz, but for television signals it is around 6 MHz.

Baseline: Line described by two stations from which GPS observations have been made simultaneously. A vector of coordinate differences between this pair of stations is one way to express a baseline.

Base Station: (*also known as* Reference Station) Known location where a static GPS receiver is set. The base station is intended to provide data with which to differentially correct the GPS measurements collected by one or more roving receivers. A base station used in differential GPS (DGPS) is used to correct pseudorange measurements collected by roving receivers to improve their accuracy. In carrier phase surveys, the base station data are combined with the other receivers' measurements to calculate double-differenced observations or used in real-time kinematic (RTK) configurations.

Beat Frequency: When two signals of different frequencies are combined, two additional frequencies are created. One is the sum of the two original frequencies. One is the difference of the two original frequencies. Either of these new frequencies can be called a beat frequency.

Between-Epochs Difference: Difference in the phase of the signal on one frequency from one satellite as measured between two epochs observed by one receiver. Because a GPS satellite and a GPS receiver are always in motion relative to one another the frequency of the signal broadcast by the satellite is not the same as the frequency received. Therefore, the fundamental Doppler observable in GPS is the measurement of the change of phase between two epochs.

Between-Receivers Single Difference: Difference in the phase measurement between two receivers simultaneously observing the signal from one satellite on one frequency. For a pair of receivers simultaneously observing the same satellite, a between-receivers single difference pseudorange or carrier phase observable can virtually eliminate errors attributable to the satellite's clock. When baselines are short, the between-receivers single difference can also greatly reduce errors attributable to orbit and atmospheric discontinuities.

Between-Satellites Single Difference: Difference in the phase measurement between signals from two satellites on one frequency simultaneously observed by one receiver. For one receiver simultaneously observing two satellites, a between-satellites single difference pseudorange or carrier phase observable can virtually eliminate errors attributable to the receiver's clock.

Bias: Systematic error. Biases affect all GPS measurements and hence the coordinates and baselines derived from them. Biases may have a physical origin, satellite orbits, atmospheric conditions, clock errors, etc. They may also originate from less than perfect control coordinates, incorrect ephemeris information, and so on. Modeling is one method used to eliminate or at least limit the effect of biases.

Binary Biphase Modulation: Method used to impress the pseudorandom noise codes onto GPS carrier waves using two states of phase modulations, since the code is binary. In binary biphase modulation, the phase changes that occur on the carrier wave are either 0° or 180°.

Bit: Unit of information. In a binary system, a bit is either a 1 or a 0.

Block I, II, IIR, and IIF Satellites: Classification of the GPS satellite's generations. Block I satellites were officially experimental. There were 11 Block I satellites. PRN4, the first GPS satellite, was launched February 22, 1978. The last Block I satellite was launched October 9, 1985, and none remain in operation. Block II satellites are operational satellites. The GPS constellation was declared operational in 1995. Twenty-seven Block II satellites have been launched. Block IIRs are replenishment satellites and have the ability to receive signals from other IIR satellites. These satellites have the capability to measure ranges between themselves. Block IIF will be the follow-on series of satellites in the new century.

Broadcast: A modulated electromagnetic wave transmitted across a large geographical area.

Broadcast Ephemeris: (*see* Ephemeris)

Byte: Sequence of eight binary digits that represents a single character, like a letter or number.

C/A Code: (*also known as* S Code) Binary code known by the names, civilian/access code, clear/access code, clear/acquisition code, Coarse/Acquisition code, and various other combinations of similar words. It is a standard spread-spectrum GPS pseudorandom noise code modulated on the L1 carrier using binary biphase modulations. The C/A code is not carried on L2. Each C/A code is unique to the particular GPS satellite broadcasting it. It has a chipping rate of 1.023 MHz, more than a million bits per second. The C/A code is a direct sequence code and a source of information for pseudorange measurements for commercial GPS receivers.

Carrier: Electromagnetic wave, usually sinusoidal, that can be modulated to carry information. Common methods of modulation are frequency modulation and amplitude modulation. However, in GPS the phase of the carrier is modulated, and there are currently two carrier waves. The two carriers in GPS are L1 and L2, which are broadcast at 1575.42 and 1227.60 MHz, respectively.

Carrier Beat Phase: (*see also* Beat Frequency; Doppler Shift) Phase of a beat frequency created when the carrier frequency generated by a GPS receiver combines with an incoming carrier from a GPS satellite. The two carriers have slightly different frequencies as a consequence of the Doppler shift of the satellite carrier compared with the nominally constant receiver generated carrier. Because the two signals have different frequencies, two beat frequencies are created. One beat is the sum of the two frequencies. One is the difference of the two frequencies.

Carrier Frequency: In GPS the frequency of the unmodulated signals broadcast by the satellites. Currently, the carrier frequencies are L1 and L2, which are broadcast at 1575.42 and 1227.60 MHz, respectively.

Carrier Phase: (*also known as* Reconstructed Carrier Phase) (1) The term is usually used to mean a GPS measurement based on the carrier signal itself, either L1 or L2, rather than measurements based on the codes modulated onto the carrier wave. (2) Carrier phase may also be used to mean a part of a full carrier wavelength. An L1 wavelength is about 19 cm, and an L2 wavelength is about 24 cm. A part of the full wavelength may be expressed in a phase angle from 0° to 360°, a fraction of a wavelength, cycle, or some other way.

Carrier Tracking Loop: (*see also* Code Tracking Loop) Feedback loop that a GPS receiver uses to generate and match the incoming carrier wave from a GPS satellite.

CEP: (*see* Circular Error Probable)

Cesium Clock: (*also known as* Cesium Frequency Standard) Atomic clock that is regulated by the element cesium. Cesium atoms, specifically atoms of the isotope Cs-133, are exposed to microwaves and the atoms vibrate at one of their resonant frequencies. By counting the corresponding cycles, time is measured.

Channel: A channel of a GPS receiver consists of the circuitry necessary to receive the signal from a single GPS satellite on one of the two carrier frequencies. A channel includes the digital hardware and software.

Check Point: Reference point from an independent source of higher accuracy used in the estimation of the positional accuracy of a data set.

Chipping Rate: In GPS the rate at which chips, binary 1's and 0's, are produced. The P code chipping rate is about 10 million bits per second. The C/A code chipping rate is about 1 million bits per second.

Circular Error Probable (CEP): Description of two-dimensional precision. The radius of a circle, with its center at the actual position, that is expected to be large enough to include half (50 percent) of the normal distribution of the scatter of points observed for the position.

Civilian Code: (*see* C/A Code)

Civilian/Access Code: (*see* C/A Code)

Class of Survey: Means to generalize and prioritize the precision of surveys. The foundation of the categories is often taken from geodetic surveying classifications supplemented with standards of higher precision applicable to GPS work. Different classifications frequently apply to horizontal and vertical surveys. The categories themselves, their notation, and accuracy tolerances are unique to each nation. A Class of Survey should reflect the quality of network design, the instruments and methods used, and processing techniques as well as the precision of the measurements.

Clear/Access Code: (*see* C/A Code)

Clear/Acquisition Code: (*see* C/A Code)

Clock Bias: Discrepancy between a moment of time per a GPS receiver's clock or a GPS satellite's clock and the same moment of time per GPS Time, or another reference, such as Coordinated Universal Time or International Atomic Time.

Clock Offset: Difference between the same moments of time as indicated by two clocks.

Coarse/Acquisition: (C/A) (*see* C/A Code)

Code Phase: Measurements based on the C/A code rather than measurements based on the carrier waves. Sometimes used to mean pseudorange measurements expressed in units of cycles.

Code Tracking Loop: Feedback loop used by a GPS receiver to generate and match the incoming codes, C/A or P codes, from a GPS satellite.

Complete Instantaneous Phase Measurement: (*see* Fractional Instantaneous Phase Measurement) Measurement including the integer number of cycles of carrier beat phase since the initial phase measurement.

Confidence Level: Probability that the true value is within a particular range of values, expressed as a percentage.

Confidence Region: Region within which the true value is expected to fall, attended by a confidence level.

Constellation: (1) The space segment, all GPS satellites in orbit. (2) The particular group of satellites used to derive a position. (3) The satellites available to a GPS receiver at a particular moment.

Continuously Operating Reference Stations (CORS): System of base stations that originated from the stations built to support air and marine navigation with real-time differential GPS correction signals.

Continuous-Tracking Receiver: (*see* Multichannel Receiver)

Control Segment: U.S. Department of Defense's network of GPS monitoring and upload stations around the world. The tracking data derived from this network are used to prepare, among other things, the broadcast ephemerides and clock corrections by the Master Control Station at Schriever Air Force Base (formerly Falcon AFB), Colorado Springs, Colorado. Then included in the Navigation message, the calculated information is uploaded to the satellites.

Coordinated Universal Time: (*also known as* UTC, Zulu Time) Universal Time systems are international, based on atomic clocks around the world. Coordinated atomic clock time called Tempes Atomique International (TAI) was established in the 1970s and it is very stable. It is more stable than the actual rotation of the Earth. TAI would drift out of alignment with the planet if leap seconds were not introduced periodically as they are in UTC. UTC is, in fact, one of several of the Universal Time standards. There are standards more refined than UTC's tenth of a second. UTC is maintained by the U.S. Naval Observatory (USNO).

Correlation-Type Channel: A channel in a GPS receiver used to shift or compare the incoming signal with an internally generated signal. The code generated by the receiver is cross-correlated with the incoming signal to find the correct delay. Once they are aligned, the delay lock loop keeps them so. Correlator designs are sometimes optimized for acquisition of signal under foliage, accuracy, multipath mitigation, etc.

Cutoff Angle: (*see* Mask Angle)

Cycle Ambiguity: (*also known as* Integer Ambiguity) The number of full wavelengths, the integer number of wavelengths, between a particular receiver and satellite is initially unknown in a carrier phase measurement. It is called the cycle or integer ambiguity. If a single-frequency receiver is tracking several satellites, there is a different ambiguity for each. If a dual-frequency receiver is tracking several satellites, there is a different ambiguity for each satellite's L1 and L2. However, in every case the ambiguity is a constant number as long as the tracking, lock, is not interrupted. However, should the signal be blocked, if a cycle slip occurs, there is a new ambiguity to be resolved. Once the initial integer ambiguity value is resolved in a fixed solution for each satellite-receiver pair, the integrated carrier phase measurements can yield very precise positions. However, the cycle ambiguity resolution processes actually utilize double-differenced carrier phase observables, not single satellite-receiver measurements. The great progress in carrier phase GPS systems has reduced the length of observation data needed, as in rapid static techniques. The integer ambiguity resolution can now occur even with the receiver in motion, the on-the-fly approach.

Cycle Slip: Discontinuity of an integer number of cycles in the carrier phase observable. A jump of a whole number of cycles in the carrier tracking loop of a GPS receiver. Usually, the result of a temporarily blocked GPS signal. A cycle slip causes the cycle ambiguity to change suddenly. The repair of a cycle slip includes the discovery of the number of missing cycles during an outage. Cycle slips must be repaired before carrier phase data can be successfully processed in double-differenced observables.

Data Logger: (*also known as* Data Recorder and Data Collector) Data entry computer, usually small, lightweight, and often handheld. A data logger stores information supplemental to the measurements of a GPS receiver.

Data Message: (*see* Navigation Message)

Data Set: Organized collection of related data compiled specifically for computer processing.

Data Transfer: Transporting data from one computer or software to another, often accompanied by a change in the format of the data.

Datum: (1) Any reference point line or surface used as a basis for calculation or measurement of other quantities. (2) A means of relating coordinates determined by any means to a well-defined reference frame. (3) The singular of data.

Decibel (dB): A tenth of a bel. The bel was named for Alexander Graham Bell. Decibel does not actually indicate the power of the antenna; it refers to a comparison. Most GPS antennas have a gain of about 3 dB (decibels). This indicates that the GPS antenna has about 50 percent of the capability of an isotropic antenna. An isotropic antenna is a hypothetical, lossless antenna that has equal capabilities in all directions.

DGPS: (*see* Differential GPS)

Differential GPS: (1) Method that improves GPS pseudorange accuracy. A GPS receiver at a base station, a known position, measures pseudoranges to the same satellites at the same time as other roving GPS receivers. Roving receivers occupy unknown positions in the same geographic area. Occupying a known position, the base station receiver finds corrective factors that can either be communicated in real-time to the roving receivers, or may be applied in post-processing. (2) The term differential GPS is sometimes used to describe relative positioning. In this context it refers to more precise carrier phase measurement to determine the relative positions of two receivers tracking the same GPS signals in contrast to absolute, or point, positioning.

Digital: Involving or using numerical digits. Information in a binary state, either a one or a zero, a plus or a minus, is digital. Computers utilize the digital form almost universally.

Dilution of Precision (DOP): In GPS positioning, an indication of geometric strength of the configuration of satellites in a particular constellation at a particular moment and hence the quality of the results that can be expected. A high DOP anticipates poorer results than a low DOP. A low DOP indicates that the satellites are widely separated. Because it is based solely on the geometry of the satellites, DOP can be computed without actual measurements. There are various categories of DOP, depending on the components of the position fix that are of most interest.

Distance Root-Mean-Square (drms): Statistical measurement that can characterize the scatter in a set of randomly varying measurements on a plane. The drms is calculated from a set of data by finding the root-mean-square value of the radial errors from the mean position. In other words, the root-mean-square value of the linear distances between each measured position and the ostensibly true location. In GPS positioning, 2 drms is more commonly used. Two drms does not mean two-dimensional rms. Two drms does mean twice the distance root mean square. In practical terms, a particular 2 drms value is the radius of a circle that is expected to contain from 95 to 98 percent of the positions a receiver collects in one occupation, depending on the nature of the particular error ellipse involved. Two drms is convenient to calculate. It can be predicted using covariance analysis by multiplying the HDOP by the standard deviation of the observed pseudoranges.

Dithering: Intentional introduction of digital noise. Dithering the satellite's clocks is the method the Department of Defense (DoD) uses to degrade the accuracy of the Standard Positioning Service. This degradation was known as Selective Availability. Selective Availability was switched off on May 2, 2000, by presidential order.

DoD: Department of Defense

DOP: (*see* Dilution of Precision)

Doppler-Aiding: Method of receiver signal processing that relies on the Doppler shift to smooth tracking.

Doppler Shift: In GPS, a systematic change in the apparent frequency of the received signal caused by the motion of the satellite and receiver relative to one other.

DOT: Department of Transportation

Double-Difference: A method of GPS data processing. In this method, simultaneous measurements made by two GPS receivers, either pseudorange or carrier phase, are combined. Both satellite and receiver clock errors are virtually eliminated from the solution. Most high-precision GPS positioning methods use double-difference processing in some form.

Dual-Capability GPS/GLONASS Receiver: (*also known as* Interoperability) A receiver that has the capability to track both GPS satellites and GLONASS satellites.

Dual-Frequency: Receivers or GPS measurements that utilize both L1 and L2. Dual-frequency implies that advantage is taken of pseudorange and/or carrier phase on both L-band frequencies. Dual-frequency allows modeling of ionospheric bias and attendant improvement in long baseline measurements particularly.

Dynamic Positioning: (*see* Kinematic Positioning)

Earth-Centered Earth-Fixed (ECEF): Cartesian system of coordinates with three axes in which the origin of the three-dimensional system is the Earth's center of mass, the geocenter. The z axis passes through the North Pole, that is, the International Reference Pole (IRP) as defined by the International Earth Rotation Service (IERS). The x axis passes through the intersection of zero longitude, near the Greenwich meridian, and the equator. The y axis extends through the geocenter along a line perpendicular from the x axis. It completes the right-handed system with the x and z axes and all three rotate with the Earth. Three coordinate reference systems (i.e., NAD83, WGS 84, and ITRS) are all similarly defined.

ECEF: (*see* Earth-Centered Earth-Fixed)

Elevation: Distance measured along the direction of gravity above a surface of constant potential. Usually, the reference surface is the geoid. Mean Sea Level (MSL) once utilized as a reference surface approximating the geoid is known to differ from the geoid up to a meter or more. The term height, sometimes considered synonymous with elevation, refers to the distance above an ellipsoid in geodesy.

Elevation Mask Angle: (*see* Mask Angle)

Ellipsoid of Revolution: (*also known as* Spheroid) Biaxial closed surface, whose planar sections are either ellipses or circles that are formed by revolving an ellipse about its minor axis. Two quantities fully define an ellipsoid of revolution, the semimajor axis, a, and the flattening, $f = (a - b)/a$, where b is the length of the semiminor axis. The computations for the North American Datum of 1983 were done with respect to the GRS80 ellipsoid, the Geodetic Reference System of 1980. The GRS80 ellipsoid is defined by

$$a = 6,738,137 \text{ m}$$

$$1/f = 298.257222101$$

Ellipsoidal Height: (*also known as* Geodetic Height) Distance from an ellipsoid of reference to a point on the Earth's surface, as measured along the perpendicular from the ellipsoid. Ellipsoidal height is not the same as elevation above Mean Sea Level or orthometric height. Ellipsoidal height also differs from geoidal height.

Ephemeris: Table of values including locations and related data from which it is possible to derive a satellite's position and velocity. In GPS, broadcast ephemerides are compiled by the Master Control Station, uploaded to the satellites by the Control Segment, and transmitted to receivers in the Navigation message. It is designed to provide orbital elements quickly and is not as accurate as the precise ephemeris. Broadcast ephemeris errors are mitigated by differential correction or in double-differenced observables over short baselines. Precise ephemerides are post-processed tables available to users via the Internet.

Epoch: Time interval. In GPS the period of each observation in seconds.

Error Budget: Summary of the magnitudes and sources of statistical errors that can help in approximating the actual errors that will accrue when observations are made.

FAA: Federal Aviation Administration

Fast-Switching Channel: (*see* Multiplexing Channel)

Federal Radionavigation Plan: Document mandated by Congress that is published every other year by the Department of Defense and the Department of Transportation. In an effort to reduce the functional overlap of federal initiatives in radionavigation, it summarizes plans for promotion, maintenance, and discontinuation of domestic and international systems.

Femtosecond: One-millionth of a nanosecond (10^{15}) of a second.

FM: Frequency modulation

Fractional Instantaneous Phase: Carrier beat phase measurement not including the integer cycle count. It is always between zero and one cycle.

Frequency: (*see also* Wavelength) Number of cycles per unit of time. In GPS the frequency of the unmodulated carrier waves are L1 at 1575.42 MHz and L2 at 1227.60 MHz.

Frequency Band: Within the electromagnetic spectrum, a particular range of frequencies.

FRP: Federal Radionavigation Plan

Gain: The gain, or gain pattern, of a GPS antenna refers to its ability to collect more energy from above the mask angle and less from below the mask angle.

General Theory of Relativity: (*see also* Special Theory of Relativity) Physical theory published by A. Einstein in 1916. In this theory, space and time are no longer viewed as separate but rather form a four-dimensional continuum called space-time that is curved in the neighborhood of massive objects. The theory of general relativity replaces the concept of absolute motion with that of relative motion between two systems or frames of reference and defines the changes that occur in length, mass, and time when a moving object or light passes through a gravitational field. General relativity predicts that as gravity weakens, the rate of clocks increase; they tick faster. Special relativity predicts that moving clocks appear to tick more slowly than stationary clocks, because the rate of a moving clock seems to decrease as its velocity increases. Therefore, for GPS satellites, general relativity predicts that the atomic clocks in orbit on GPS satellites tick faster than the atomic clocks on Earth by about 45,900 ns/day. Special relativity predicts that the velocity of atomic clocks moving at GPS orbital speeds tick slower by about 7200 ns/day than clocks on Earth. The rates of the clocks in GPS satellites are reset before launch to compensate for these predicted effects.

Geodesy: Science concerned with the size and shape of the Earth. Geodesy also involves the determination of positions on the Earth's surface and the description of variations of the planet's gravity field.

Geodetic Datum: Model defined by an ellipsoid and the relationship between the ellipsoid and the surface of the Earth, including a Cartesian coordinate system. In modern usage, eight constants are used to form the coordinate system used for geodetic control. Two constants are required to define the dimensions of the reference ellipsoid. Three constants are needed to specify the location of the origin of the coordinate system, and three more constants are needed to specify the orientation of the coordinate system. In the past, a geodetic datum was defined by five quantities: the latitude and longitude of an initial point, the azimuth of a line from this point, and the two constants to define the reference ellipsoid.

Geodetic Survey: Survey that takes into account the size and shape of the Earth. A geodetic survey may be performed using terrestrial or satellite positioning techniques.

Geographic Information System (GIS): Computer system used to acquire, store, manipulate, analyze, and display spatial data. A GIS can also be used to conduct analysis and display the results of queries.

Geoid: Equipotential surface of the Earth's gravity field that approximates Mean Sea Level more closely. The geoid surface is everywhere perpendicular to gravity. Several sources have contributed to models of the Geoid, including ocean gravity anomalies derived from satellite altimetry, satellite-derived potential models, and land surface gravity observations.

Geoidal Height: Distance from the ellipsoid of reference to the geoid measured along a perpendicular to the ellipsoid of reference.

Geometrical Dilution of Precision (GDOP): (*see* Dilution of Precision)

Gigahertz (GHz): One billion cycles per second.

Global Navigation Satellite System (GNUS): A satellite-based positioning system. In Europe, GNUS-1 is a reference to a combination of GPS and GLONASS. However, GNUS-2 is a reference to a proposed combination of GPS, GLONASS, and other systems both space and ground based.

Global Orbiting Navigation Satellite System (GLONASS): Satellite radio navigation system financed by the Soviet Commonwealth. It is comprised of 21 satellites and 3 active spares arranged in three orbital rings approximately 11,232 nautical miles above the Earth, though the number of functioning satellites may vary because of funding. Frequencies are broadcast in the ranges of 1597 to 1617 MHz and 1240 to 1260 MHz. GLONASS positions reference the datum PZ90 rather than WGS84.

Global Positioning System (GPS): Radio navigation system for providing the location of GPS receivers with great accuracy. Receivers may be stationary on the surface of the Earth or in vehicles: aircraft, ships, or in Earth-orbiting satellites. The Global Positioning System is comprised of three segments; the user segment, the space segment, and the Control Segment.

GPSIC: GPS Information Center, U.S. Coast Guard

GPS Receiver: Apparatus that captures modulated GPS satellite signals to derive measurements of time, position, and velocity.

GPS Time: GPS Time is the time given by all GPS Monitoring Stations and satellite clocks. GPS Time is regulated by Coordinated Universal Time (UTC). GPS Time and Coordinated Universal Time were the same at midnight UT on January 6, 1980. Since then, GPS Time has not been adjusted as leap seconds were inserted into UTC approximately every 18 months. These leap seconds keep UTC approximately synchronized with the Earth's rotation. GPS Time has no leap seconds and is offset from UTC by an integer number of seconds; the number of seconds is in the Navigation message and most GPS receivers apply the correction automatically. The exact difference is contained in two constants, within the Navigation message, A0 and A1, providing both the time difference and rate of system time relative to UTC. Disregarding the leap second offset, GPS Time is mandated to stay within one microsecond of UTC, but over several years, it has actually remained within a few hundred nanoseconds.

Ground Antennas: S band antennas used to upload information to the GPS satellites, including broadcast ephemeris data and broadcast clock corrections.

GRS80 (Geocentric Reference System 1980): An ellipsoid adopted by the International Association of Geodesy.

$$\text{Major semiaxis } (a) = 6{,}378{,}137 \text{ m}$$

$$\text{Flattening } (1/f) = 298.257222101$$

Hand-Over Word: Information used to transfer tracking from the C/A code to the P code. This time synchronization information is in the Navigation message.

Height, Ellipsoidal: (*see* Ellipsoidal Height)

Height, Orthometric: (*see* Orthometric Height)

Hertz: One cycle per second.

Heuristic Approach: Fancy words for trial and error.

Hydrogen Maser: Atomic clock. A device that uses microwave amplification by stimulated emission of radiation is called a maser. Actually, the microwave designation is not entirely accurate because masers have been developed to operate at many wavelengths. In any case, a maser is an oscillator whose frequency is derived from atomic resonance. One of the most useful types of maser is based on transitions in hydrogen, which occurs at 1421 MHz. The hydrogen maser provides a very sharp, constant oscillating signal and thus serves as a time standard for an atomic clock. The active hydrogen maser provides the best known frequency stability for a frequency generator commercially available. At a 1 hour averaging time the active maser exceeds the stability of the best known cesium oscillators by a factor of at least 100, and the hydrogen maser is also extremely environmentally rugged.

Independent Baselines: (*also known as* Nontrivial Baselines) Baselines observed using GPS Relative Positioning techniques. When more than two receivers are observing at the same time, both independent (nontrivial) and trivial baselines are generated. For example, where r is the number of receivers, every complete static session yields $r - 1$ independent (nontrivial) baselines. If four receivers are used simultaneously, six baselines are created. However, three of those lines will fully define the position of each occupied station in relation to the others. Therefore, the user can consider three $(r - 1)$ of the six lines independent (nontrivial), but once the decision is made, only those three baselines are included in the network.

Inmarsat: International Maritime Satellite

Integrity: Quality measure of GPS performance including a system to provide a warning when the system should not be used for navigation because of some inadequacy.

Interferometry: (*see* Relative Positioning)

International Earth Rotation Service (IERS): Created in 1988 by the International Union of Geodesy and Geophysics (IUGG) and the International Astronomical Union (IAU). It is an interdisciplinary service organization that includes astronomy, geodesy, and geophysics. IERS maintains the International Celestial Reference System and the International Terrestrial Reference System.

International GPS Service (IGS): Comprised of many civilian agencies that operate a worldwide GPS tracking network. IGS produces postmission ephemerides, tracking station coordinates, Earth orientation parameters, satellite clock corrections, and tropospheric and ionospheric models. It is an initiative of the International Association of Geodesy and other scientific organizations, established in 1994 and originally intended to serve precise surveys for monitoring crustal motion. The range of users has since expanded.

International Terrestrial Reference System (ITRS): Very precise, geocentric coordinate system. The ITRS is geocentric, including the oceans and the atmosphere. Its length unit is the meter. Its axes are consistent with the Bureau International de l'Heuer (BIH) System at 1984.0 within +/–3 milliarcseconds. The IERS Reference Pole (IRP) and Reference Meridian (IRM) are consistent with the corresponding directions in the BIH directions within +/–0.005". The BIH reference pole was adjusted to the Conventional International Origin (CIO) in 1967 and kept stable independently until 1987. The uncertainty between IRP and CIO is +/–0.03". The ITRS is realized by estimates of the coordinates and velocities of a set of stations, some of which are satellite laser ranging (SLR) stations, or very long baseline interferometry (VLBI) stations, but the vast majority are GPS tracking stations of the IGS network.

Ionosphere: Layer of atmosphere extending from about 50 to 1000 km above the Earth's surface in which gas molecules are ionized by ultraviolet radiation from the Sun. The apparent speed, polarization, and direction of GPS signals are affected by the density of free electrons in this nonhomogeneous and dispersive band of atmosphere.

Ionospheric Delay: (*also known as* Ionospheric Refraction) Difference in the propagation time for a signal passing through the ionosphere compared with the propagation time for the same signal passing through a vacuum. The magnitude of the ionospheric delay changes with the time of day, latitude, season, solar activity, and observing direction. For example, it is usually least at the zenith and increases as a satellite gets closer to the horizon. The ionospheric delay is frequency dependent and, unlike the troposphere, affects the L1 and L2 carriers differently. There are two categories of the ionospheric delay, phase and group. Group delay affects the codes, the modulations on the carriers; phase advance affects the carriers themselves. Group delay and phase advance have the same magnitude but opposite signs. Because the ionospheric delay is frequency dependent it can be nearly eliminated by combination of pseudorange or carrier phase observations on both the L1 and L2 carriers. Still, even with dual-frequency observation and relative positioning methods in place, over long baselines the residual ionospheric delay can remain a substantial bias for high precision GPS. Single-frequency receivers cannot significantly mitigate the error at all and must depend on the ionospheric correction available in the Navigation message to remove even 50 percent of the effect.

Isotropic Antenna: Hypothetical, lossless antenna that has equal capabilities in all directions.

Iteration: Converging on a solution by repetitive operations.

IVHS: Intelligent Vehicle Highway System.

Joint Program Office (JPO): Office responsible for the management of the GPS system in the U.S. Department of Defense.

Kalman Filter: In GPS a numerical data combiner used in determining an instantaneous position estimate from multiple statistical measurements on a time-varying signal in the presence of noise. The Kalman filter is a set of mathematical equations that provides an estimation technique based on least squares. This recursive solution to the discrete-data linear filtering problem was proposed by R.E. Kalman in 1960. Since then, the Kalman filter has been applied to radionavigation, in general, and GPS in particular, among other methods of measurement.

Kinematic Positioning: (*also known as* Stop & Go Positioning and Real-Time Kinematic Positioning) (1) Version of relative positioning in which one receiver is a stationary reference and at least one other roving receiver coordinates unknown positions with short occupation times while both track the same satellites and maintain constant lock. If lock is lost, reinitialization is necessary to fix the integer ambiguity. (2) GPS applications in which receivers on vehicles are in continuous motion.

L-Band: Radio frequencies from 390 to 1550 MHz.

L1: GPS signal with the C/A code, the P code, and the Navigation message modulated onto a carrier with the frequency 1575.42 MHz.

L2: A GPS signal with the P code and the Navigation message modulated onto a carrier with the frequency 1227.60 MHz.

Latency: Time taken for a system to compute corrections and transmit them to users in real-time GPS.

Latitude: Angular coordinate, the angle measured from an equatorial plane to a line. On Earth the angle measured northward from the equator is positive, southward is negative. The geodetic latitude of a point is the angle between the equatorial plane of the ellipsoid and a normal to the ellipsoid through the point. At that point the astronomic latitude differs from the geodetic latitude by the meridional component of the deflection of the vertical.

Local Accuracy: Uncertainty of a position relative to other positions nearby. In other words, local accuracy would be useful in knowing the accuracy of adjacent points in relation to each other. Local Accuracy is also known as relative accuracy. In other words, local horizontal and vertical accuracies represent the averaged uncertainty in points relative to adjacent points to which they are directly connected. Within a well-defined geographical area local accuracy may be the most immediate concern.

Longitude: An angular coordinate. On Earth, the dihedral angle from the plane of reference, $0°$ meridian, to a plane through a point of concern, and both planes perpendicular to the plane of the equator. In 1884, the Greenwich Meridian was designated the initial meridian for longitudes. The geodetic longitude of a point is the angle between the plane of the geodetic meridian through the point and the plane of the $0°$ meridian. At that point the astronomic longitude differs from the geodetic longitude by the amount of the component in the prime vertical of the local deflection of the vertical divided by the cosine of the latitude.

Loop Closure: A procedure by which the internal consistency of a GPS network is discovered. A series of baseline vector components from more than one GPS session, forming a loop or closed figure, is added together. The closure error is the ratio of the length of the line representing the combined errors of all the vector's components to the length of the perimeter of the figure.

Mask Angle: Elevation angle below which satellites are not tracked. The technique is used to mitigate atmospheric, multipath, and attenuation errors. A usual mask angle is $15°$.

Master Control Station: Facility manned by the 2nd Space Operations Squadron at Schriever Air Force Base (formerly Falcon AFB) in Colorado Springs, Colorado. The Master Control Station is the central facility in a network of worldwide tracking and upload stations that comprise the GPS Control Segment.

MCS: GPS Master Control Station.

Megahertz (MHz): One million cycles per second.

Microsecond (μs or p sec): One-millionth (10^6) of a second.

Millisecond (ms or msec): One-thousandth of a second.

Minimally Constrained: (*see* Network Adjustment) A least squares adjustment of all observations in a network with only the constraints necessary to achieve a meaningful solution. For example, the adjustment of a GPS network with the coordinates of only one station fixed.

Modem (Modulator/Demodulator): Device that converts digital signals to analog signals and analog signals to digital signals. Computers sharing data usually require a modem at each end of a phone line to perform the conversion.

Monitor Stations: Stations used in the GPS Control Segment to track satellite clock and orbital parameters. Data collected at monitor stations are linked to a master control station at which corrections are calculated and from which correction data are uploaded to the satellites as needed.

Multichannel Receiver: (*also known as* Parallel Receiver) A receiver with many independent channels. Each channel can be dedicated to tracking one satellite continuously.

Multipath: (*also known as* Multipath Error) Error that results when a portion of the GPS signal is reflected. When the signal reaches the receiver by two or more different paths, the reflected paths are longer and cause incorrect pseudoranges or carrier phase measurements and subsequent positioning errors. Multipath is mitigated with various preventative antenna designs and filtering algorithms.

Multiplexing Channel: (*also known as* Fast-Switching, Fast-Sequencing, and Fast-Multiplexing) A channel of a GPS receiver that tracks through a series of a satellite's signals, from one signal to the next in a rapid sequence.

Multiplexing Receiver: GPS receiver that tracks a satellite's signals sequentially and differs from a multichannel receiver in which individual channels are dedicated to each satellite signal.

NAD83 (North American Datum, 1983): Horizontal control datum for positioning in Canada, the United States, Mexico, and Central America based on a geocentric origin and the Geodetic Reference System 1980 (GRS80) ellipsoid. The values for GRS 80 adopted by the International Union of Geodesy and Geophysics in 1979 are $a = 6,378,137$ m and reciprocal of flattening = $1/f = 298.257222101$. It was designed to be compatible with the Bureau International de l'Heuer (BIH) Terrestrial System BTS84. The ellipsoid is geocentric. The origin was defined by satellite laser ranging (SLR), orientation by astronomic observations, and scale by both SLR and very long baseline interferometry (VLBI). NAD83 was actually realized through Doppler observations using internationally accepted transformations from the Doppler reference frame to BTS84 and adjustment of some 250,000 points. VLBI stations were also included to provide an accurate connection to other reference frames. While NAD83 is similar to the World Geodetic System of 1984 (WGS 84), it is not the same as WGS 84. Defined and maintained by the U.S. Department of Defense, WGS84 is a global geodetic datum used by the GPS Control Segment. Access to NAD83 is through a national network of horizontal control monuments established mainly by conventional horizontal control methods, triangulation, trilateration, and astronomic azimuths. Some GPS baselines were used. The adjustment of these horizontal observations, together with several hundred observed Doppler positions, provides a practical realization of NAD83.

Nanosecond (ns or nsec): One-billionth, (10^9) of a second.

NANU: Notice Advisory to NAVSTAR Users.

Navigation Message: (*also known as* Data Message) A message modulated on L1 and L2 of the GPS signal that includes an ionospheric model, the satellite's broadcast ephemeris, broadcast clock correction, constellation almanac, and health, among other things.

NAVSAT: A European radionavigation system under development.

NAVSTAR (NAVigation Satellite Timing and Ranging): GPS satellite system.

Network Accuracy: Uncertainty of a position relative to a datum. Network accuracy is not about the accuracy of the positions at each end of the line relative to each other but rather relative to the whole datum. Network accuracy is also known as absolute accuracy. In other words, network accuracy represents the accuracy of a point with respect to the reference system. Those tasked with constructing a control network that embraces a wide geographical scope will most often need to know the positions relationship to the realization of the datum on which they are working. Typically network horizontal and vertical accuracies require that a point's accuracy be specified with respect to an appropriate national geodetic datum. In the United States as a practical matter this most often means that the work is tied to at least one of the more continuously operating reference stations (CORS), which represent the most accessible realization of the National Spatial Reference System in the nation.

Network Adjustment: Least squares solution in which baselines vectors are treated as observations. It may be minimally constrained. It may be constrained by more than one known coordinate, as is usual in a GPS survey to densify previously established control of a geodetic framework.

North American Vertical Datum of 1988 (NAVD88): A minimally constrained adjustment of U.S., Canadian, and Mexican leveling observations holding fixed the height of the primary tidal benchmark of the new International Great Lakes Datum of 1985 (IGLD85) at Father Point/Rimouski, Quebec, Canada. NAVD88 and IGLD85 are now the same. Between NAVD88 orthometric heights and those referred to the National Geodetic Vertical Datum of 1929 (NGVD29) there are differences ranging from −40 cm to +150 cm within the lower 48 states. The differences range from +94 cm to +240 cm in Alaska. GPS-derived orthometric heights estimated using the precise geoid models now available are compatible with NAVD88. NAVD88 includes 81,500 km of new leveling data never before adjusted to NGVD29. The principal impetus for NAVD88 was minimizing the recompilation of national mapping products. The NAVD88 datum does not correspond exactly to the theoretical level surface defined by the GRS80 definitions.

Observing Session: (*also known as* Observation) Continuous and simultaneous collection of GPS data by two or more receivers.

OMEGA: Radio navigation system that can provide global coverage with only eight ground-based transmitting stations.

On-the-Fly (OTF): Method of resolving the carrier phase ambiguity very quickly. The method requires dual-frequency GPS receivers capable of making both carrier phase and precise pseudorange measurements. The receiver is not required to remain stationary, making the technique useful for initializing in carrier phase kinematic GPS.

Orthometric Height: Vertical distance from the geoid to the surface of the Earth. GPS heights are ellipsoidal. Ellipsoidal heights are the vertical distance from an ellipsoid of reference to the Earth's surface. Ellipsoidal heights differ from leveled, orthometric heights, and conversion from ellipsoidal to orthometric heights requires the vertical distance from the ellipsoid of reference to the geoid (i.e., the geoidal height). The vertical distance from the ellipsoid of reference to the geoid around the world varies from +75 to −100 m; in the coterminous United States it varies from −8 to −53 m. The geoidal heights are negative because the geoid is beneath the ellipsoid. In other words, the ellipsoid is overhead. The relationship between these three heights is

$$h = H + N$$

where h is the ellipsoid height, H is the orthometric height, and N is the geoidal height.

Outage: GPS positioning service is unavailable. Possible reasons for an outage include sufficient number of satellites are not visible, the dilution of precision value is too large, or the signal-to-noise ratio value is too small.

P Code: (*also known as* Protected Code) A binary code known by the names P code, precise code, and protected code. It is a standard spread-spectrum GPS pseudorandom noise code. It is modulated on the L1 and the L2 carrier using binary biphase modulations. Each week long segment of the P code is unique to a single GPS satellite and is repeated each week. It has a chipping rate 10.23 MHz, more than ten million bits per second. The P code is sometimes replaced with the more secure Y code in a process known as anti-spoofing.

Parallel Receiver: (*see* Multichannel Receiver)

PDOP-Position Dilution of Precision: (*see* Dilution of Precision)

Perturbation: Deviation in the path of an object in orbit. Deviation being departure of the actual orbit from the predicted Keplerian orbit. Perturbing forces of Earth-orbiting satellites are caused by atmospheric drag, radiation pressure from the Sun, the gravity of the Moon and the Sun, the geomagnetic field, and the noncentral aspect of Earth's gravity.

Phase Lock: Adjustment of the phase of an oscillator signal to match the phase of a reference signal. First, the receiver compares the phases of the two signals. Next, using the resulting phase difference signal. the reference oscillator frequency is adjusted. When the two signals are next compared, the phase difference between them is eliminated.

Phase Observable: (*see* Carrier Phase)

Phase Smoothed Pseudorange: Pseudorange measurement with its random errors reduced by combination with carrier phase information.

Picosecond: One-trillionth (10^{12}) of a second, one-millionth of a microsecond.

Point Positioning: (*see* Absolute Positioning)

Position: Description, frequently by coordinates, of the location and orientation of a point or object.

Post-Processed GPS: Method of deriving positions from GPS observations in which base and roving receivers do not communicate as they do in real-time kinematic (RTK) GPS. Each receiver records the satellite observations independently. Their collections are combined later. The method can be applied to pseudoranges to be differential corrected or carrier phase measurements to be processed by double-differencing.

Precise Code: (*see* P Code)

Precise Ephemeris: (*see* Ephemeris)

Precise Positioning Service (PPS): GPS positioning for the military at a higher level of absolute positioning accuracy than is available to C/A code receivers, which relies on SPS (Standard Positioning Service). PPS is based on the dual-frequency P code.

Precision: Agreement among measurements of the same quantity; widely scattered results are less precise than those that are closely grouped. The higher the precision, the smaller the random errors in a series of measurements. The precision of a GPS survey depends on the network design, surveying methods, processing procedures, and equipment.

Protected Code: (*see* P Code)

Pseudolite: (*also known as* Pseudo Satellite) Ground-based differential station that simulates the signal of a GPS satellite with a typical maximum range of 50 km. Pseudolites can enhance the accuracy and extend the coverage of the GPS constellation. Pseudolite signals are designed to minimize their interference with the GPS signal.

Pseudorandom Noise: (*also known as* PRN) A sequence of digital 1's and 0's that appear to be randomly distributed but can be reproduced exactly. Binary signals with noise-like properties are modulated on the GPS carrier waves as the C/A codes and the P codes. Each GPS satellite has unique C/A and P codes. A satellite may be identified according to its PRN number. Thirty-two GPS satellite pseudorandom noise codes are currently defined.

Pseudorange: In GPS a time-biased distance measurement. It is based on code transmitted by a GPS satellite, collected by a GPS receiver and then correlated with a replica of the same code generated in the receiver. However, there is no account for errors in synchronization between the satellite's clock and the receiver's clock in a pseudorange. The precision of the measurement is a function of the resolution of the code; therefore, a C/A code pseudorange is less precise than a P code pseudorange.

Quartz Crystal Controlled Oscillator: GPS receivers rely on a quartz crystal oscillator to provide a stable reference so that other frequencies of the system can be compared with or generated from this reference. The fundamental component is the quartz crystal resonator. It utilizes the piezoelectric effect. When an electrical signal is applied, the quartz resonates at a frequency unique to its shape, size, and cut. The first study of the use of quartz crystal resonators to control the frequency of vacuum tube oscillators was made by Walter G. Cady in 1921. Important contributions were made by G. W. Pierce, who showed that plates of quartz cut in a certain way could be made to vibrate so as to control frequencies proportional to their thickness.

R95: Representation of positional accuracy. The radius of a theoretical circle centered at the true position that would enclose 95 percent of the other positions.

Radio Navigation: Determination of position, direction, and distance using the properties of transmitted radio waves.

Radio Technical Commission on Maritime Services (RTCM): In DGPS the abbreviation RTCM has come to mean the correction messages transmitted by some reference stations using a protocol developed by the Radio Technical Commission on Maritime services Special Committee 104. These corrections can be collected and decoded by DGPS receivers designed to accept the signal. The corrections allow the receiver to generate corrected coordinates in real time. There are several sources of RTCM broadcasts. One source is the U.S. Coast Guard system of beacons in coastal areas. Other sources are commercial services, some of which broadcast RTCM corrections by satellite; some use FM subcarriers.

Range: (*also known as* Geometric Range) Distance between two points, particularly the distance between a GPS receiver and satellite.

Range Rate: Rate at which the range between a GPS receiver and satellite changes. Usually measured by tracking the variation in the Doppler shift.

Real-Time DGPS: Method of determining relative positions between known control and unknown positions using pseudorange measurements. A base station or satellite at the known position transmits corrections to the roving receiver or receivers. The procedure offers less accuracy than RTK. However, the results are immediately available, in real-time and need not be post-processed.

Real-Time Kinematic (RTK) Positioning: Method of determining relative positions between known control and unknown positions using carrier phase measurements. A base station at the known position transmits corrections to the roving receiver or receivers. The procedure offers high accuracy immediately, in real-time. The results need not be post-processed. In the earliest use of GPS, kinematic and rapid static positioning were not frequently used because ambiguity resolution methods were still inefficient. Later when ambiguity resolution such as on-the-fly (OTF) became available, real-time kinematic and similar surveying methods became more widely used.

Receiver Channel: (*see* Channel)

Receiver Independent Exchange Format (RINEX): Package of GPS data formats and definitions that allow interchangeable use of data from dissimilar receiver models and post-processing software developed by the Astronomical Institute of the University of Berne in 1989. More than 60 receivers from 4 different manufacturers were used in the GPS survey EUREF 89. RINEX was developed for the exchange of the GPS data collected in that project.

Reconstructed Carrier Phase: (*see* Carrier Phase)

Reference Station: (*see* Base Station)

Relative Positioning: GPS surveying method that improves the precision of carrier phase measurements. One or more GPS receivers occupy a base station, a known position. They collect the signals from the same satellites at the same time as other GPS receivers that may be stationary or moving. The other receivers occupy unknown positions in the same geographic area. Occupying known positions, the base station receivers find corrective factors that can either be communicated in real-time to the other receivers, as in

RTK, or may be applied in post-processing, as in static positioning. Relative positioning is in contrast to absolute, or point, positioning. In relative positioning, errors that are common to both receivers—such as satellite clock biases, ephemeris errors, and propagation delays—are mitigated.

Root Mean Square (RMS): Square root of the mean of squared errors for a sample.

Rover: (*also known as* Mobile Receiver) GPS receiver that is in motion relative to a stationary base station during a session.

Rubidium Clock: Environmentally tolerant and very accurate atomic clock whose working element is gaseous rubidium. The resonant transition frequency of the Rb-87 atom (6,834,682,614 Hz) is used as a reference. Rubidium frequency standards are small, light, and have low power consumption.

SA: (*see* Selective Availability)

S Code: (*see* C/A Code)

Satellite Clocks: Two rubidium (Rb) and two cesium (Cs) atomic clocks are aboard Block II/IIA satellites. Three rubidium clocks are on Block IIR satellites. Hydrogen maser time standards may be used in future satellites.

Satellite Constellation: In GPS, four satellites in each of six orbital planes. In GLONASS, eight satellites are in each of three orbital planes.

Second Base: Unit of time in the International System of Units. The duration of 9,192,631,770 periods of the radiation corresponding to the transition between the two hyperfine levels of the ground state of the cesium 133 atom is undisturbed by external fields.

Selective Availability (SA): By dithering the timing and ephermides data in the satellites, standard positioning service (SPS), users are denied access to the full GPS accuracy. The intentional degradation of the signals transmitted from the satellites may be removed by precise positioning service (PPS) users. It is also virtually eliminated in carrier phase relative positioning techniques and double-differencing processing. Selective Availability was begun in 1990 and was removed from August 10, 1990, to July 1, 1991. It was enabled on November 15, 1991, and finally switched off by presidential order on May 2, 2000.

Single Difference: (*see* Between-Receivers Difference) (*See also* Between-Satellites Single Difference)

Space Segment: Portion of the GPS system in space, including the satellites and their signals.

Spatial Data: (*also known as* Geospatial Data) Information that identifies the geographic location and characteristics of natural or constructed features and boundaries of Earth. This information may be derived from, among other things, remote sensing, mapping, and surveying technologies.

Special Theory of Relativity: (*see also* General Theory of Relativity) A theory developed by Albert Einstein, predicting, among other things, the changes that occur in length, mass, and time at speeds approaching the speed of light. General relativity predicts that as gravity weakens the rate of clocks increase and they tick faster. Special relativity predicts that moving clocks appear to tick more slowly than stationary clocks because the rate of a moving clock seems to decrease as its velocity increases. Therefore, for GPS satellites, general relativity predicts that the atomic clocks in orbit on GPS satellites tick faster than the atomic clocks on Earth by about 45,900 ns/day. Special relativity predicts that the velocity of atomic clocks moving at GPS orbital speeds tick slower by about 7200 ns/day than clocks on earth. The rates of the clocks in GPS satellites are reset before launch to compensate for these predicted effects.

Spherical Error Probable (SEP): Description of three-dimensional precision. The radius of a sphere with its center at the actual position that is expected to be large enough to include half (50 percent) of the normal distribution of the scatter of points observed for the position.

Spread-Spectrum Signal: A signal spread over a frequency band wider than needed to carry its information. In GPS a spread-spectrum signal is used to prevent jamming, mitigate multipath, and allow unambiguous satellite tracking.

SPS: (*see* Standard Positioning Service)

Standard Deviation: (*also known as* 1 sigma, 1σ) An indication of the dispersion of random errors in a series of measurements of the same quantity. The more tightly grouped the measurements around their average (mean), the smaller the standard deviation. Approximately 68 percent of the individual measurements will be within the range expressed by the standard deviation.

Standard Positioning Service (SPS): Civilian absolute positioning accuracy using pseudorange measurements from a single frequency C/A code receiver, ±20 to ±40 m 95 percent of the time (2 drms) with Selective Availability switched off.

Standard Varieties of Dilution of Precision:

    PDOP: Position (three coordinates)

    GDOP: Geometric (three coordinates and clock offset)

    RDOP: Relative (normalized to 60 s)

    HDOP: Horizontal (two horizontal coordinates)

    VDOP: Vertical (height)

    TDOP: Time (clock offset)

Static Positioning: A relative, differential, surveying method in which at least two stationary GPS receivers collect signals simultaneously from the same constellation of satellites during long observation sessions. Generally, static GPS measurements are post-processed and the relative position of the two units can be determined to a very high accuracy. Static positioning is in contrast to kinematic and real-time kinematic positioning where one or more receivers track satellites while in motion. Static positioning is in contrast to absolute, or point, positioning, which has no relative positioning component. Static positioning is in contrast to rapid-static positioning where the observation sessions are short.

Stop-and-Go Positioning: (*see* Kinematic Positioning)

SV: Space Vehicle

Time Dilation: (*also known as* Relativistic Time Dilation) (*see* Special Theory of Relativity and General Theory of Relativity) Systematic variation in time's rate on an orbiting GPS satellite relative to time's rate on Earth. The variation is predicted by the special theory of relativity and the general theory of relativity as presented by A. Einstein.

Transit Navigation System: Satellite-based Doppler positioning system funded by the U.S. Navy.

Triple-Difference: (*also known as* Receiver-Satellite-Time Triple Difference) Combination of two double differences. Each of the double differences involves two satellites and two receivers. The triple difference is derived between two epochs. In other words, a triple difference involves two satellites, two receivers, and time. Triple differences ease the detection of cycle slips.

Trivial Baseline: Baselines observed using GPS relative positioning techniques. When more than two receivers are observing at the same time, both independent (nontrivial) and trivial baselines are generated. For example, where r is the number of receivers, every complete static session yields $r - 1$ independent (nontrivial) baselines and the remaining baselines are trivial. For example, if four receivers are used simultaneously, six baselines are created. Three are trivial, and three are independent. The three independent baselines will fully define the position of each occupied station in relation to the others. The three trivial baselines may be processed, but the observational data used for them has already produced the independent baselines. Therefore, only the independent baseline results should be used for the network adjustment or quality control.

Tropospheric Effect: (*also known as* Tropospheric Delay) The troposphere comprises approximately 9 km of the atmosphere over the poles and 16 km over the equator. The tropospheric effect is nondispersive for frequencies under 30 GHz. Therefore, it affects both L1 and L2 equally. Refraction in the troposphere has a dry component and a wet component. The dry component is related to the atmospheric pressure and contributes about 90 percent of the effect. It is more easily modeled than the wet component. The GPS signal that travels the shortest path through the atmosphere will be the least affected by it. Therefore, the tropospheric delay is least at the zenith and most near the horizon. GPS receivers at the ends of short baselines collect signals that pass through substantially the same atmosphere and the tropospheric delay may not be troublesome. However, the atmosphere may be very different at the ends of long baselines.

UKOOA: Member companies of the U.K. Offshore Operators Association (UKOOA) are licensed by the government to search for and produce oil and gas in U.K. waters. It publishes, among other things "The Use of Differential GPS in Offshore Surveying." These guidelines cover installation and operation, quality measures, minimum training standards, receiver outputs, and data exchange format.

Universal Transverse Mercator (UTM): Represents ellipsoidal positions, latitude and longitude, as grid coordinates, northing and easting, on a cylindrical surface that can be developed into a flat surface. Universal Transverse Mercator is a particular type of transverse Mercator projection. It was adopted by the U.S. Army for large-scale military maps and is shown on all 7.5 min quadrangle maps and 15 min quadrangle maps prepared by the U.S. Geological Survey. The Earth is divided into 60 zones between 84°N latitude and 84°S latitude, most of which are 6° of longitude wide. Each of these UTM zones has a unique central meridian, and the scale varies by 1 part in 1000 from true scale at equator.

USCG: U.S. Coast Guard

User-Equivalent Range Error (UERE): Contribution in range units of individual uncorrelated biases to the range measurement error.

User Interface: Software and hardware that activate displays and controls that are the means of communication between a GPS receiver and the receiver's operator.

User Segment: Component of the GPS system that includes the user equipment, applications, and operational procedures. The part of the whole GPS system that includes the receivers of GPS signals.

UTC: (*see* Coordinated Universal Time)

UTM: (*see* Universal Transverse Mercator)

Very Long Baseline Interferometry (VLBI): By measuring the arrival time of the wave front emitted by a distant quasar at two widely separated Earth-based antennas, the relative position of the antennas is determined. Because the time difference measurements are precise to a few picoseconds, the relative positions of the antennas are accurate within a few millimeters and the quasar positions to fractions of a milliarcsecond.

Voltage-Controlled Quartz Crystal Oscillator: (*see also* Quartz Crystal Controlled Oscillator) Quartz crystal oscillator with a voltage controlled frequency.

Wavelength: (*see also* Frequency) Along a sine wave the distance between adjacent points of equal phase. The distance required for one complete cycle.

Waypoint: Two-dimensional coordinate to be reached by GPS navigation.

Wide Area Augmentation System (WAAS): U.S. Federal Aviation Authority (FAA) system that augments GPS accuracy, availability, and integrity. It provides a satellite signal for users to support en route and precision approach aircraft navigation. Similar systems are Europe's European Geostationary Navigation Overlay System and Japan's MT-SAT.

World Geodetic System 1984 (WGS84): A world geodetic Earth-centered, Earth-fixed terrestrial reference system. The origin of the WGS 84 reference frame is the center of mass of the Earth. The GPS Control Segment has worked in WGS84 since January 1987, and therefore, GPS positions are said to be in this datum.

Y Code: When anti-spoofing on the P code is encrypted into the Y code and transmitted on L1 and L2.

Z-Count Word: GPS satellite clock time at the leading edge of the next data subframe of the transmitted GPS message (usually expressed as an integer number of 1.5 s periods) [van Dierendonck et al., 1978].

Zero Baseline Test: Setup using two GPS receivers connected to one antenna. Nearly all biases are identical for both receivers, and only random observation errors attributable to both receivers remain. The baseline measured should be zero if the receivers' calibration was ideal.

Zulu Time: (*see* Coordinated Universal Time)

References

Brunner, F.K., and W.M. Welsch. 1993. Effect of the troposphere on GPS measurements. *GPS World* 4(1):42–51. Advanstar Communications.

Federal Geographic Data Committee (FGDC). 1998. *Geospatial Positioning Accuracy Standards, Part 2: Standards for Geodetic Networks*. National Ocean and Atmospheric Administration (NOAA). Available at https://www.fgdc.gov/standards/projects/FGDC -standards-projects/accuracy/part2/index_html (accessed Oct. 21, 2014).

Fotopoulos, G. 2000. *Parameterization of DGPS Carrier Phase Errors over a Regional Network of Reference Stations*. MS thesis, University of Calgary, Department of Geomatic Engineering (UCGE Reports, Number 20142), Calgary, Alberta, Canada.

Lazar, S. 2002. Modernization and the move to GPS III. *Crosslink* 3(2):42–46.

van Dierendonck, A.J., S.S. Russell, E.R. Kopitzke, and M. Birnbaum. 1978. The GPS navigation message. *Navigation* 25:147–165.

Wells, D., Ed. 1986. *Guide to GPS Positioning*. Canadian GPS Associates, Fredericton, NB, Canada.

Index

Page numbers followed by f and t indicate figures and tables, respectively.